GW00776205

Engineering With Mathcad

Using Mathcad to Create and Organize Your Engineering Calculations

Engineering With Mathcad

Using Mathcad to Create and Organize Your Engineering Calculations

Brent Maxfield

ELSEVIER

AMSTERDAM • BOSTON • HEIDELBERG • LONDON • NEW YORK • OXFORD
PARIS • SAN DIEGO • SAN FRANCISCO • SINGAPORE • SYDNEY • TOKYO

Butterworth-Heinemann is an imprint of Elsevier

Butterworth-Heinemann is an imprint of Elsevier
Linacre House, Jordan Hill, Oxford OX2 8DP
30 Corporate Drive, Suite 400, Burlington, MA 01803

First published 2006

British Library Cataloguing in Publication Data
A catalogue record for this book is available from the British Library

Library of Congress Cataloguing in Publication Data
A catalogue record for this book is available from the Library of Congress

ISBN-13: 978-0-7506-6702-9
ISBN-10: 0-7506-6702-8

For information on all Butterworth-Heinemann publications
visit our web site at http://books.elsevier.com

Printed and bound in Great Britain

06 07 08 09 10 10 9 8 7 6 5 4 3 2 1

Acknowledgment

Grateful acknowledgment is given to Mathsoft Engineering and Education, Inc. for permission to use information from the *Mathcad Users Guide*, *Mathcad Help*, *Mathcad Advisor Newsletters*, and the *Mathsoft Technical Support webpage*.

Contents

Preface

I first began using Mathcad back in the early 1990s, starting with Mathcad Version 4. Once I started using it, I was hooked. Since that time, I have essentially thrown away my calculation pads. Where I formerly used a pencil, paper, and a calculator, I now use Mathcad.

I have written thousands of pages of Mathcad calculations. They include one-page worksheets to several hundred-page worksheets. Even with all of this experience, I am still learning useful things about Mathcad—thing that I wish I had learned years ago. As I talk with other engineers who use Mathcad, I often show them things that they have never used. Mathcad has so many wonderful functions and features, it is very difficult to learn them all, and it is difficult to discern which of them will be useful to your particular application. My purpose in writing this book is to share with other engineers the things that I have learned through my years of using Mathcad.

As practicing engineers, students, or professionals, our lives are so focused on getting-the-job-done that we do not have the luxury of spending the time necessary to thoroughly study and understand a computer program as complex as Mathcad. I wrote this book for you. It filters out the complex functions and focuses on the functions and features that will be most useful to engineering practice.

Many books provide a comprehensive study of all the Mathcad functions. If you are interested in learning about the very mathematical and complex Mathcad functions, you will want to study one of these books. In *Engineering with Mathcad*, I focus on the application of simple, yet powerful Mathcad functions, and applying these to building comprehensive engineering calculations.

If you are new to Mathcad, I hope you enjoy learning about this great engineering program. If you are an experienced Mathcad user, I hope that you are able to learn new concepts, which you can apply to your engineering calculations.

I wish to thank the many people at Mathsoft Engineering & Education, Inc. who have assisted with this book. In particular, I want to thank Leslie Bondaryk, who has supported me every step of the way. She provided valuable insight into many different aspects of this book, and provided valuable feedback to improve the book.

Many engineers have improved my career. Indirectly, they have helped create this book. There are too many to mention by name, but I wish to express thanks to the engineers in the Structural Engineers Association of Utah. I wish also to thank my colleagues who provide continual improvement to our use of Mathcad at work. It is a pleasure to work with them.

This list of thanks would not be complete without special thanks to my wife Cherie, who put up with the late nights, early mornings, and Saturday's spent working on this book. She has been a stalwart supporter of this effort. I could not have done it without her support. I also want to thank my five sons who are the joy of my life. They have been very understanding and patient with the time spent away from them. Let's go play!

Brent Maxfield

Introduction

Welcome to the wonderful world of Mathcad. If this is your first time using Mathcad, then you are in for a real treat. You will soon see how easy it is to start using Mathcad anytime you need to do simple or complex calculations. If you are an experienced Mathcad user, this book is intended to give you new perspectives to increase your Mathcad productivity.

New users will benefit by quickly getting up-to-speed, and by learning easy-to-use Mathcad functions. The more advanced and complicated Mathcad elements are discussed in later chapters. Experienced Mathcad users will benefit by learning how to use Mathcad to better organize and create engineering calculations.

Mathcad is a very powerful interactive worksheet that utilizes standard math notation and equation entry to solve problems. It is easy to learn, and no programming is required to use it. It immediately returns and updates results.

This book is directly focused toward engineers, scientists, engineering and science students, and others performing technical calculations. Its purpose is to quickly teach basic Mathcad skills, teach some very useful and powerful features and then teach how to apply these features to create and organize comprehensive technical calculations—even calculations created from other software programs.

Mathcad has hundreds of features and functions, many of which are very complex and high-power math functions. In general, practicing engineers use only a small percentage of these functions. It takes a considerable amount of time to sort through all the Mathcad features and determine which ones are going to be useful to your specific needs. If you are a busy engineer or student, who wants to get up-to-speed quickly, then this book is for you. It ignores the complex Mathcad features and emphasizes the essential engineering features.

This book uses the terms engineer and engineering quite often. If you are involved in any technical calculations as a scientist or technician this book will still be of benefit to you. The principles discussed for engineers are still applicable to many scientific and technical calculations.

Why use Mathcad

Mathcad is perfectly suited for engineering calculations. You work in Mathcad just as you would on a piece of calculation paper. The equations you write on a piece of paper can be entered into Mathcad just as they look on the paper, and just as quickly as you write them on the paper. The benefit of Mathcad is that your calculations are now electronic and can be archived, shared with coworkers, reused on other projects and updated as variables change. Another advantage of Mathcad is that your results can be used in further calculations. If the variables for the first result change, then all the calculations based on that first result are also immediately updated.

Mathcad is easy to learn. There is no programming language that you need to learn. With just a few simple instructions, you can begin using Mathcad.

With Mathcad all your calculations are presented on the worksheet. They are not hidden in cells as they are in spreadsheets. Also, all variables are named. You do not need to remember which cells the variables are stored in. This makes checking the Mathcad calculations much simpler.

These are all the great benefits and reasons for using Mathcad, but there is one more significant advantage to using Mathcad—Units! You can attach units to every number or variable used in Mathcad. Once a unit is attached to a number or variable, Mathcad remembers the unit. You can then display this variable in any unit system. You do not need to convert it. Mathcad does it for you. No more dividing by 12 to convert inches to feet, or remembering how many millimeters per inch. Mathcad does all of this internally. You can even mix unit systems. Units are discussed in great detail in Chapter 4—Units!

Mathcad can even replace your calculator. If you are doing your calculations the old fashioned way—using a paper and pencil, you can still use Mathcad to help with your calculations—such as adding or multiplying strings of numbers, evaluating equations, etc. The advantage is that the math stays on the screen until you close the worksheet. With a calculator all your numbers quickly disappear.

With Mathcad, both the input and results stay on the screen. This way you can look back at previous calculations to check input or results. You can even change the input and get new results.

Don't give up your favorite engineering software

If you are like many engineers, you probably have many Microsoft Excel spreadsheet programs customized for your specific applications. You also have specialized software programs that meet your specific engineering needs. You don't need to give up these programs when you begin using Mathcad. Mathcad supplements these programs.

Mathcad can be used as a vehicle to create, organize and assemble all the pieces of your engineering calculations. Mathcad can be used to calculate the input to be used in other software programs. In the case of Microsoft Excel, the input and output can even be dynamic with Mathcad. Mathcad can pass input information into Excel. Excel will perform its calculations, and then give the output back to Mathcad. This is a fantastic feature. You can also insert the results from your other software programs into Mathcad. Mathcad can even store the input and output files from these software programs within the Mathcad worksheet so that when you double-click on the files, the software program opens.

Book overview

Engineering with Mathcad uses an analogy of teaching you how to build a house. If you were to learn how to build a house, the final goal would be the completed house. Learning how to use the tools would be a necessary step, but the tools are just a means to help you complete the house. It is the same with this book. The ultimate goal is to teach you how to apply Mathcad to build comprehensive engineering calculations.

In order to begin building, you need to learn a little about the tools. You also need to have a toolbox where you can put the tools. When building a house, there are simple hand tools and more powerful power tools. It is the same with Mathcad. We will learn to use the simple tools before learning about the power tools. After learning about the tools, we will learn to build.

This book is divided into four parts:

Part I—Building your Mathcad Toolbox. This is where you build your Mathcad toolbox—your basic understanding of Mathcad. It teaches the basics of the

Mathcad program. The chapters in this part create a solid foundation upon which to build.

Part II—Hand Tools for your Mathcad Toolbox. The chapters in this part will focus on simple features to get you comfortable with Mathcad.

Part III—Power Tools for your Mathcad Toolbox. This part addresses more complex and powerful Mathcad features.

Part IV—Creating and Organizing your Engineering Calculations with Mathcad. This is where you start using the tools in your toolbox to build something—engineering calculations. This part discusses embedding other programs into Mathcad. It also discusses how to assemble calculations from multiple Mathcad files, and files from other programs.

Before you begin

If you don't already have Mathcad installed on your computer, take a few minutes and install the included Academic Evaluation version of Mathcad. This will allow you to follow along and practice the concepts discussed is this book. It will also give you access to the Mathcad Help and Mathcad Tutorials.

Engineering with Mathcad is based on the U.S. version of Mathcad. It is also based on the U.S. Keyboard. There may be slight differences in Mathcad versions sold outside of the United States.

Mathcad basics

Whenever you open Mathcad, a blank worksheet opens. You can liken this worksheet to a clean sheet of calculation paper waiting for you to add information to it.

Let's begin with some simple math. Type `5+3=`. You should get the following:

$$5 + 3 = 8$$

Now type `(2+3)*2=`. You should get the following:

$$(2 + 3)\cdot 2 = 10$$

You can also assign variable names to these equations. To assign a value to a variable, type the variable name and then type the colon `:` key. For example type `a1:5+3`.

$$a1 := 5 + 3$$

Now type `a1=`. This displays the value of variable "a1."

$$a1 = 8$$

Let's assign another variable. Type `b1:(2+3)*2`.

$$b1 := (2 + 3)\cdot 2$$

Now type `b1=`. This displays the value of variable "b1."

$$b1 = 10$$

Now that values are assigned to variable "a1" and variable "b1," you can use these variables in equations. Type `c1:a1+b1`.

$$c1 := a1 + b1$$

Now type `c1=`. You should get the following result:

$$c1 = 18$$

As you can see, there is no programming involved. Simply type the equations as you would write them on paper.

Additional resources

This book is written as a supplement to the Mathcad Help and the Mathcad User's Guide. It adds insights not contained in these resources. You should become familiar with the use of both these resources prior to beginning an earnest study of this book. To access the Mathcad Help, click **Mathcad Help** from the **Help** menu, or press the `F1` key. The Mathcad User's Guide is a PDF file located in the Mathcad program directory in the "doc" folder.

In addition to the Mathcad Help and the Mathcad User's Guide, the Mathcad Tutorials provide an excellent resource to learn Mathcad. The Mathcad Tutorials are accessed by clicking **Tutorials** from the **Help** menu. Take the opportunity to review some of the topics covered by the tutorials.

Terminology

There are a few terms we need to discuss in order to communicate effectively.

The terms, "click," "clicking," or "Select" will mean to click with the left mouse button.

The terms "expression" and "equation" are sometimes used interchangeably. The term "equation" is a subset of the term "expression." When we use the term "equation," it generally means some type of algebraic math equation that is being defined on the right side of the definition symbol "=:". The term "expression" is broader. It usually means anything located to the right of the definition symbol. It can mean "equation," or it can mean a Mathcad program, a user-defined function, a matrix or vector or any number of other Mathcad elements.

Part I

Building your Mathcad Toolbox

Just as you store tools in a toolbox, your Mathcad tools are stored in your Mathcad toolbox. Your Mathcad toolbox is the place where you will store your Mathcad skills—the tools that will be discussed in Parts II and III. You build your Mathcad toolbox by learning about the basics of the Mathcad program and the Mathcad worksheet. The chapters in Part I teach about variables, expression editing, user-defined functions, units, Mathcad settings, and customizing Mathcad. These chapters create a foundation upon which to build. They create your Mathcad Toolbox.

1 Variables

Variables

Variables are one of the most important features of Mathcad. As in Algebra, variables define constants and create relationships. Your Mathcad worksheet will be full of variables. It is therefore important to quickly gain a solid foundation in their use.

Chapter 1 will:

- Discuss types of variables.
- Show how to define variables.
- Give rules for naming variables.
- List characters that can be used in variable names.
- Introduce string variables.
- Provide guidelines for naming variables.

Types of variables

Variables can consist of numbers or constants such as: A := 1, or B := 67. They can consist of equations such as: C := A + B, or D := A + 3. You can set one variable equal to another such as E := A. Variables can also consist of strings of characters such as F := "This is an example of a string variable." Variables can even have logic programs associated with them so that the value of the variable depends on the outcome of Boolean logic. As you go through this book you will see that variables can be very simple or be very complex. For the purpose of this chapter we will stay with simple examples. More detailed examples will follow in later chapters.

Defining variables

To define a variable, type the variable name followed by pressing the colon [:] key. This inserts the definition symbol, ":=" and a placeholder (G := ∎). The placeholder is a small black box that indicates that the expression is incomplete. Click on the placeholder and complete the definition by typing in a number or an equation. To view the value assigned to a variable, type the variable name followed by the equal sign.

Rules for naming variables

Case and font

The first important thing to remember about variable names is that they are case, font, size, and style sensitive. Thus the variable "ANT" is different than the variable "ant" (UPPER case versus lower case), and the variable "Bat" is different than the variable "**Bat**" (Normal font versus bold font). The variable "Cat" is different than the variable "Cat" (Different font size), and the variable "Dog" is different than the variable "`Dog`" (Different font style). If at some point in your worksheet Mathcad isn't recognizing your variable, check to make sure that your variables are exactly the same in case, font, size, and style.

Characters that can be used in variable names

There are some rules for naming variables. These are listed below:

- Variable names can consist of upper and lower case letters.
- The digits 0 through 9 can be used in a variable name, except that the leading character in a variable name cannot be a digit. Mathcad interprets anything beginning with a digit to be a number and not a variable.
- Variable names may consist of Greek letters. The easiest way to insert Greek letters is to use the Greek letter toolbar. Find this toolbar by highlighting **Toolbars** from the **View** menu and then clicking on **Greek**. Select the desired Greek letter from the toolbar, and it will be inserted into your worksheet. Another way to insert Greek letters is to type the equivalent

roman letter and then type **CTRL+G**. See Figure 1.1 for a table of equivalent Greek letters. Search "Greek toolbar" in the Index of Mathcad Help for this table of Greek equivalent letters.

α	a	η	h	ο	o	ϖ	v
β	b	ι	i	π	p	ω	w
χ	c	φ	j	θ	q	ξ	x
δ	d	κ	k	ρ	r	ψ	y
ε	e	λ	l	σ	s	ζ	z
ϕ	f	μ	m	τ	t		
γ	g	ν	n	υ	u		
Α	A	Η	H	Ο	O	ς	V
Β	B	Ι	I	Π	P	Ω	W
Χ	C	ϑ	J	Θ	Q	Ξ	X
Δ	D	Κ	K	Ρ	R	Ψ	Y
Ε	E	Λ	L	Σ	S	Ζ	Z
Φ	F	Μ	M	Τ	T		
Γ	G	Ν	N	ϒ	U		

Figure 1.1 Table of equivalent Greek letters

- The infinity symbol ∞ may be used only as the beginning character in a variable name. To insert the infinity symbol, type **CTRL+SHIFT+z**.

Subscripts

Subscripts can be used. However, once a subscript is used, all remaining characters remain as part of the subscript (See Chemistry Notation below for an exception). You cannot use a subscript and then return to normal text. To type a subscript, type the first part of the variable name, and then type a period. The insertion point will drop down half of a line. All characters typed after this point will be part of the subscript. This subscript is called a literal subscript. It looks very similar to an array subscript, but it behaves much differently. Array subscripts will be discussed in Chapter 9. See Figure 1.2 for an example of variable names using subscripts.

Example_1

Sample_2

f_y

f'_c

Figure 1.2 Example of variable names using subscripts

Special text mode

Keyboard symbols may be used in variable names, but Mathcad operators may not. However, most keyboard symbols are also Mathcad shortcuts that insert a Mathcad operator or perform another Mathcad function. (See "Operator Keystrokes" in the Index of Mathcad Help for a list of keyboard shortcuts.) This prevents you from using most keyboard symbols in your variable names. When you try to use a symbol that is also a Mathcad shortcut, Mathcad inserts the operator or executes the command referenced by the shortcut. For example, if you type **A $ B**, then Mathcad inserts the range sum symbol because the $ symbol is the keyboard shortcut for the range sum. A variable name cannot use the addition operator. If you type **A + B : 6**, Mathcad will not recognize the variable name, and will give an error. See Figure 1.3.

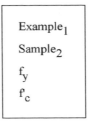

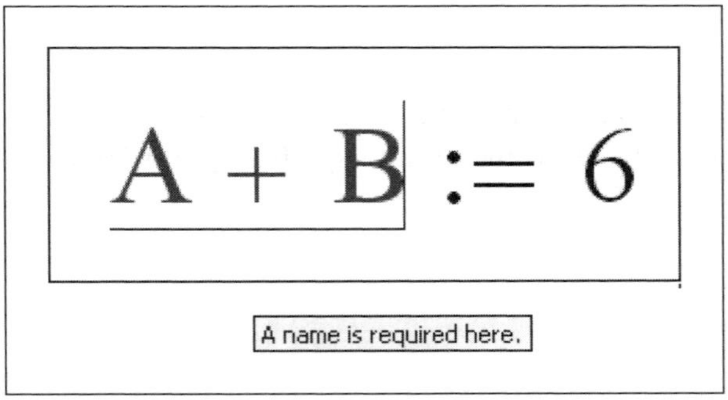

Figure 1.3 Operators may not be used in a variable name

Mathcad provides a way to use both symbols and operators in variable names by providing a special text mode. To activate the special text mode, begin the variable name by typing a letter, then type **CTRL+SHIFT+k**. Once the special text mode is entered, the editing lines turn from blue to red. You are now free to enter any keyboard symbols. If you want your variable name to begin with a symbol, then move the cursor back to the beginning of the variable and type the symbol. When you are done entering symbols, type **CTRL+SHIFT+k** again to return to normal math mode. See Figure 1.4 for examples of variable names using keyboard symbols.

A+B := 1	A+B = 1
B$C: := 2	B$C: = 2
x^2 := 4	x^2 = 4
{Test} := 100	{Test} = 100

Figure 1.4 Examples of variable names using the special text mode

Chemistry notation

Mathcad provides a means to have your variable names look like an expression or an equation. It is a special mode called Chemistry Notation. To activate the Chemistry Notation type **CTRL+SHIFT+j**. This inserts a pair of brackets with a placeholder between them. You are now free to insert whatever letters, numbers, and operators you want between the brackets. Using this mode you can make your variable name look like an equation. You are not limited to staying within subscripts like you are when you name a normal variable. Chemistry Notation is useful when you have a long equation with many parts. You may want to separate the equation into smaller parts. In order to do this you need to give each part of the equation a variable name. Sometimes it is difficult to determine what to name each part. With Chemistry Notation, each part can be given the variable name to match the part of the equation. See Figure 1.5.

$a := 1 \quad b := 5 \quad c := 6$

$\left[\sqrt{b^2 - 4 \cdot a \cdot c}\right] := \sqrt{b^2 - 4 \cdot a \cdot c}$

This variable name was entered using Chemistry Notation. The variable name looks just like the defined equation.

$\left[\sqrt{b^2 - 4 \cdot a \cdot c}\right] = 1$

$\left[\frac{NaOH}{Na_2 \cdot CO_3}\right] := \frac{80}{106}$

An example of a chemistry ratio entered using Chemistry Notation.

$\left[\frac{NaOH}{Na_2 \cdot CO_3}\right] = 0.755$

Figure 1.5 Variable names using Chemistry Notation

The following chapter will show you how to create and edit these Mathcad expressions. These examples illustrate the power and versatility that Mathcad has to name variables.

String variables

A string is a sequence of characters between double quotes. It has no numeric value, but it can be defined as a variable. To create a string variable, type the variable name followed by pressing the colon [:] key. This inserts the definition symbol ":=" and a placeholder. Type the double quotes key ["] in the placeholder. You will see an insertion line between a pair of double quotes. You can then type any combination of letters, numbers or other characters. When you are finished with the string, press enter.

String variables are useful to use as error messages. If you need a certain input to be positive, you can assign a string variable to have the value, "Input must be positive." If a number less than zero is entered, Mathcad can display this string variable as an error message. String variables are also useful to use as a means of displaying whether or not a certain condition is met. You can assign one variable to have the value, "Yes" and another variable to have the value "No." If a specific condition is met Mathcad can display the string variable associated with "Yes." If the specific condition is not met Mathcad can display the string variable associated with "No." These simple logic programs will be discussed in Chapter 12, "Simple Logic Programming." See Figure 1.6 for some examples of text strings.

$TextString_1 := \text{"This is a text string"}$

$TextString_1 = \text{"This is a text string"}$

$TextString_2 := \text{"Yes"}$ $\quad$ $TextString_2 = \text{"Yes"}$

$TextString_3 := \text{"No"}$ $\quad$ $TextString_3 = \text{"No"}$

Figure 1.6 Examples of text strings

Why use variables

Figure 1.7 shows three different ways of getting similar results. The first method shown is very direct. If you were not going to be saving the worksheet and you needed a quick answer, the first method works fine. Just type in the numbers to

The following example shows three different methods of getting the same results.

1. Use direct numbers

$5 + 7 = 12$

$12 \cdot 3 = 36$

$36 - 8 = 28$

2. Use intermediate results

$Answer_1 := 5 + 7 \qquad Answer_1 = 12$

$Answer_2 := Answer_1 \cdot 3 \qquad Answer_2 = 36$

$Answer_3 := Answer_2 - 8 \qquad Answer_3 = 28$

3. Assign variable names to input and output values

$Input_1 := 5 \quad Input_2 := 7 \quad Input_3 := 3 \quad Input_4 := 8$

$Answer_4 := Input_1 + Input_2 \qquad Answer_4 = 12$

$Answer_5 := Answer_4 \cdot Input_3 \qquad Answer_5 = 36$

$Answer_6 := Answer_5 - Input_4 \qquad Answer_6 = 28$

Figure 1.7 Using variables in engineering calculations

get an answer. Use the result of the first equation, and type it into the second equation. Use the result of the second equation, and type it into the third equation.

The second method shown is to assign a variable name to the intermediate answers. The benefit of this method is that you will always have the result of each expression available to use in other expressions in your worksheet. The equations shown are very simple and basic, but in your engineering calculations the equations or expressions can be very complex. Once the result of an expression is calculated by Mathcad, you want to capture it for future use. You do this by assigning the result to a variable.

The third method shown is to assign all values to variable names. There are four input values and three output results. You may be saying to yourself, "Why would I type all those extra key strokes. It is much more time consuming to type $Input_1$+$Input_2$ than just typing 5+7." Well, let's assume that the numbers 5, 7, 3, and 8 represent some type of engineering input. You now use these numbers over and over in your calculations. If you keep using just the numbers 5, 7, 3, and 8 in your many different equations, then what happens if at some point the input value 7 is changed to 9? If you have used the variable $Input_2$=7, then you just change the value of $Input_2$ from 7 to 9 and you are done. Mathcad does the rest. If you didn't use the input variable, then you will need to go through your worksheet and change (or attempt to change) every instance where the number 7 represented the input variable, and change it from 7 to 9. This could be an impossible task if you have a complex worksheet.

Remember that the goal of this book is to teach you how to use Mathcad as a tool for creating engineering calculations. Because of this, it is recommended that you get into the habit of using the third method illustrated in Figure 1.7. Most of the examples used in this book will use this method. We will assign the input values to variable names; assign a variable name to the expression, and then display the results of the expression.

Hopefully you understand how important variables are to Mathcad. The next section discusses how to best use variables in your engineering calculations.

Guidelines for naming variables

If your worksheets are only a few pages long, or if your worksheets are not reused, then it is not critical to worry about how your variables are named. However, if you use Mathcad to create comprehensive calculations, then the naming of variables becomes critical. Therefore, the following naming guidelines are given to assist in choosing variable names that will be used in your engineering calculations.

We will first establish a few useful naming guidelines to assist you in choosing variable names. These are not strict rules that you must always follow. There will be exceptions to each of these guidelines, but if they are consistently followed, then your worksheets will be better organized. These guidelines will also prevent problems that will be discussed in Part IV.

Naming guideline 1

Make your variable names more than just a single letter

Single letters are appropriate if you are just doing a quick calculation for something that won't be saved. They are also appropriate if the single letter represents a universal constant such as "e." They are also useful if you have an equation that uses single letter variables. But, generally, it is better to use variable names with more than one letter. This gives you the ability to make the variable name more descriptive.

Naming guideline 2

Make your variable names mean something

Sometimes longer variable names are more useful because they can be descriptive of the use of the variable. If your calculations are more than one page long and include many variable names, it is much easier to remember a descriptive variable name. It also makes it easier to check printed calculations.

Naming guideline 3

Use a combination of upper case and lower case letters to help make your variable names easier to read

Unless there is some specific reason not to, your variable names should begin with an upper case letter. If your variable name is more than one word, then use a combination of upper case and lower case letters to make the variable name easier to read. For example instead of naming a variable "vesselpressure," name it "VesselPressure." See Figure 1.8 for some examples.

Buildingwidth	BuildingWidth
Maximumacceleration	MaximumAcceleration
Momentofinertia	MomentOfInertia
Transversewavevelocity	TransverseWaveVelocity
Areaofcircle	AreaOfCircle

Figure 1.8 Using uppercase letters to separate words in variable names

Naming guideline 4

Use underscore to separate different names in your variable names

In the previous guideline we discussed using a combination of upper case and lower case letters to make variable names easier to read. Another way to do this is to use the underscore to separate names. For example, instead of naming a variable "VesselPressure," you could name it "Vessel_Pressure." See Figure 1.9 for some examples.

BuildingWidth	Building_Width
MaximumAcceleration	Maximum_Acceleration
MomentOfInertia	Moment_Of_Inertia
TransverseWaveVelocity	Transverse_Wave_Velocity
AreaOfCircle	Area_Of_Circle

Figure 1.9 Using underscore to separate words in variable names

Naming guideline 5

Make good use of subscripts in your variable names

Subscripts are created by typing a period "." within your variable name. It lowers the text into a subscript font. Subscripts are useful to distinguish variables which are very similar or related. Figure 1.10 shows a few examples of using subscripts.

Area1	Area_1
Volume1	Volume_1
Velocity1	$\mathrm{Velocity}_1$
Integral1	$\mathrm{Integral}_1$

Figure 1.10 Using subscripts in variable names

These subscripts are referred to as literal subscripts. It is important to understand the difference between the use of literal subscripts and the array subscripts that we will talk about in Chapter 9. Remember that once you use a subscript in a variable name, you cannot return to normal text size.

Naming guideline 6

Use the (`) key if you need to use a "prime" (single apostrophe) in your variable name

Sometimes your variable name has an apostrophe in it, such as the strength of concrete, f'c. If you type the single quote key ('), then it will put parenthesis around your variable name. It does not type a single quote. In order to use the prime symbol in your calculations, use the (\`) key in the upper left hand corner of your keyboard. It is usually located on the same key as the (~) tilda key. See Figure 1.11.

(f)	If you type f followed by typing the single quote key ' , you get (f).
f'	The \` mark is usually located next to the number 1 key.
f'	If you really want to use a single quote mark, then you can use the special text feature mentioned previously to include the single quote.

Figure 1.11 Options for using a prime symbol

Summary

Variables are an important part of engineering calculations in Mathcad. Learning how to use variables effectively will make it much easier to create engineering calculations.

In Chapter 1 we:

- Learned what a variable is.
- Showed how to define a variable.
- Learned that variables are case, font, size and style sensitive.
- Discussed which characters can be used in variable names.
- Discussed string variables.
- Emphasized the importance of defining variables in Mathcad calculations.
- Gave several guidelines for naming variables.

Practice

1. Open a new Mathcad worksheet.
2. Add 10 simple variable definitions from your field of study, include some subscript names. Follow the guidelines suggested in this chapter.

3. Add 10 variable definitions that are two words or more. Use two different methods for differentiating between words.
4. Add 10 variable definitions that include characters requiring the use of the special text mode.
5. Add 10 variable definitions that use the Chemistry Notation.
6. Add 10 string variable definitions.

2
Creating and editing Mathcad expressions

The ability to quickly create and edit complicated engineering expressions is critical to effectively using Mathcad. This chapter teaches techniques to create and edit Mathcad expressions. It also discusses characteristics associated with text regions.

Chapter 2 will:

- Introduce and define regions.
- Discuss the worksheet ruler and tabs.
- Tell how to move, align and resize regions.
- Explain how to create and edit text regions.
- Explain how to create and edit math expressions.
- Introduce the editing cursor and the different forms it takes.
- Discusses the use of operators.
- Demonstrate how to wrap a math region.
- Discuss the use of Find and Replace.
- Provide several Mathcad examples to emphasize each of the items discussed.
- Encourage completing several Mathcad Tutorials.

Introduction

The purpose of this chapter is to provide you with a quick overview of how to create, edit and arrange math and text regions in your worksheet. It will reinforce the information contained in the Mathcad tutorial.

It is suggested that you read and do the exercises in the Mathcad tutorial before or just after reading this chapter. You can open the Mathcad tutorial by clicking **Tutorials** from the **Help** menu. This opens a new window called the Mathcad Resources window. In this window you will see a list of Mathcad tutorials. Click on the **Getting Started Primers**. Each of these primers is excellent. You may choose to do them all, but for the purpose of this chapter, focus on the following topics: Entering Math Expressions, Building Math Expressions, Editing Math Expressions, First Things First, and Adding Text and Images. This chapter cannot replace the experience gained by completing the Mathcad tutorials.

Regions

Definition of region

Your entire Mathcad worksheet will be comprised of individual regions. A region is a location where information is stored on the worksheet. Each variable expression you create comprises a separate region. Any text you type will be contained in a region. There are two easy ways to view the regions in your worksheet. The first is to click **Regions** from the **View** menu. This will turn the background from white to gray and the regions will be white. You will be able to clearly see each region and see how large it is. Another way to view the regions is to click with your left mouse button outside of a region. Hold the left mouse button and drag it across several regions and release the mouse button. All regions within this area will now be selected, and each will have a dashed line surrounding the region. This method is referred to as "drag select."

There may be times when one region is placed on top of another region. These methods of viewing regions will help identify when one region is overlapping another region.

Using the worksheet ruler

The worksheet ruler at the top of your worksheet can help you align regions and set tabs. To make the ruler appear, click **Ruler** from the **View** menu. Repeat the procedure for hiding the ruler.

You can change the measurement system used on the ruler by right clicking the ruler and selecting from the list of measurements: inches, centimeters, points or picas. Remember that there are 72 points per inch and 6 picas per inch. When you change the ruler measurements, some of the dialog box measurement systems change to the ruler measurement system.

Tabs

You can use tabs to help align regions in your worksheet. If you press the `tab` key prior to creating a math or text region, the region will be left aligned with a tab stop.

Mathcad defaults to tab stops of one-half inch. You can set tab stops on the worksheet ruler by clicking on the worksheet ruler at the location where you want to set the tab stop. Once a tab stop is shown on the ruler, you can adjust the tab by clicking on the tab and dragging it along the ruler. You can clear the tab stop by clicking on the tab and dragging it off the worksheet toolbar. Another way to set tab stops is by choosing **Tabs** from the **Format** menu. This opens the Tabs dialog box. From this dialog box you can clear all tab stops and set new tab stops at exact tab stop locations.

Selecting and moving regions

You can select and move a single region or multiple regions. To move a single region, drag select the region or click within the region, and then place your cursor near the perimeter of the region until the cursor changes from an arrow to a hand. Now left click and hold the mouse button. Drag the region to where you want it. Once a region is selected, you may also use the arrow keys to move the region.

To select adjacent regions, drag select the regions. To select non-adjacent regions, hold the `ctrl` key and click within each desired region. To move the selected regions, place the cursor in one of the regions, left click and hold the mouse button. Drag the regions to a new location. You may also use the arrow keys to move the regions.

Aligning regions and alignment guidelines

There will be times when you want to align different regions either vertically or horizontally. Aligned regions appear much more professional. To align regions, highlight **Align Regions** from the **Format** menu, and then select either **Across** or **Down**.

If your selected regions are roughly aligned in a horizontal row, the Across alignment will place the top of each region in a horizontal line. If your selected regions are roughly aligned in a vertical column, the Down alignment will place the left side of each region in a vertical line.

Using the **Align Regions** feature may cause regions to overlap. If a vertical line will pass through more than one of your selected regions, then the Across alignment will cause these regions to overlap. If a horizontal line will pass through more than one of your selected regions, then the Down alignment will cause these regions to overlap. In order to prevent regions from overlapping, it is important to only select regions in a roughly horizontal or roughly vertical layout. If this is not possible, move some of the regions prior to using the alignment feature.

Mathcad warns you to check your regions prior to executing the requested alignment. If you accidentally align regions that cause an overlap, you can undo the alignment, or you can select the overlapping regions and click **Separate Regions** from the **Format** menu. This will separate the regions vertically.

It was mentioned earlier that if you press the `tab` key prior to creating a region, the new region will be left aligned to the tab stop. If you did not use the tab stop when creating regions, you can still align your regions to the tab stop. Mathcad has a feature called alignment guidelines. These are green lines that extend down from the tab stops. See Figure 2.1.

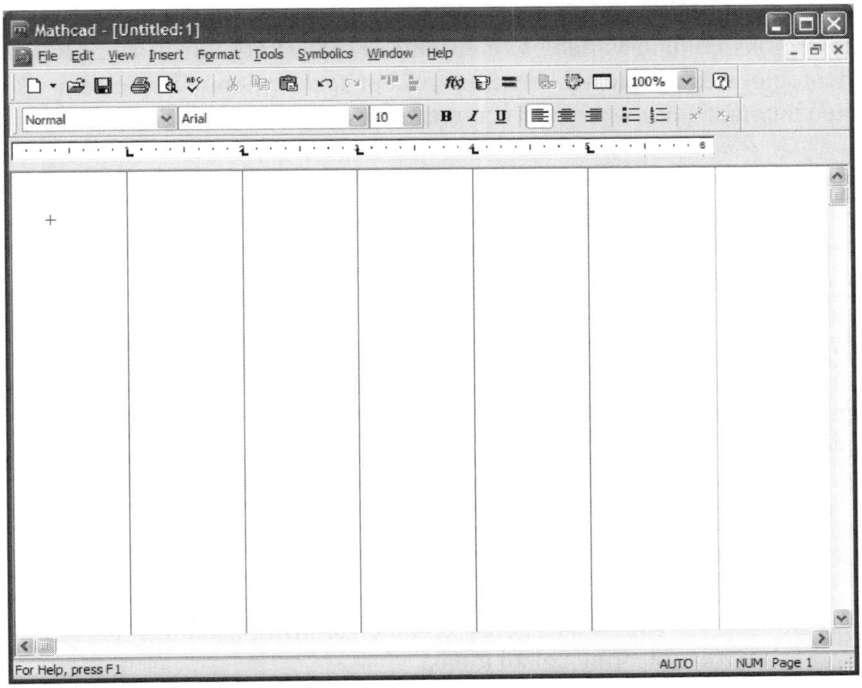

Figure 2.1 Alignment guide lines

You can move your regions to align with these alignment guidelines. To turn on guidelines for all the tab stops, open the Tabs dialog box by clicking **Tabs** from the **Format** menu. Then place a check in the "Show guide lines for all tabs" box. This will place the green guidelines at all existing tab stops. If you add additional tab stops after checking this box, you will need to repeat the procedure. To set a guideline for an individual tab stop, right click on the tab stop and select **Show Guideline**. To remove an existing guideline, right click on the tab stop and select **Show Guideline** (there should be a check next to it). You can also remove all guidelines at once by unchecking the "Show guide lines for all tabs" box in the Tabs dialog box.

After you have done a Down alignment of a selected group of regions, you can move this group of regions to align them with one of the guidelines.

Math regions

Math regions contain variables, constants, expressions, functions, plots, etc. These regions are basically anything except text regions. These regions are created automatically whenever you create any expression or definition.

When calculating worksheets, Mathcad reads from top to bottom and left to right. This means that once you define a variable such as Length := 3, you can compute with it anywhere below and to the right of the definition. This is important to remember!

Math regions can be very simple or very complex. The majority of this book discusses various math regions, so we will not discuss them in detail in this chapter. We will discuss how to create and modify math regions later in this chapter.

Text regions

Text regions allow you to add notes, comments, titles, headings, and items of interest to your engineering calculations.

Creating text regions

There are several ways to create a text region. The simplest way to create a text region is to start typing text. As soon as you use the spacebar, Mathcad converts the math region into a text region. This is a handy feature, unless you type the spacebar by accident when you are entering a variable name. Once a math region is converted to a text region, it cannot be changed back to a math region. (You may use the undo command, if you immediately catch the mistake.)

Other ways to create text regions are to use the double quote (") key, or choose **Text Region** from the **Insert** menu.

Changing font characteristics

Once you create a text region you can type text just as you would in a word processor. You can also use tabs or change font characteristics such as font type, font size or font color. You can also use such things as bold, italic, underline, strikeout, subscript, and superscript. To change the font characteristics while in a text region, highlight the text and choose **Text** from the **Format** menu.

Inserting Greek symbols

To insert Greek letters, use the Greek Symbol toolbar. You can open this toolbar by choosing **Toolbars** from the **View** menu and then selecting **Greek**. If the Math toolbar is open, you can click on the icon representing the Greek letters. You can also type a Roman letter and immediately type CTRL+g. This converts the alphabetic character to its Greek symbol equivalent.

Ending a text region

When you are finished typing the text, if you press the enter key, Mathcad inserts a new paragraph in the same text region. In order to exit a text region, click outside the region. This closes the text region. You may also type CTRL+SHIFT+ENTER, or you may use the arrow keys to move the cursor outside the text region.

Controlling the width of a text region

When you start typing in a text region, the region grows to the right until it reaches the right margin. At that point the text wraps to a new line. There are times when you do not want the text region to grow all the way to the right margin. To force the region to wrap before it reaches the right margin, type CTRL+ENTER at the point where you want the text region to wrap. The text may not immediately wrap, but when you begin typing the next word, the cursor will move to the next line. Do not use the ENTER key to change the width of text. Only use the ENTER key to add a new paragraph to a text box.

To change the width of an existing text box, place the cursor at the point where you want the text box to wrap and press CTRL+ENTER. You may also click within the text box and then move the handle on the right side of the text box. The text will wrap according to the new text box width. If you used the ENTER key to change the width of the original text region, the text will not wrap to the new width.

Moving regions below the text region

As you add text to a new text region, the region grows. You may find that the growing text region begins to overlap on top of other regions. There is a way to prevent this from occurring. To do this, right click inside the text region, click on **Properties**, select the Text tab, and place a check in the box adjacent to "Push Regions Down As You Type." Now as you type new text, or modify the width of existing text, any regions below the text region will move down or up depending on how you size the text box. See Figure 2.2.

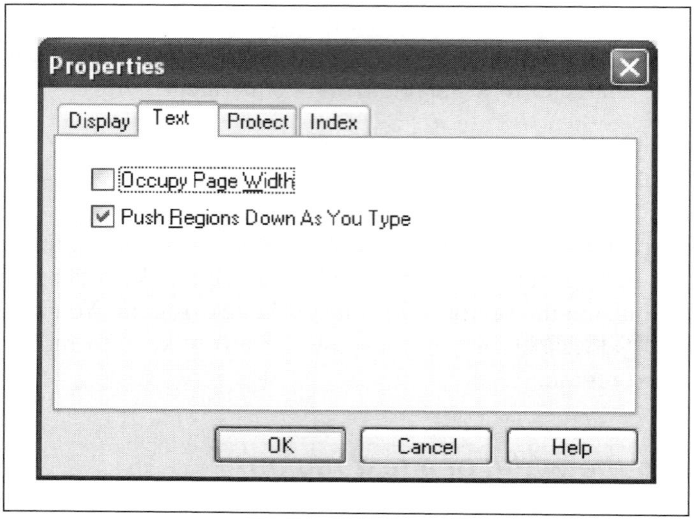

Figure 2.2 Push Regions Down As You Type check box

This feature must be set for every text region. There is not a way to set it globally. Be cautious about using this feature. If it is set, some of your math regions may be moved downward. This could cause some of your variable definitions to change or to not be recognized. This would occur if a variable definition is moved downward and an adjacent expression (to the right) uses the values from the variable definition.

Paragraph properties

A text region is similar to a simple word processor. You are able to format the text region in much the same way as you would in a word processor. We discussed earlier how to change the font characteristics of text in the text region. You can also set many paragraph characteristics such as: margins, alignment, first line indent, hanging indent, bullets, automatic numbering, tabs, etc.

To set the paragraph characteristics, click in the text box and select **Paragraph** from the **Format** menu. You may also right click in a text region and select **Paragraph** from the dropdown menu. See Figure 2.3.

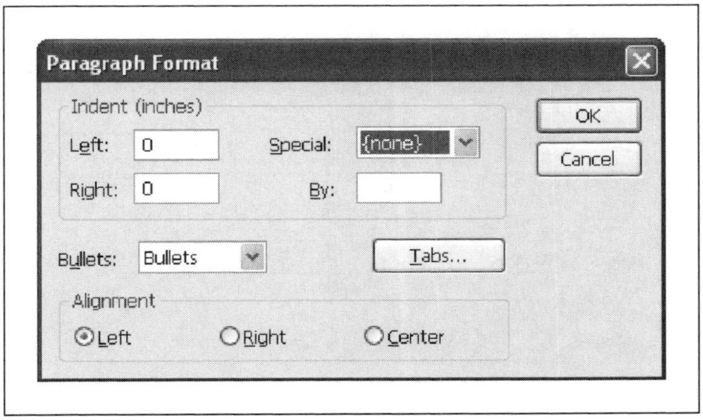

Figure 2.3 Paragraph Format dialog box

The indent boxes for left and right are based on the edges of the text region, not the page margins of your worksheet. If you set the margins to be one inch from both left and right, then as you change the size of your text region, the text will always remain one inch from the edges of your text region.

Clicking "Special" will allow you to indent the first line or allow you to have a hanging indent on your first line. After selecting **First Line** or **Hanging** Indent, then tell Mathcad how much you want to indent by changing the number in the **By** box.

The "Bullets" box allows you to use bullets or automatic numbering to each paragraph in your text region.

The **Tabs** button allows you to set tab locations just as in a word processing program. The tabs are measured from the left edge of the text region, not from the left edge of the page or page margin.

Each paragraph in the text region may have different paragraph settings. See Figure 2.4 for an example of how a text region will look with specific settings. The first paragraph has different features than the last three paragraphs.

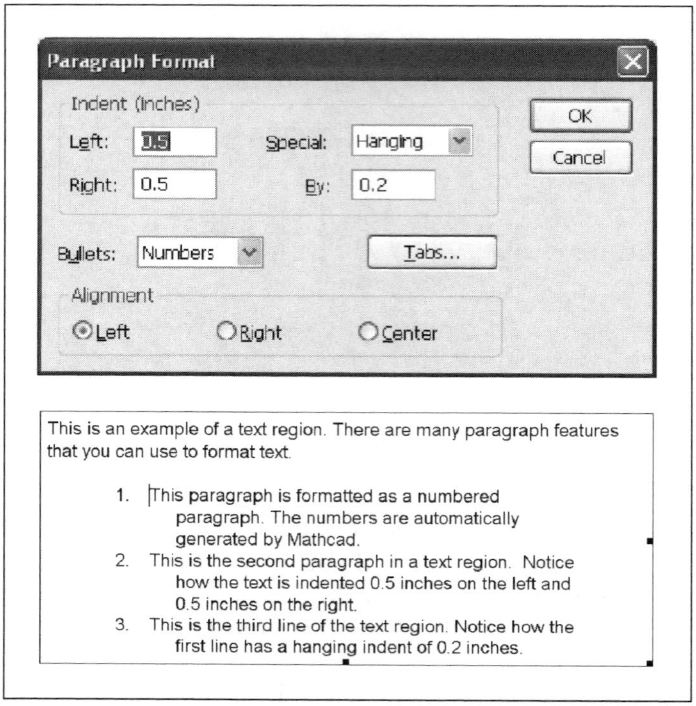

Figure 2.4 Paragraph formatting for bottom three paragraphs of text region

Text ruler

When the worksheet ruler is showing, you will also have a ruler when you are working in a text region. This text ruler changes width to match the width of the text region. The ruler begins at the left edge of the text region and extends to the right edge of the text region. From the ruler, you can set left and right margins, indents, hanging indents, and tabs. To do this, slide the left and right indent markers to the desired positions. You can also add tabs to the text ruler by clicking on the ruler. See Figure 2.5.

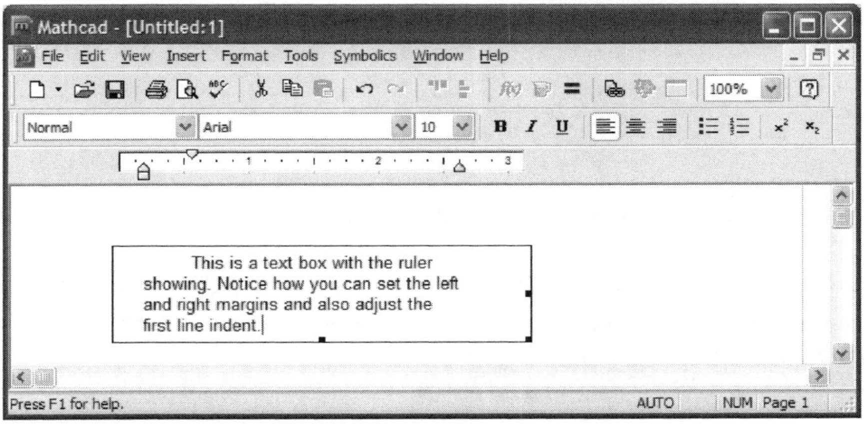

Figure 2.5 Text box with ruler turned on

Spell check

Mathcad has a built-in spell checker. The spell checker only checks the spelling in text regions, not math regions. To activate the spell checker, click **Spelling** on the **Tools** menu. If a misspelled word is found, you have the option to change to one of the suggested replacement words, ignore the suggestions, add the word to your personal dictionary, or have Mathcad offer additional suggestions.

Mathcad can check several different languages. It can also check several different dialects. For example, you can tell Mathcad to use the British English instead of the American English. To select a different language or dialect, select **Preferences** from the **Tools** menu, and then click on the Language tab. From this tab, under the "Spell Check Options," you can select a specific language, and some languages will allow you to select a specific dialect.

Creating simple math expressions

There are two ways to create a simple expression. The first way is to type an operator such as +, −, *, or /. This will create empty placeholders that you can then click to fill in the numbers or *operands*. For example, if you press the [+] key anywhere in your worksheet, you will get the following:

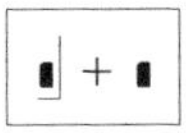

Click in the first placeholder and type 2, then press tab or click in the second placeholder and type 5. Your expression should now look like this:

$$2 + 5$$

In this example 2 and 5 are operands of the + operator.

You can use the above procedure with any operator. Let's try the exponent operator. Press ^ to create the exponent operator. You can also click x^Y on the calculator toolbar. You should have the following:

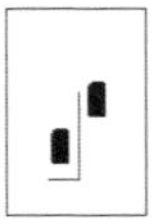

Click in the lower placeholder and type 2, then press tab or click in the upper placeholder and type 4. Your expression should now look like this:

$$2^4$$

A second way to create a simple expression is to just type as you would say the expression. For example you would say 2 plus 5, so you would type the following **2 + 5**. You would say 2 to the 4th power, so you would type **2 ^ 4**.

These methods work very well for creating simple expressions. As your expressions become more complex, there are a few things that we must learn.

Creating more complex expressions

Creating more complex math expressions is very easy once you learn the concept of the editing lines. These are similar to a two-dimensional cursor with a vertical and a horizontal component. There is a vertical editing line and a horizontal editing line. As an expression gets larger, the editing lines can grow larger to contain the expanding expression. Notice how in the previous examples the editing lines just contained a single operand. Pressing the spacebar will cause the editing lines to grow to hold more of the expression. For example, if you type `2+5 spacebar`, you get the following:

$$2 + 5$$

Whatever is held between the editing lines becomes the operand for the next operator. So, if you type `2+5 spacebar ^ 3`, you get the following:

$$(2 + 5)^{3}$$

In this case (2+5) is the x operand for the operator x to the power of y. Notice how the editing lines now only contain the number 3. This means that if you type any operator, the number 3 is the operand for the operator. Thus, if you type `+ 4`, then you get the following:

$$(2 + 5)^{3+4}$$

But, if you type the `spacebar` first, the editing lines expand to enclose the whole expression. This expression becomes the operand for the next operator. Thus, if you now type `+ 4`, you get the following:

$$(2 + 5)^{3} + 4$$

The whole expression became the operand for the addition operator.

It is very important to understand this concept of using the editing lines to determine what the operand is of your next operator. You can also use parenthesis to set the operand for operators. Pressing the single quote (`'`) adds a pair of opposing parenthesis.

The following example will help reinforce these concepts.

Let's create the following expression:

$$\frac{\left(\frac{1}{2}-\frac{1}{3}\right)^2}{\sqrt{\frac{4}{5}+\frac{2}{7}}}$$

To create this expression, use the following steps:

1. Type `1 / 2 spacebar`. The editing lines now hold the fraction 1/2. This becomes the operand for the subtraction operator.

$$\frac{1}{2}$$

2. Type `- 1 / 3 spacebar spacebar`. The editing lines should now hold both fractions. This becomes the operand for the power operator.

$$\frac{1}{2}-\frac{1}{3}$$

3. Type `^ 2 spacebar`. The editing lines should now hold the entire numerator. This becomes the operand for the division operator.

$$\left(\frac{1}{2}-\frac{1}{3}\right)^2$$

4. Type `/ \(or use the square root icon on the math toolbar) 4 / 5 spacebar spacebar`. This makes everything under the radical the operand for the addition operator.

$$\frac{\left(\frac{1}{2}-\frac{1}{3}\right)^2}{\sqrt{\frac{4}{5}}}$$

5. Type `+ 2 / 7`. This completes the example.

$$\frac{\left(\frac{1}{2}-\frac{1}{3}\right)^2}{\sqrt{\frac{4}{5}+\frac{2}{7}}}$$

Notice how during each step, the spacebar was used to enlarge the editing lines to include the operand for the following operator.

The Mathcad tutorial has additional examples that provide worthwhile practice.

Editing expressions

Another important concept to know is how to edit existing expressions.

In order to understand this concept, it is important to understand how to move the vertical editing line. This vertical editing line can be moved left and right using the left and right arrow keys. You can also toggle the vertical editing line from the right side to the left side and back by pressing the `Insert` key. For expressions that are more complex you can also use the up and down arrows to move both editing lines.

Selecting characters

If you click anywhere in an expression and then press the spacebar, the editing lines expand to include more and more of the expression. How the editing lines expand depend on where you begin and on what side the vertical editing line is on. The editing lines work differently in different versions of Mathcad. The best way to understand how they work is to experiment and to follow the examples in the Mathcad Tutorial.

Tip! I have found that if you begin with the vertical editing line on the right side of the horizontal editing line, the expansion of the editing lines makes more sense. The general rule is that as the editing lines expand and cross an operator, the operand for that operator is then included within the lines.

Deleting characters

You can delete characters in your expressions by moving the vertical editing line adjacent to the character. If the vertical editing line is to the left of the character, press the `Delete` key. If the vertical editing line is to the right of the character, press the `Backspace` key.

To delete multiple characters, drag-select the portion of the expression you want to delete. If the vertical editing line is to the left of the highlighted area, press the `Delete` key. If the vertical editing line is to the right of the highlighted area, press the `Backspace` key.

Deleting and replacing operators

To replace an operator, place the editing lines so that the vertical editing line is just to the left of the operator. Next, press the `Delete` key. This will delete the operator usually leaving a hollow box symbol where the operator used to be. Now, type a new operator, and it will replace the symbol. See Figure 2.6.

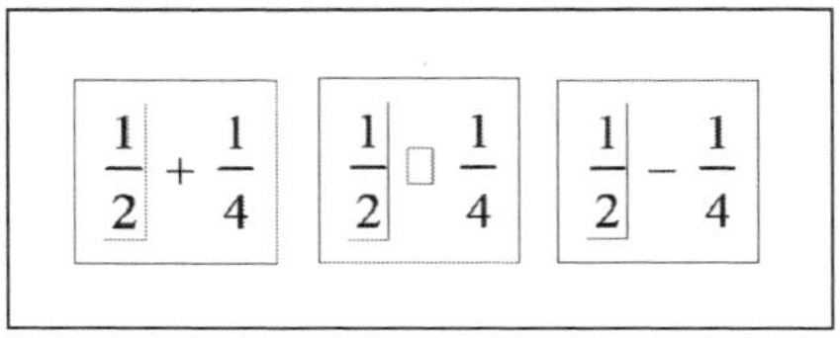

Figure 2.6 Replacing an operator

You may also have the vertical editing line to the right of the operator and use the backspace key to delete and replace the operator.

The best way to understand this concept is to experiment with it.

Wrapping equations

There are times when a very long expression may extend beyond the right margin. If this is the case, then the entire expression will not print on the same sheet of paper.

There is a way to wrap your equations so that they are contained on two or more lines; however, you are only able to wrap equations at an addition operator.

To wrap an equation, type **CTRL+ENTER** just prior to an addition operator. Mathcad inserts three dots indicating that the expression is to be continued on a following line, and on the following line, Mathcad inserts the addition operator with a placeholder box. Because Mathcad automatically inserts the addition operator, you are not able to wrap an equation at other operators.

> **Tip!** You may wrap an equation at a subtraction operator by making the following operand a negative number (in essence adding a negative number).

See Figure 2.7 for examples of wrapping equations.

$$\text{Example}_1 := 1 + 2 + 3 + 4 + 5 + 6 \ldots \\ + 7 + 8 + 9$$

Wrapped Equation

$$\text{Example}_1 = 45$$

$$\text{Example}_2 := 1 + 5 + 3 + 4 \ldots \\ + -3 + 2$$

Wrapped Equation using negative number.

$$\text{Example}_2 = 12$$

$$\text{Example}_3 := \begin{pmatrix} 1 + 2 + 3 + 4 + 5 + 6 \ldots \\ + 7 + 8 + 9 \end{pmatrix} \cdot 2$$

Multiplication using wrapped equation.

$$\text{Example}_3 = 90$$

Figure 2.7 Wrapping equations

Find and replace

Find

The Find and Replace features can easily help you to either find or replace variables or text in your worksheets. To use these features click on the **Edit** menu.

Let's first look at the Find dialog box. See Figure 2.8.

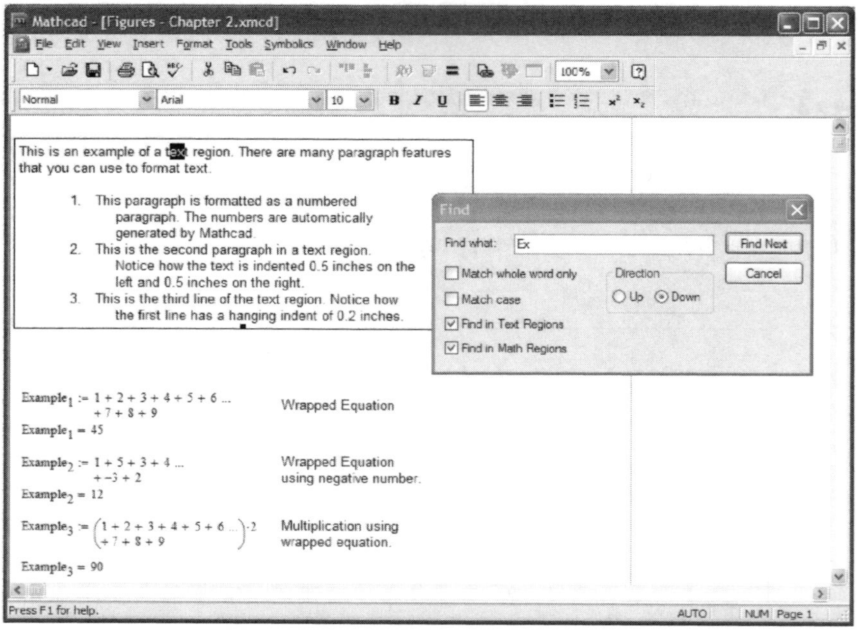

Figure 2.8 Using the Find dialog box

Type what you want to find in the "Find what" box. If you are searching for Greek letters, type \ followed by the Roman equivalent letter. If you are searching for a tab, use ^t. If you are searching for a return, use ^p.

In Figure 2.8, we typed "Ex." Now let's look at how the different check boxes affect how Mathcad finds things.

You can make Mathcad look in only text regions or only math regions. In order to look in both text regions and math regions it is important to place a check in both the "Find in Text Regions" and "Find in Math Regions" check boxes. The other check boxes tell Mathcad to match the whole word or to match the case.

In Figure 2.8 we checked only the bottom two check boxes. This means that Mathcad will find all instances of "Ex"—uppercase and lowercase—in both the text regions and math regions. In Figure 2.8, Mathcad found the "ex" in "text." It will also find the "ex" in "example" and "Ex" in "Example."

If we checked the top box, "Match whole word only," Mathcad will not find any instances of "Ex" because all instances of ex are contained in other words. If we uncheck the top box and check the next box, "Match case," Mathcad will only find the variables using the word "Example" in the math regions.

If we change the "Find what" from "Ex" to "ex," and leave the "Match case" box checked, then Mathcad will find the instances of "ex" in the text regions, but will not find the variables using the word "Example" in the math regions.

Tip! The "Find in Math Regions" is unchecked by default. Make sure to check this box if you are searching for variables.

Replace

The Replace dialog box is identical to the Find dialog box, except that there is a new line "Replace with." In Figure 2.9, we will find all instances of "Example" and replace them with "WrappingExample." It is important to make sure that the top box, "Match whole word only" is not checked, otherwise Mathcad will not

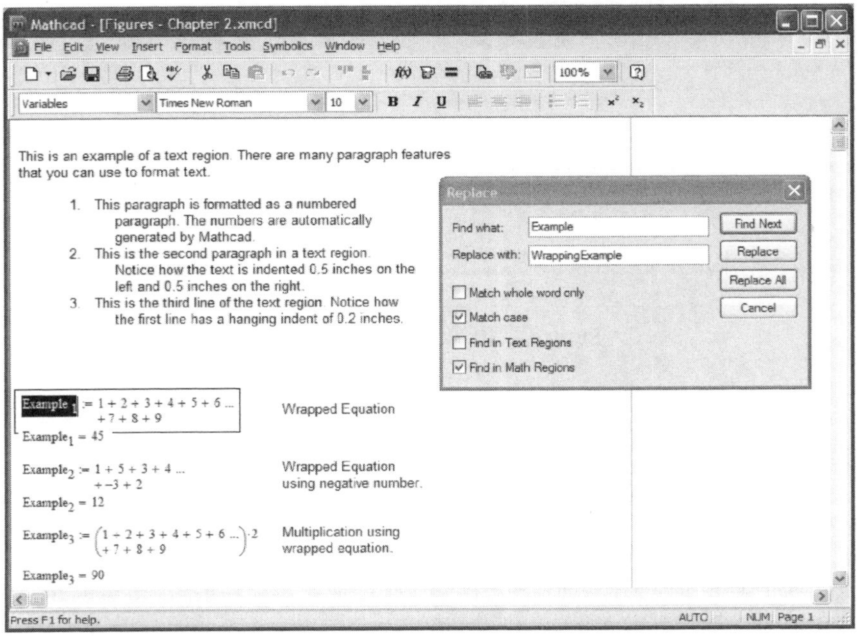

Figure 2.9 Using the Replace dialog box

find the variables such as "Example$_1$." We also checked the "Match case" box, but in our case, it really would not matter because the "Find in Text Regions" box is unchecked.

Figure 2.10 shows what happened after we replaced all instances of the variable "Example" with "WrappingExample," Mathcad quickly replaced these. Because we replaced the variable name with a longer variable name, the math regions are now overlapping the text regions. If this happens, you can quickly select the text regions and move them to the right as we discussed earlier in the chapter.

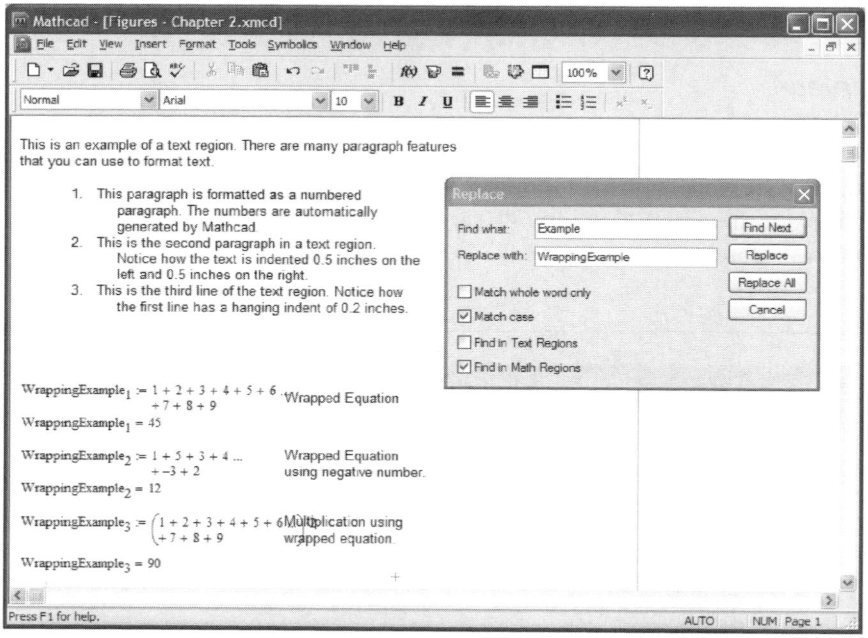

Figure 2.10 After using Replace

Inserting and deleting lines

It is possible to move regions down and create more blank space for new regions. You can do this by holding down the **ENTER** key, or you can right click above the region you want to move and select "Insert Lines." This opens the Insert Lines dialog box, where you can input the number of lines to insert. When pasting new information into Mathcad, it is wise to insert blank spaces between the area where you are pasting the information.

If you have extra space in between regions, you can delete the blank lines. To do this, place your cursor at the top of the blank space and use the **DELETE** key to delete the blank lines. You can also right-click and select "Delete Lines." From here, use the up arrow button. Keep clicking the up arrow until the line numbers stop increasing. Mathcad knows how many blank lines are between the cursor and the next region. Clicking OK will delete the blank spaces and move the regions together.

If you want to force a page break in your worksheet, press **CTRL+ENTER**. This places a horizontal line in your worksheet indicating the location of the page break. You can drag this page break indicator up or down in your worksheet.

Summary

The intent of this chapter was to introduce you to text regions and math regions. It is not a complete discussion of all associated concepts. The best way to gain an understanding of these concepts is to practice. If you have not done so already, open the Mathcad tutorials and go through the **Getting Started Primers** mentioned at the beginning of this chapter.

In Chapter 2 we:

- Introduced the concept of regions.
- Discussed moving and aligning regions.
- Described the attributes of text regions.
- Explained paragraph properties in text regions.
- Illustrated creating and editing math expressions.
- Encouraged the use of the Mathcad Tutorials to practice the concepts discussed in this chapter.

Practice

Enter the following equations into a Mathcad worksheet.

$$\frac{-b+\sqrt{b^2-4\cdot a\cdot c}}{2\cdot a}$$

$$\frac{F_0}{\sqrt{m^2\cdot\left(\omega_0{}^2-\omega^2\right)^2+b^2\cdot\omega^2}}$$

$$\left(\frac{2}{3}-\frac{F_y\left(\frac{1}{r_T}\right)^2}{1530x10^3\cdot C_b}\right)\cdot F_y$$

$$\frac{-1}{2}d_2+\sqrt{\frac{2\cdot v_2{}^2\cdot d_2}{g}+\frac{d_2{}^2}{4}}$$

1. Give each of the above equations a variable name. Assign variable names and a value to the variables used in the equations. These variable assignments will need to be made above the equation definition. Show the result. Change some of the input variable values and see the impact they have on the results.
2. Choose 10 equations from your field of study (or from a physics book) and enter them into a Mathcad worksheet. Assign the variables that the equation needs prior to entering the equation. Select appropriate variable names. Don't select easy equations. Pick long complicated formulas that will give you some practice entering equations.
3. Choose some of the above equations and change some of the operators in the equation.
4. Chose some of the above equations and make the equation wrap at an addition or subtraction operator.
5. Create a text region and write two or three paragraphs about the things you learned in this chapter. After creating the paragraphs, change some of the paragraph characteristics and font characteristics. Use the spell checker to check for spelling errors.
6. Use the Find and Replace features to search for and replace certain text and math characters.

3
Simple functions

Mathcad has hundreds of built-in functions. This chapter introduces these built-in functions and describes their basic use. It also introduces the use of user-defined functions.

The power of using user-defined functions is not realized even by many long-time Mathcad users. In some instances, user-defined functions can be confusing and complicated, causing many users to ignore using them. After briefly discussing built-in functions, this chapter will focus on simple user-defined functions. The goal of this chapter is to make you comfortable with the concept and use of simple user-defined functions.

Chapter 3 will:

- Introduce built-in functions.
- Show how to insert built-in functions into your worksheet.
- Discuss what an "argument" is.
- Introduce user-defined functions.
- Show different types of arguments.
- Give examples of different function names and argument names.
- Tell when to use a user-defined function.
- Describe how to use variables in user-defined functions.
- Give examples of user-defined functions in technical calculations.
- Provide warnings about the use of functions.

Built-in functions

Built-in functions range from very simple to very complex. Examples of simple built-in Mathcad functions are: ***sin()***, ***cos()***, ***ln()***, and ***max()***. Every Mathcad function is set-up in a similar way. The function name is given, followed by a pair of parenthesis. The information that is typed within the parenthesis is called the argument. Every function has a name and an argument. The function takes the information from the argument (contained within the parenthesis) and processes the information based on rules that are defined for the specific function, returning a result.

To see a list of all the built-in functions that Mathcad has, select **Function** from the **Insert** menu. A dialog box will appear that lists all of the built-in functions of Mathcad. See Figure 3.1. The functions are grouped by category in the left column. The functions assigned to the highlighted category appear on the right. If you know the function name, you can search for the function by clicking on **All** in the left column and then clicking in the right column and typing the function name. The function will then be highlighted.

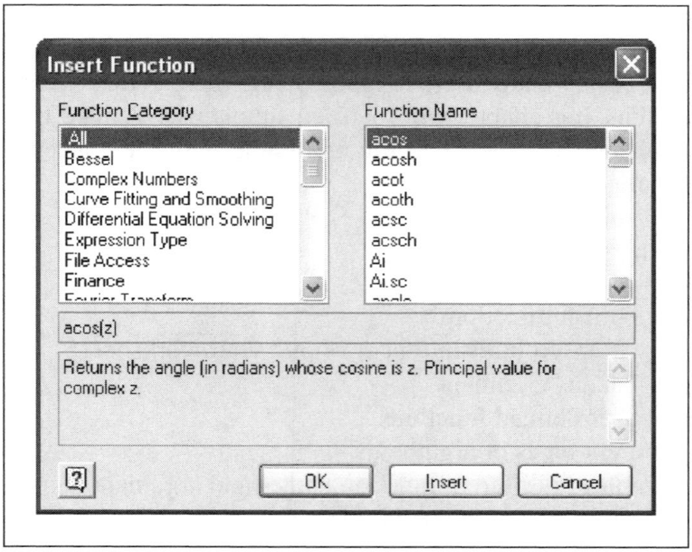

Figure 3.1 Insert Function dialog box showing the Function Categories and Function Names

Once a function is selected in the Insert Function dialog box, you will see some useful information in boxes at the bottom of the dialog box. The upper box shows a list of arguments (within the parenthesis) that the function is expecting.

Some functions expect only a single argument. Other functions expect two arguments. Some functions require multiple arguments. There are some functions that can have a variable number of arguments. These will be indicated by three dots following the listed arguments. Figure 3.2 shows four functions with different numbers of arguments.

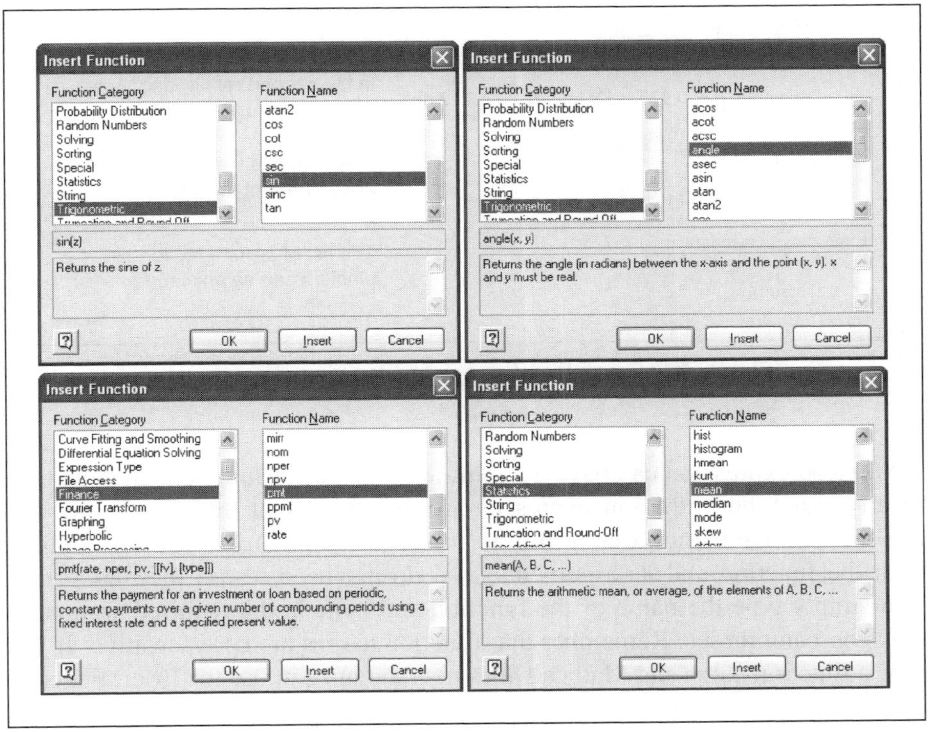

Figure 3.2 Note the two boxes below the Function Name. The first box shows the arguments. The second box describes the function. Note that some functions require only one argument. Other functions require multiple arguments. Some arguments allow unlimited arguments

The lower box contains a description of what the function does. This describes what Mathcad will return when the arguments are included in the function list. It also describes what type of information the function is expecting (such as if the argument must be in radians, or whether it must be an integer).

Take a moment to scan the complete list of Mathcad's built-in functions from the Insert Function dialog box. In later chapters, we will discuss selected functions. If you are interested in knowing about a specific function refer to Mathcad Help.

Figure 3.3 shows examples of using four different Mathcad functions requiring various argument lengths.

$\text{Example}_{1a} := \sin\left(\frac{\pi}{4}\right)$

$\text{Example}_{1a} = 0.707$

$\text{Example}_{1b} := \text{angle}(3,4)$

$\text{Example}_{1b} = 0.927 \quad \text{Example}_{1b} = 53.13\,\text{deg}$

In the second result "deg" was typed in the placeholder

$\text{Example}_{1c} := \text{pmt}\left(\frac{10\%}{12}, 36, 10000, 0, 0\right)$

See Mathcad Help for information about the pmt function

$\text{Example}_{1c} = -322.672$

$\text{Example}_{1d} := \text{mean}(1,3,5,7,10)$

The list of arguments for this function can be any length.

$\text{Example}_{1d} = 5.2$

Figure 3.3 Examples of using various types of arguments

A Mathcad expression can have an unlimited number of functions included. To insert a function within an expression, simply highlight the placeholder and use the Insert Function dialog box to insert the function. If you are familiar with the function and its arguments, you do not need to use the dialog box. You simply type the name of the function and include the required arguments between parentheses. Remember that function names are case sensitive. If you type a function name and Mathcad does not recognize it, use the Insert Function dialog box to see if the first letter is upper case or lower case. Figure 3.4 shows an expression using multiple built-in functions.

$\text{Example}_2 := 1 + \cos\left(\frac{\pi}{2} + 7\tan\left(\frac{\pi}{4}\right)\right) + \sin(\pi)$

$\text{Example}_2 = 0.343$ This is the default result

$\text{Example}_2 = 0.343\,\text{rad}$ Result with "rad" typed in placeholder

$\text{Example}_2 = 19.653\,\text{deg}$ Result with "deg" typed in placeholder

Figure 3.4 Example of using built-in functions within an expression

User-defined functions

User-defined functions are very similar to built-in functions. They consist of a name, a list of arguments, and a definition giving the relationship between the arguments. The name of the user-defined function is simply a variable name. The same rules that apply to naming variables also apply to naming functions. Refer to Chapter 1 for a listing of the naming rules. The arguments used in the user-defined function do not need to be defined previously in your worksheet. The function is simply telling Mathcad what to do with the function arguments, thus the arguments are not defined prior to using the function.

The following is a simple user-defined function: $\mathrm{SampleFunction}(x) := x^2$

In the above example, "SampleFunction" is the name of the function, and "x" is the argument. When you type: `SampleFunction(2)=`, Mathcad takes the value of the argument (2), and applies it everywhere there is an occurrence of the argument in the definition. So in the above example, Mathcad replaces "x" with the number "2" and squares the number, returning the value of 4. See Figure 3.5 for examples of this function with various arguments. Notice that the argument may also be the result of an expression. Any expression is allowed as long as the result of the expression is a value that is expected by the function. The beauty of this is that you don't need to calculate the value of the function argument if it is the result of another equation.

$\mathrm{SampleFunction}(x) := x^2$	Function definition
$\mathrm{SampleFunction}(2) = 4$	Simple positive argument
$\mathrm{SampleFunction}(-4) = 16$	Simple negative argument
$\mathrm{SampleFunction}(\sqrt{16}) = 16$	Argument using the result of a built-in function
$\mathrm{SampleFunction}\left(\frac{1}{2}\right) = 0.25$	Argument using the result of an expression
$\mathrm{SampleFunction}(2\cdot 4) = 64$	Argument using the result of an expression
$\mathrm{SampleFunction}\left(4\cdot\sin\left(\frac{\pi}{4}\right)\right) = 8$	Argument using the result of an expression with a built-in function.

Figure 3.5 Sample Function with various arguments

Let us look at some sample user-defined functions. See Figure 3.6. Notice the different types of function names and function arguments. It doesn't matter what letter or combination of letters you use for the argument.

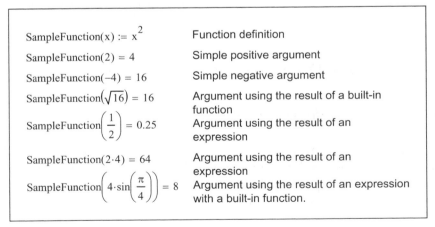

$f(a) := a^{\frac{1}{3}}$ $f(27) = 3$	Function names can be a single letter.
$F_2(a) := 2 \cdot a^2 + a + 3$ $F_2(3) = 24$	Note that the "a" in this function is totally independent of the "a" in the function above. Each argument applies only to the defined function.
$\text{ExampleFunction}_2(\text{Dog}) := 2^{\text{Dog}}$ $\text{ExampleFunction}_2(4) = 16$	Arguments are not limited to single letters
$\text{AnotherFunction}(x2) := \sqrt{x2 + 8}$ $\text{AnotherFunction}(8) = 4$	Arguments may also consist of letters and numbers. This is not recommended as it appears that the argument is "x" multiplied by 2, rather than "x2".

Figure 3.6 Examples of various function names and arguments

You can even use characters as arguments if you switch to the special text mode discussed in Chapter 1 (**CTRL+SHIFT+k**). See Figure 3.7.

$\text{SampleFunction}_3(@) := @^2 + 2$ $\text{SampleFunction}_3(10) = 102$	Type the following to get this function: SampleFunction.3 (**[Shift]+[Ctrl]+k** @ **[Shift]+[Ctrl]+k**): **[Shift]+[Ctrl]+k** @ **[Shift]+[Ctrl]+k** ^2 + 2" This takes many key strokes. The advantage of using a symbol as an argument must outweigh the extra effort to add the symbol.

Figure 3.7 Example of using a symbol as the argument

Why use user-defined functions?

Once a user-defined function is defined, it can be used over and over again. This makes user-defined functions very useful and powerful. User-defined functions are useful if you are repeatedly doing the same steps over and over again in your calculations. This is illustrated in Figure 3.8.

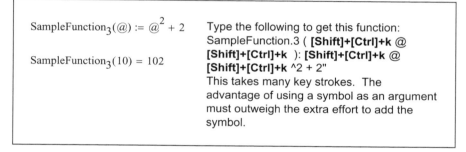

Suppose that you need to find the result of $x^3 + 2 \cdot x^2 - x + 3$ for many different values of x. You can setup this expression: $y := x^3 + 2 \cdot x^2 - x + 3$. You can then define x:=1 and then type y=. You could then redefine x to be x:=5 ,and then type y= to get another value of y. This method works, but it takes time and is not convenient because you must continuously redefine "x".

$x := 1$

$y := x^3 + 2 \cdot x^2 - x + 3 \qquad y = 5$

You will need to redefine "x" to be another number in order to get a result for a different value of x. Your original answer using x:=1 will also be lost.

(In later chapters we will discuss range variables and arrays which will make the above scenario easier.)
Another way to solve the above situation is to define a function. After the function is defined, you can type the function using numerous values for the argument.

$y(x) := x^3 + 2 \cdot x^2 - x + 3$

$y(1) = 5 \qquad y(10) = 1.193 \times 10^3$

$y(5) = 173 \qquad y(100) = 1.02 \times 10^6$

Note the wavy line beneath the "y" indicating that the variable "y" is being redefined. In this case "y" is changed from an expression to a function.

Figure 3.8 Using a Function

Using multiple arguments

Until now, we have been using a single argument in the argument list. Mathcad allows you to have several arguments in the argument list. These are illustrated in Figure 3.9.

$\text{MultipleArgument}_1(a,b) := a^2 + b^2$

$\text{MultipleArgument}_1(2,3) = 13$

$\text{MultipleArgument}_2(dog,cat,goat) := \dfrac{\sqrt{dog} + 2^{cat}}{goat}$

$\text{MultipleArgument}_2(4,3,2) = 5$

Figure 3.9 Using multiple arguments

Variables in user-defined functions

You are also allowed to use a variable in your function definition that is not a part of your argument list; however, this variable must be defined prior to using it in your user-defined function. The value of the variable—at the time you define your user-defined function—becomes a permanent part of the function.

If you redefine the variable below this point in your worksheet, the function still uses the value at the time the function was defined. This case is illustrated in Figures 3.10 and 3.11.

$Input_1 := 4$

$VariableExample(V) := Input_1 \cdot V$

$VariableExample(3) = 12$

The value of "Input1" in the above function remains the value at the point at which the function was defined (4).

$Input_1 := 8$ Redefine the value of "$Input_1$".

$VariableExample(3) = 12$

In the above example, "$Input_1$" was redefined from 4 to 8. This did not affect the result of the function. The value of "$Input_1$" in the function depends on the value of "$Input_1$" at the point at which the function was defined (4), not the new value of "$Input_1$" (8).

Figure 3.10 Using variables in user-defined functions

$Input_1 := 4$ (Note: The wavy lines are because the variable names were used previously in Figure 3.10.)

$Input_1 := 8$ Redefine the value of "$Input_1$".

$VariableExample(V) := Input_1 \cdot V$

$VariableExample(3) = 24$

In this example, the value of "$Input_1$" in the function IS changed. This is because the value of "$Input_1$" was changed from 4 to 8 PRIOR to the definition of the user-defined function.

Figure 3.11 Changing the value of a variable in a user-defined function

When you evaluate a previously defined user-defined function, the values you put between the parentheses in the argument list may also be previously defined variables. This is illustrated in Figures 3.12–3.14.

$\text{Distance}(a,b) := \sqrt{a^2 + b^2}$

$\text{Distance}(3,4) = 5$

$x_1 := 3 \quad y_1 := 4$

$x_2 := 5 \quad y_2 := 12$

$x_3 := 20 \quad y_3 := 25$

$\text{Distance}(x_1, y_1) = 5$

$\text{Distance}(x_2, y_2) = 13$

$\text{Distance}(x_3, y_3) = 32.016$

The argument list can use previous defined variables

Figure 3.12 Using variables in the argument list

$B := 10$	This defines variable "B".
$\text{Function}_1(B) := B^2$	The argument "B" in this function is independent of the value of variable "B" above. In this case, "B" defines the argument of the function "Function_1". It does not depend on the value of variable "B" above.
$\text{Function}_1(3) = 9$	Uses 3 as the argument of "Function_1". Result of "Function_1" is 3^2.
$\text{Function}_1(B) = 100$	Uses variable "B" as the argument of "Function_1" Result of "Function_1" is 10^2
$B := 12$	Redefines variable "B"
$\text{Function}_1(B) = 144$	Result of the function changes because the value of variable "B" has changed. Result of "Function_1" is 12^2

Figure 3.13 Note the difference between variable "B" and argument "B"

$C_1 := 2$	This defines variable "C_1".
$\text{Function}_2(D) := C_1^{D}$	In this function "C_1" is a fixed variable from outside the function, and "D" is the argument of the function (taken from the argument list).
$\text{Function}_2(3) = 8$	Result of "Function_2" is 2^3. C_1=2, Argument D=3.
$\text{Function}_2(C_1) = 4$	Result of "Function_2" is 2^2. The variable "C_1" becomes the value of the argument in "Function_2". C_1=2, Argument D=2 (Value of variable "C_1").
$C_1 := 4$	Redefine variable "C_1".
$\text{Function}_2(C_1) = 16$	Result of "Function_2" is 2^4. C_1=2 (for the function), Argument D=4 (New value of variable "C_1"). Remember that for the function, "C_1" remains the original value at the time that the function was defined, not the redefined value of "C_1".

Figure 3.14 Note how "C_1" is captured in the function and how it also becomes the argument "D"

Examples of user-defined functions

Let's now look at two examples of how user-defined functions can be used in technical calculations. See Figures 3.15 and 3.16.

Write a function to calculate the area of a trapezoid.

$$\text{Trap}_{\text{Area}}(B1,B2,h) := \frac{1}{2}\cdot h\cdot(B1+B2) \qquad \text{Trap}_{\text{Area}}(5,10,18) = 135 \qquad \text{Trap}_{\text{Area}}(8,0,9) = 36$$

Write a function to calculate the center of gravity of a trapezoid.

$$\text{Trap}_{\text{CG}}(B1,B2,h) := \frac{h\cdot(2\cdot B2+B1)}{3\cdot(B1+B2)} \qquad \text{Trap}_{\text{CG}}(5,10,18) = 10 \qquad \text{Trap}_{\text{CG}}(8,0,9) = 3$$

In order to use these functions and results in technical calculations, you may want to assign the input values and results to variable names such as these:

$T1_{B1} := 5 \quad T1_{B2} := 10 \quad T1_{\text{Height}} := 18$

$T2_{B1} := 8 \quad T2_{B2} := 0 \quad T2_{\text{Height}} := 9$

Input values to be used in functions.

$T1_{\text{Area}} := \text{Trap}_{\text{Area}}(T1_{B1}, T1_{B2}, T1_{\text{Height}}) \qquad T1_{\text{Area}} = 135$

$T1_{\text{CG}} := \text{Trap}_{\text{CG}}(T1_{B1}, T1_{B2}, T1_{\text{Height}}) \qquad T1_{\text{CG}} = 10$

$T2_{\text{Area}} := \text{Trap}_{\text{Area}}(T2_{B1}, T2_{B2}, T2_{\text{Height}}) \qquad T2_{\text{Area}} = 36$

$T2_{\text{CG}} := \text{Trap}_{\text{CG}}(T2_{B1}, T2_{B2}, T2_{\text{Height}}) \qquad T2_{\text{CG}} = 3$

Results are assigned to variable names. These results can now be used later in the calculations.

Figure 3.15 Finding the area and center of gravity of a trapezoid

Write a function to calculate the surface area of a right circular cylinder.

$\text{Cylinder}_{\text{Area}}(r,h) := 2\cdot\pi\cdot r\cdot h + 2\cdot\pi\cdot r^2$

$\text{Cylinder}_{\text{Area}}(4,12) = 402.124$ $\quad\quad$ $\text{Cylinder}_{\text{Area}}(3,10) = 245.044$

Write a function to find the volume of a right circular cylinder.

$\text{Cylinder}_{\text{Volume}}(r,h) := \pi\cdot h\cdot r^2$

$\text{Cylinder}_{\text{Volume}}(4,12) = 603.186$ $\quad\quad$ $\text{Cylinder}_{\text{Volume}}(3,10) = 282.743$

In order to use these functions and results in technical calculations, you may want to assign the input values and results to variable names such as these:

$r_1 := 4$ $\quad\quad$ $h_1 := 12$ $\quad\quad$ Input values to be used in functions.

$r_2 := 3$ $\quad\quad$ $h_2 := 10$

$\text{Area}_1 := \text{Cylinder}_{\text{Area}}(r_1,h_1)$ $\quad\quad$ $\text{Area}_1 = 402.124$

$\text{Volume}_1 := \text{Cylinder}_{\text{Volume}}(r_1,h_1)$ $\quad\quad$ $\text{Volume}_1 = 603.186$

$\text{Area}_2 := \text{Cylinder}_{\text{Area}}(r_2,h_2)$ $\quad\quad$ $\text{Area}_2 = 245.044$

$\text{Volume}_2 := \text{Cylinder}_{\text{Volume}}(r_2,h_2)$ $\quad\quad$ $\text{Volume}_2 = 282.743$

Results are assigned to variable names. These results can now be used later in the calculations.

Figure 3.16 Finding the surface area and volume of a right circular cylinder

Warnings

Functions are very useful in technical calculations. You need to become familiar with their use. There are several warnings that will help make the use of functions more effective.

- Be careful not to redefine a user-defined function. User-defined function names are similar to variables. If you redefine a function, it no longer works. See Figure 3.17.

$h(x) := x^2 + x + 1$ $\quad\quad$ Defines function "h".

$h(2) = 7$

$h := 3$ $\quad\quad$ Redefines "h" as a variable

$h(2) = ■$ $\quad\quad$ Function "h" is no longer recognized, because it was redefined as variable "h".

This value must be a function, but has the form: Unitless.

Figure 3.17 Be careful not to overwrite user-defined functions

- Be careful not to redefine a built-in function. See Figure 3.18.

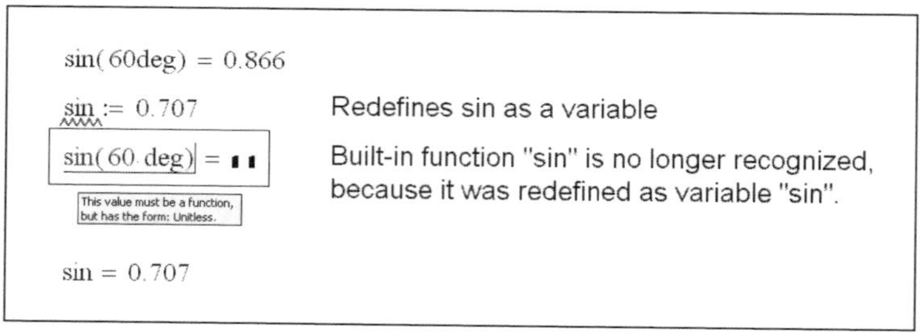

Figure 3.18 Be careful not to overwrite built-in functions

- Remember that if you use a variable in your function definition, the value of the variable does not change, even if you later rename the variable.
- Once you define your user-defined function, you will not be able to display the function definition again in your worksheet. If you type the name of the function and an equal sign, Mathcad does not display the function. See Figure 3.19. This may make it difficult to remember exactly what the function definition was. You must also remember what order the arguments go in. One option for displaying the user-defined function definition later in your worksheet is to copy the function definition and paste the definition where you want it displayed. This essentially redefines the function to the same definition. Now right-click within the pasted version of the definition and click on **Disable Evaluation**. This will disable the redefinition of the user-defined function. You may wonder, why worry about it? You get the same answer as before because it is the same definition. The problem comes if you later redefine the original user-defined function. Mathcad will use the value of the changed function until it comes to the point in your worksheet where you pasted the original function. From this point on, Mathcad will use the old function definition. By disabling the function definition, Mathcad continues to use the new version of the function. However, Mathcad still displays an old version of the function. This may lead to some confusion in your calculation, but not an incorrect answer.

$g_1(x) := x^2$ Function definition

g_1 = f(any1) => any1^2 Mathcad does not display the original definition

In order to display the function definition at a later point in your worksheet, copy the origial definition and paste it where you want it displayed. Right click within the definition and click "Disable Evaluation." WARNING: BE SURE TO DISABLE THE PASTED FUNCTION! You do not want to have two definitions of the same function in your worksheet. If you change the original function, you may forget to change the other copies of the function. This will result in incorrect results in your calculations.

$g_1(x) := x^2$ ■ This is a disabled definition. Mathcad ignores the definition. If you want to display the function after it is defined be sure to disable the definition. WARNING: If the original function is changed, this display will be WRONG.

$g_1(3) = 9$ The result of this fuction is based on the original definition, not the disabled definition. If the original function is changed, this result will also change.

Figure 3.19 How to display a function definition later in your calculations

Summary

We have just scratched the surface of user-defined functions. The intent of this chapter was to introduce the concepts of simple built-in functions, and to get you comfortable with the concept and use of user-defined functions. There is much more to learn. User-defined functions will be discussed in much more detail in later chapters of this book.

In Chapter 3 we:

- Discussed Mathcad's built-in functions.
- Introduced user-defined functions.
- Explained when a user-defined function can be used.
- Showed how to include multiple arguments and variables in user-defined functions.
- Issued several warning about using built-in and user-defined functions.

Practice

Note: Save your worksheet with the following user-defined functions for use with the practice exercises in Chapter 4.

1. Write user-defined functions to calculate the following. Choose your own descriptive function names. These functions will have a single argument. Evaluate each function for two different input arguments.

   [a] The volume of a circular sphere with a radius of R1.
   [b] The surface area of a sphere with a radius of R1.
   [c] Converting degree Celsius to degree Fahrenheit.
   [d] Converting degree Fahrenheit to degree Celsius.
   [e] The surface area of a square box with length L1.

2. Write user-defined functions to calculate the following. Choose your own descriptive function names. These functions will have multiple arguments. Evaluate each function for two sets of input arguments.

   [a] The inside area of a pipe with an outside diameter of D1 and thickness of T1.
   [b] The material area of a pipe with an outside diameter of D1 and thickness of T1.
   [c] The volume of a box with sides: L1, L2, and L3.
   [d] The surface area of a box with sides: L1, L2, and L3.
   [e] The distance traveled by a free falling object (neglecting air resistance) with an initial velocity, acceleration, and time: V0, a, and T.

3. From your area of study (or from a physics book) write 10 user-defined functions. At least 5 of these should be with multiple arguments. Evaluate each function for two sets of input arguments.

4
Units!

One of the most powerful features of Mathcad is its ability to attach units to numbers. Mathcad units will be one of your best friends as an engineer. Not only will Mathcad units simplify your life, they are one of the best means you have of catching mistakes. You may remember the 1999 Mars Climate Orbiter that ended in disaster by burning-up in the atmosphere of Mars. Why? Because engineers failed to convert English measures of rocket thrust into metric measures. Using Mathcad units will help prevent problems like this from occurring.

Chapter 4 will:

- Introduce units.
- Discuss unit dimensions, units, and the default unit systems.
- Show how to assign units to numbers.
- Discuss units of force and units of mass.
- Explain how to create custom units.
- Describe how to use units in equations and functions.
- Present ways of dealing with units in empirical formulas.
- Demonstrate how to use custom scaling units.
- Illustrate the use of custom dimensionless units.

Introduction

Once a unit is attached to a variable, Mathcad keeps track of it internally and displays the unit automatically. You will never need to remember the conversion factors for various units. You will never need to convert it from one unit system to another. Mathcad does it all for you. All you need to do is tell Mathcad how you want the unit displayed. For example, you can attach the unit of meters (m) to a variable. You will then be able to tell Mathcad to display this variable in any unit of length such as: millimeters (mm), centimeter (cm), kilometers (km), inches (in), yards (yd), or miles (mi). Mathcad does the conversion for you. If Mathcad does not have the unit of measurement built-in, you can define it, and use it over and over.

If you use units consistently, Mathcad will alert you to problems in your expressions. Mathcad will not allow you to manipulate numbers if the unit dimensions do not match. For example, Mathcad will not allow you to add units of length to units of force. However, you can multiply units of length and units of force to get units of torque or energy.

If the units in your result do not match the units you expected, then there is probably something wrong with your expression or your input numbers. For example, if you expected the results to be in lbf/ft^2, but they are in lbf/ft, then something is missing from your expression or from your input.

Definitions

For our discussion of units, we will use the following terminology:

- Unit dimension: A physical quantity, such as mass, time, or pressure, which can be measured.
- Unit: A means of measuring the quantity of a unit dimension.
- Base unit dimension: One of seven basic unit dimensions: length, mass, time, temperature, luminous intensity, substance, and current or charge.
- Base unit: A default unit measuring one of the seven base unit dimensions. For example, the following are base units: meter (m), kilogram (kg), second (s), Kelvin (K), candela (cd), mole (mol), or Ampere (A).
- Derived unit dimension: A unit dimension derived from a combination of any of the seven base unit dimensions. For example, the following are derived unit dimensions: area (length^2), pressure (mass/(length*time^2), energy (mass*length^2/time^2) and power (mass*length^2/time^3).
- Derived unit: A unit measuring a derived unit dimension. For example, the following are derived units: pounds per square inch (psi), Joules (J), British Thermal Unit (BTU), and Watts (W).
- Unit system: A group of units used to measure base unit dimensions and derived unit dimensions. There are four unit systems available in Mathcad:
  - SI, with base units of meter (m), kilogram (kg), second (s), Kelvin (K), candela (cd), mole (mol), and Ampere (A).

- MKS, with base units of meter (m), kilogram (kg), second (s), Kelvin (K), candela (cd), mole (mol), and Coulomb (C).
- CGS, with base units of centimeter (cm), gram (gm), second (s), Kelvin (K), candela (cd), mole (mol), and statampere (statamp).
- U.S., with base units of feet (ft), pound mass (lb), second (s), Kelvin (K), candela (cd), mole (mol), and Ampere (A).

- Default unit system: The unit system you tell Mathcad to use. Mathcad defaults to the SI unit system, unless you change it.
- Custom unit system: Beginning with Mathcad version 13, you can choose a unit system and then change the default base units (i.e. from feet to inches). You can also select which derived units to use.

Changing the default unit system

When you open Mathcad, the SI unit system is the default unit system. To change to another default unit system in your current worksheet, select **Worksheet Options** from the **Tools** menu. Click on the Unit System tab, and select the desired default unit system. See Figure 4.1. You can also choose "None" as the default unit system. However, if you select "None," then Mathcad will not recognize any of the built-in units. The custom unit system will be discussed shortly.

Tip! Do not choose "None" as the default unit system. In engineering calculations, units are critical.

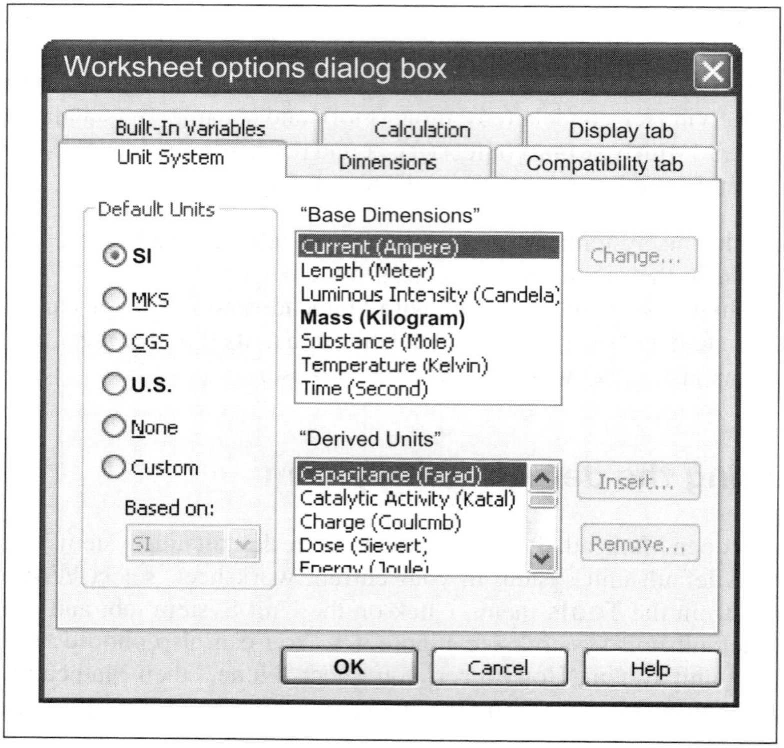

Figure 4.1 Setting default unit system

Assigning units to numbers

To assign units to a number, simply multiply the number by the name of the unit. If you cannot remember the name of the unit, you can select from a list of over one-hundred built-in Mathcad units. These are found in the Insert Unit dialog box. See Figure 4.2.

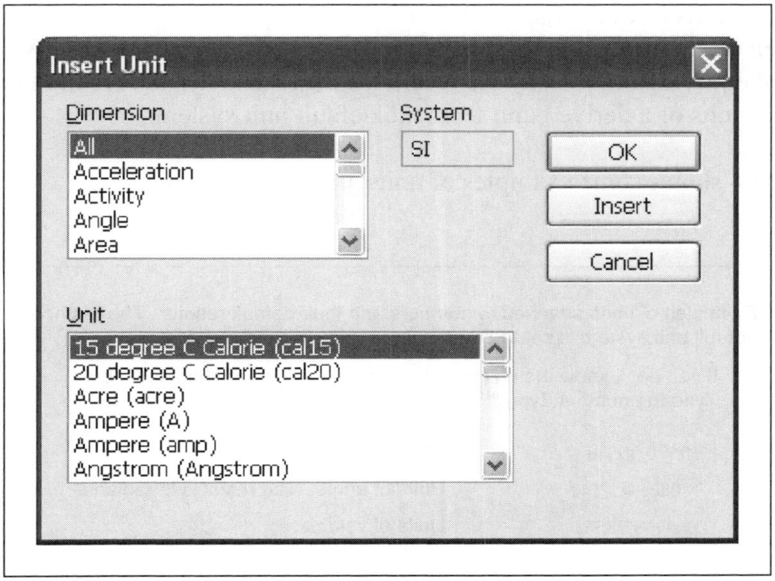

Figure 4.2 Insert Unit dialog box

To open the Insert Unit dialog box, select **Unit** from the **Insert** menu. You can also click on the measuring cup icon in the Standard Toolbar, or you can use the shortcut **CTRL+u**. See Figure 4.3. The "System" shown in the Insert Unit dialog box is the default unit system selected from the Worksheet Options dialog box. If you have selected "Custom" for your default unit system, then the "System" displayed here is the system that you based your units on. (The Custom unit system will be discussed shortly.) If you select **All** from the "Dimension" box, then all the built-in units available for that system will be shown in the "Unit" box. Note that some units will only be available for some specific unit systems.

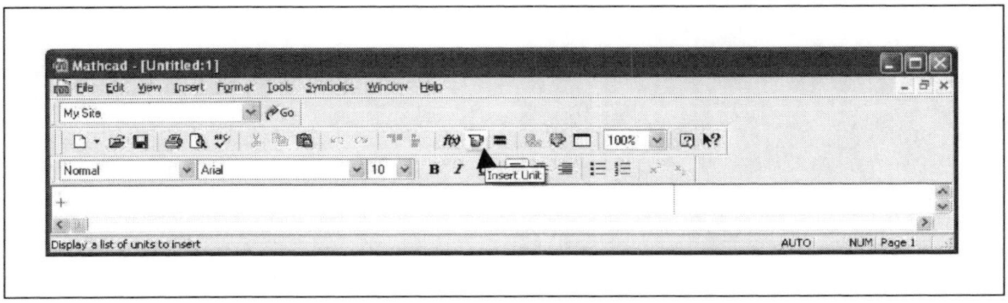

Figure 4.3 Icon to insert units

To assign units from the Insert Unit dialog box, type a number, type the asterisk and then select the desired unit from the "Unit" box, then click **OK**. When you type the equal sign to display the result, Mathcad displays the result in terms of the base units or a derived unit from the default unit system.

Figure 4.4 shows some examples of units attached to numbers.

Examples of units attached to numbers and their default results. The Mathcad default unit system is set to SI, so units display in the SI system.

If you don't know the unit names, then use the Insert Unit dialog box. Type the number, type "*", and then click "Insert."

$1\text{ft} = 0.305\ \text{m}$	Units of length
$180\text{deg} = 3.142$	Units of angle. The result is in radians.
$1\text{gal} = 3.785\ \text{L}$	Units of volume
$1\text{min} = 60\ \text{s}$	Units of time
$1\text{gm} = 1 \times 10^{-3}\ \text{kg}$	Units of mass

Figure 4.4 Examples of units attached to numbers

To add numbers with units attached, the units must all be from the same unit dimension. For example, you can add any units of length, or you can add any units of time, but you cannot add units of length to units of time. Figure 4.5 shows some examples of adding simple units. If you attempt to add different unit dimensions, Mathcad warns that the unit dimensions do not match.

Examples of unit addition.

Attached units can be from any unit system as long as the unit dimensions are the same (such as length, force, etc).

If you don't know the unit names then use the Insert Units dialog box.

$Units_A := 6m + 4mm$ — Type: 6*m+4*mm:

$Units_A = 6.004\ m$

$Units_B := 6ft + 3in$ — Type 6*ft+3*in:

$Units_B = 1.905\ m$ — Mathcad's default unit system is set to SI, so Mathcad displays units of length in meters.

$Units_C := 6N + 1kip$

$Units_C = 4.454 \times 10^3\ N$

$Units_D := 1day + 2hr + 25s$

$Units_D = 9.363 \times 10^4\ s$

$Units_E := 5m + 3N$

$Units_E = \blacksquare$

Mathcad warns that the fundamental units do not match. The variables are in red, and there is no result shown.

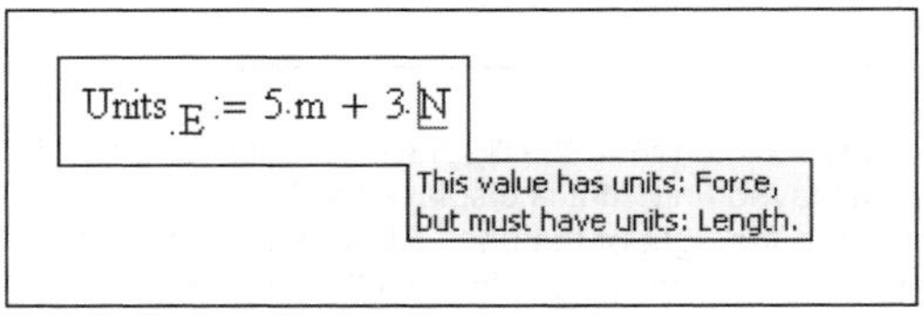

Figure 4.5 Unit addition

The units Mathcad displays by default are based on the chosen default unit system (or custom unit system), but you can change the displayed units. Let's now look at how to display different units. See Figures 4.6, 4.7, and 4.8.

$Units_F := 3ft + 1m + 33mm + 4yd$

$Units_F = 5.605\ m$

Mathcad displays in the default unit system (in this case SI). To display the results in inches, do the following:

Type: Units.F=[tab]in[enter]

$Units_F = 220.669\ in$

To display the results in mm, do the following:

Type: Units.F=[tab]mm[enter]

$Units_F = 5.605 \times 10^3\ mm$

You can display the results in many different units:

$Units_F = 5.605 \times 10^{-3}\ km$

$Units_F = 3.483 \times 10^{-3}\ mi$

$Units_F = 0.028\ furlong$

$Units_F = 6.13\ yd$

$Units_F = 18.389\ ft$

Figure 4.6 Displaying results in different units

If you want to change the displayed results for a number already displayed, then click on the displayed unit. If it is displayed in the default unit system, then a placeholder will appear to the right of the unit. Click on the placeholder, and type the unit you want displayed. You may also double click on the displayed unit and the Insert Unit dialog box will appear.

$Units_F = 5.605\ m$ Type the desired display unit in the placeholder

$Units_F = 5.605 \cdot m$ ■

Figure 4.7 Unit placeholder

If the result has already had a different unit attached to the result, then the placeholder will not appear. In this case delete the displayed unit, and type a new unit. You may also double click on the displayed unit and select a new unit from the Insert Unit dialog box.

$Units_F = 18.389\ ft$ To display a different unit, delete ft and type a new unit.

$Units_F = 3.483 \times 10^{-3}\ mi$ $Units_F = 18.389\ ft$

Figure 4.8 Changing displayed units

If you type inconsistent units in the unit placeholder, Mathcad will add units that make the result consistent. See Figure 4.9.

$Units_G := 5ft + 4in$	
$Units_G = 1.626\ m$	Mathcad defaults to m in the SI default unit system.
$Units_G = 1.626\ \frac{m}{s}\ sec$	If you type "sec" in the unit placeholder, Mathcad adds an "s" in the denominator in order for the final units to be units of length.

Figure 4.9 Balancing displayed units

Mathcad can combine units. Look at the examples in Figure 4.10.

$Units_H := 40mm \cdot 5mm$	
$Units_H = 2 \times 10^{-4}\ m^2$	
$Units_H = 200\ mm^2$	Hint: To get the superscript, use the "^" character above the number 6 key. Type mm^2.
$Units_I := \frac{10ft}{2\ sec}$	
$Units_I = 1.524\ \frac{m}{s}$	Default display
$Units_I = 5\ \frac{ft}{s}$	Displayed as feet per second
$Units_I = 3.409mph$	Displayed as miles per hour

Figure 4.10 Combining units

Tip! To attach units of area or volume, use the "^" symbol (which is above the number 6 key) to raise the unit to a power. For area type `m^2` or `ft^2`. For volume type `m^3` or `ft^3`.

Displaying derived units

A derived unit dimension is derived from combinations of any of the seven base unit dimensions. A derived unit measures a derived unit dimension. Some examples of derived unit dimensions include: acceleration, area, conductance, permeability, permittivity, pressure, viscosity, etc. Some examples of derived unit names are: atmosphere, hectare, farad, joule, newton, watt, etc.

Mathcad can display derived unit dimensions by the derived unit name or as a combination of the base unit names. The default is to use the derived unit names. To display derived unit dimensions as a combination of base unit names, open the Result Format dialog box by selecting **Result** from the **Format** menu. Then select the Unit Display tab. See Figure 4.11. This tab has three check boxes that control how units are displayed. The "Simplify units when possible" box is checked by default. If it is checked, Mathcad displays a derived unit name—if it is available. If a derived unit is not available, Mathcad displays a combination of base unit names. When the "Simplify units when possible" box is unchecked, derived unit dimensions are displayed in a combination of base unit names. The "Format units" check box is checked by default. If it is checked, Mathcad displays combinations of base units as mixed fractions with both numerators and denominators. Figure 4.12 illustrates the use of these boxes.

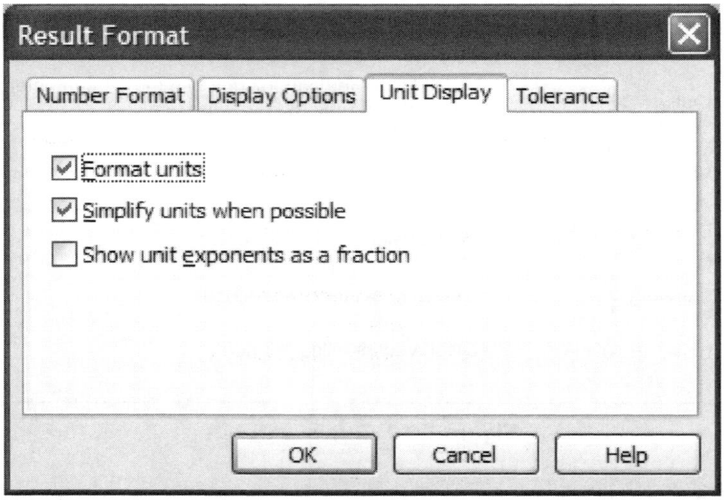

Figure 4.11 Unit display

Compare the display of derived units using the check boxes on the "Unit Display" tab of the Result Format dialog box.

Pressure	Energy	Work	Volume
Press := 1Pa	Energy := 1J	Work := 1J	Vol := $1m^3$

"Simplify units when possible" is checked.
"Format units is unchecked."

Press = 1 Pa	Energy = 1 J	Work = 1 J	Vol = 1×10^3 L

"Simplify units when possible" is checked.
"Format units" is checked."

Press = 1 Pa	Energy = 1 J	Work = 1 J	Vol = 1×10^3 L

"Simplify units when possible" is unchecked.
"Format units" is checked.

$\text{Press} = 1\,\frac{kg}{m \cdot s^2}$	$\text{Energy} = 1\,\frac{m^2 \cdot kg}{s^2}$	$\text{Work} = 1\,\frac{m^2 \cdot kg}{s^2}$	Vol = $1m^3$

"Simplify units when possible" is unchecked.
"Format units" is unchecked.

Press = $1\ m^{-1} \cdot kg \cdot s^{-2}$	Energy = $1\ m^2 \cdot kg \cdot s^{-2}$	Work = $1\ m^2 \cdot kg \cdot s^{-2}$	Vol = $1\ m^3$

Figure 4.12 Compare results of units formatting

Custom default unit system

When you choose one of the four default unit systems, you get pre-selected base units and pre-selected derived units. The addition of the custom default unit system in Mathcad version 13 is a great improvement because it allows you to tell Mathcad what base units to use and what derived units to use. For example, if you are using the U.S. default unit system, you can change the base unit of length from feet (ft) to inches (in). If you are using the SI system, you can change the derived unit of force from Pascal (Pa) to kilogram force (kgf).

To create a custom default unit system, open the Worksheet Options dialog box from the **Format** menu. On the Unit Systems tab, select **Custom** and select a "Based on" unit system. You can now change the base units from any of the seven base dimensions. Next, add or remove any of the derived units. If you see a derived unit that you do not want to use, select it and click **Remove**. For example, the derived unit of volume for SI is Liter (L) and for U.S., it is gallons (gal). If you do not like it, then remove it. If you leave the derived unit of volume, then when you have a unit of length cubed you will get a display of liters (L) in SI and gallons (gal) in U.S. If you delete the derived unit of volume, you will get m^3 or ft^3. If you changed the base unit of length from feet to inches then you will get in^3. You can add additional derived units by clicking on the **Insert** button. This will give you a choice of all the Mathcad derived units. Select the desired unit, and click **OK**. Unfortunately, you must choose one of the built-in Mathcad units. You cannot choose units that you have defined.

Tip! I always use the custom default unit system because it allows me to select the derived units to display. I highly recommend it.

Units of force and units of mass

It is important to understand how Mathcad considers units of force and units of mass. In the U.S. unit system, "lbf" (pound force) is a unit of force, and "lb" (pound mass) and "slug" are units of mass. In the SI unit system, "N" (Newton) and "kgf" (kilogram force) are units of force, and "kg" is a unit of mass. Figure 4.13 shows the relationship between various units of force and mass in the U.S. unit system. Figure 4.14 shows the relationship between various units of force and mass in the SI unit system.

Note: For this example, the default unit system was changed to US.

$g = 32.17 \frac{ft}{s^2}$ g is a built-in Mathcad unit for the acceleration of gravity.

$Units_J := 1 lb \cdot g$ lbf (pound force) = lb (mass) * acceleration of gravity

$Units_J := 1\ lbf$

$1\ lbf = 32.17 \frac{lb \cdot ft}{sec^2}$ Shows the relationship between lbf (pound force) and lb (pound mass).

$1\ slug = 32.17\ lb$ 1 slug = 32.174 pound mass

Note: Later in the chapter we will define a new unit "lbm" for pound mass. This helps to eliminate confusion as to whether lb mean mass or force. It is suggested to always use either lbf or lbm.

Figure 4.13 Relationship between various units of mass and force in the U.S. system

$g = 9.807\ \frac{m}{s^2}$	g is a built-in Mathcad unit for the acceleration of gravity.
$Units_K := 1kg \cdot g$	mass * acceleration of gravity = Force
$Units_K = 9.807\ N$	This is the way Mathcad defaults in the SI unit system.
$Units_K = 1\ kgf$	Force displayed as kgf. 1 kgf =1 kg * g
$1kgf = 9.807\ N$	Shows relationship between kgf and N.
$1N = 0.102kgf$	

Note: Sometimes it is easier to display force as kgf rather than N because it eliminates the 9.807 factor.

Figure 4.14 Relationship between various units of mass and force in the SI system

Tip! In order to avoid confusion in the U.S. unit system, create a new unit "lbm"-lbm:=lb. Now you can use lbm as a unit of mass and lbf as a unit of force in the U.S. system. See Figure 4.15.

Creating custom units

Even though Mathcad has over one-hundred built-in units, you will still need to create your own custom units from time to time. This is very easy to do. You define custom units the same way you define variables. For example, if you want to define a unit of "cfs" for cubic feet per second, then follow the steps in Figure 4.15.

$cfs := \frac{ft^3}{sec}$	This defines "cfs" as a custom unit of ft^3 divided by second.
$Units_L := \frac{1000ft^3}{1min}$	
$Units_L = 0.472\ \frac{m^3}{s}$	Mathcad default in SI units.
$Units_L = 16.667\ cfs$	Type: Units.L=[tab]cfs[enter] The custom unit "cfs" is attached to the result.

Figure 4.15 Creating custom units

In order to have the custom unit available anywhere in your worksheet, place the definition at the top of your worksheet. You may also use the global definition symbol when creating custom units. The global definition symbol appears as a triple equal sign. All global definitions in the worksheet are scanned by Mathcad prior to scanning the normal definitions. This way the unit definition does not need to be at the top of your worksheet. To define a global custom unit, type the name of the unit, press the tilde key [~], and then type the definition. See Figure 4.16.

> **Tip!** Mathcad discourages the use of global definitions because they do not participate in redefinition warnings, and they can create confusing redefinition chains if used in the middle of a document. I have had good experience using them for unit definitions. It is still a good idea to include unit definitions at the top of your worksheet.

$\text{lbm} \equiv 1\text{lb}$	This is a global definition. The unit will now be available anywhere in the worksheet rather than just below and to the right. Use the tilde key (~) to get the global definition symbol. You may also choose the global definition equal sign from the Evaluation Toolbar.

Figure 4.16 Global definition of custom unit

Figure 4.17 gives some more examples of custom units. Notice how easily Mathcad deals with the mixing of U.S. and SI units.

Custom Defined Unit		Arbitrary Usage: Mathcad Default Display (SI)	Arbitrary Usage: Using Custom Unit
$\text{cfm} := \frac{\text{ft}^3}{\text{min}}$	$\text{Custom}_1 := \frac{2\text{m}^3}{45\text{sec}}$	$\text{Custom}_1 = 0.044\,\frac{\text{m}^3}{\text{s}}$	$\text{Custom}_1 = 94.172\,\text{cfm}$
$\text{MGD} := \frac{1000000\text{gal}}{\text{day}}$		$\text{Custom}_1 = 0.044\,\frac{\text{m}^3}{\text{s}}$	$\text{Custom}_1 = 1.014\,\text{MGD}$
$\text{rpm} := \frac{360\text{deg}}{1\text{min}}$	$\text{Custom}_2 := \frac{360\text{deg}}{0.2\text{sec}}$	$\text{Custom}_2 = 31.416\,\frac{1}{\text{s}}$	$\text{Custom}_2 = 300\,\text{rpm}$
$\text{kPa} := 1000\text{Pa}$	$\text{Custom}_3 := \frac{5\text{kip}}{4\text{ft}^2}$	$\text{Custom}_3 = 5.985 \times 10^4\,\text{Pa}$	$\text{Custom}_3 = 59.85\,\text{kPa}$
$\text{cup} := 8\text{fl\_oz}$	$\text{Custom}_4 := 1\text{gal}$	$\text{Custom}_4 = 3.785\,\text{L}$	$\text{Custom}_4 = 16\,\text{cup}$

Figure 4.17 Examples of custom units

Units in equations

Now that you understand the concept of units, let's explore the use of units in equations and functions. Units are almost always a part of any engineering equation. In order to take advantage of Mathcad's wonderful unit system, it is important to always attach units to your variables. Even for simple calculations, you should get in the habit of using units. There are a few cases where this is not possible. These cases will be noted as they occur.

Figure 4.18 shows the formula for kinetic energy. Notice how the dimension of mass, length, and time combine to form a unit of energy (Joules).

The formula for kinetic energy is: $\frac{Mass \cdot Velocity^2}{2}$

$Mass := 2kg$ $\qquad$ $Velocity := 200\,\frac{cm}{s}$

$$KineticEnergy := \frac{Mass\ Velocity^2}{2}$$

$KineticEnergy_1 = 4\ J$

Figure 4.18 Example of units in an equation

Do not redefine built-in units

It is important not to redefine built-in Mathcad units. Figure 4.19 shows what will happen if you define m:=2 kg. Note the squiggle line below the "m." This means that the value of "m" is being redefined. The variable "m" is a built-in variable for meter. If you use "m" for meters below this point in your worksheet, Mathcad uses m=2 kg (mass) not m=1 meter.

The formula for kinetic energy is:

$$\frac{m \cdot v^2}{2}$$

$$v := 200 \, \frac{cm}{s}$$

$m := 2kg$

$$KineticEnergy_2 := \frac{m \cdot v^2}{2}$$

$KineticEnergy_2 := 4 \, J$

If the same equation as used in Figure 4.18 is rewritten with only single letter variables, the variable "m" redefines a built-in variable for meter. This is indicated by the squiggle line below "m."

$m := 2 \cdot kg$

This expression redefines a Mathcad built-in unit.

The variable "m" now represents 2 kg instead of 1000mm. Notice what will happen when you try to use the variable "m" to represent a length of 1 meter.

$2ft = 0.61 \, m$ Meters is still a unit and Mathcad will still display "m" for meters.

$2 \cdot m = 4 \, kg$ If you try to attach units of meter to a number, Mathcad thinks you want to attach 2 kg, because that is how you have now defined "m."

Be very careful. Do not redefine built-in Mathcad units.

Figure 4.19 Redefinition warning

Figure 4.20 shows the final velocity of an object based on initial velocity (v_0), acceleration (a), and distance (s). For this example the default unit system was changed to U.S.

Note: For this example, the default unit system was changed to US.
The formula for final velocity based on initial velocity (V_0), acceleration (a), and distance (s) is: $\sqrt{v_0^{\,2} + 2 \cdot a \cdot s}$

$InitialVelocity_1 := 80 \, \frac{ft}{sec}$ $Acceleration_1 : 10 \, \frac{ft}{sec^2}$ $Distance_1 := 300ft$

$$FinalVelocity_1 := \sqrt{InitialVelocity_1^{\,2} + 2 \cdot Acceleration_1 \cdot Distance_1}$$

$FinalVelocity_1 = 111.36 \, \frac{ft}{s}$ Mathcad defaults to ft/s in US default unit system.

$FinalVelocity_1 = 75.92 \, mph$ Result after attaching mph to the unit placeholder.

$FinalVelocity_1 = 33.94 \, \frac{m}{s}$ Result after attaching m/s to the unit placeholder.

Figure 4.20 Example of units in an equation

Let's look at the same equation, but using units from different unit systems. Notice how Mathcad does all the conversions for you. See Figure 4.21.

Note: For this example, the default unit system was changed to US.
The following input quantities are the same as in Figure 4.20. They are just input using different units.

$\text{InitialVelocity}_2 := 54.545\text{mph}$ $\text{Acceleration}_2 := 3.048 \dfrac{\text{m}}{\text{sec}^2}$ $\text{Distance}_2 := 100\text{yd}$

$$\text{FinalVelocity}_2 := \sqrt{\text{InitialVelocity}_2{}^2 + 2\cdot\text{Acceleration}_2\cdot\text{Distance}_2}$$

$\text{FinalVelocity}_2 = 111.35 \dfrac{\text{ft}}{\text{s}}$ Mathcad defaults to ft/s in US default unit system.

$\text{FinalVelocity}_2 = 75.92\text{mph}$ Result after attaching mph to the unit placeholder.

$\text{FinalVelocity}_2 = 33.94 \dfrac{\text{m}}{\text{s}}$ Result after attaching m/s to the unit placeholder.

Notice how the input units of initial velocity do not need to be in ft/sec. The input units of acceleration do not need to be in ft/sec^2, nor does the input distance need to be in feet. They can be in any units of length and time. Mathcad does all the conversion you! The result is exactly the same as in Figure 4.20, even though the input units were all different.

Figure 4.21 Same example as Figure 4.20, but using mixed units

Units in user-defined functions

Using units in user-defined functions is very similar to using units in equations, except that you need to include the units in the arguments and not the function. Figure 4.22 uses a function similar to the equation used in Figure 4.20.

$\text{FinalVelocity}(v_0,a,s) := \sqrt{v_0^2 + 2 \cdot a \cdot s}$

Creates a user-defined function based on initial velocity (v_0), acceleration (a), and distance (s).

$\text{FinalVelocity}\left(80\frac{ft}{s}, 10\frac{ft}{sec^2}, 300ft\right) = 111.36\frac{ft}{s}$

Units must be attached to each argument.

$\text{FinalVelocity}(80,10,300) = 111.36$

In this example, the numbers used for the arguments are the same as above, but no units are attached to the numbers. The numeric result is the same as above, but no units are attached to the result. For engineering calculations you want units attached to all results. Therefore, make sure that units are attached to all the input information.

$\text{FinalVelocity}\left(54.545mph, 3.048\frac{m}{sec^2}, 100yd\right) = 111.35\frac{ft}{s}$

Input arguments can be mixed units.

$\text{FinalVelocity}(54.545,3.048,100) = 59.87$

If no units are attached to the arguments, and the units do not match, then the numeric result is incorrect. BE SURE TO ATTACH UNITS TO ALL THE INPUT ARGUMENTS.

Figure 4.22 Units in user-defined functions

Units in empirical formulas

Many engineering equations have empirical formulas. There are times when units may not work with the empirical equations. This can occur when the empirical formula raises a number to a power that is not an integer such as $x^{1/2}$ or $x^{2/3}$. If units are attached to "x" then the units of the result will not be accurate. In order to resolve this problem, divide the variable by the units expected of the equation, and then multiply the results by the same unit. For example, the shear strength of concrete (lbf/in^2-psi) is based on the square root of the concrete strength (psi). See Figures 4.23, 4.24, and 4.25 to see how to resolve the use of units in empirical formulas.

The Figures 4.23, 4.24, and 4.25 illustrate the following when dealing with units in empirical equations.

- Don't stop using units if they appear to not work with your equation.
- Divide the affected variables by the units in the system expected in the equation.
- It doesn't matter what input units you used as long as you divide by the units in the system expected to be used in the equation.
- After you divide by the units, then you may need to multiply again at some point in the equation by the same units.

Figure 4.26 illustrates another empirical equation using units.

$\phi := 0.85$ $\quad f'_c := 4000psi$ — Input variables: phi and strength of concrete.

$ShearStrength_1 := 2 \cdot \phi \cdot \sqrt{f'_c}$ — Empirical Formula for shear strength of concrete based on f'_c. Result should be in psi.

$ShearStrength_1 = 7318.35 \, \frac{lb^{0.5}}{ft^{0.5} \cdot s}$ — Incorrect result because of the units under the square root.

The result needs to be in psi. The empirical formula takes the square root of f'_c (in psi) and expects the result in psi, but this did not happen in the above equation.

In order to resolve this problem, divide f'_c by psi to make it unitless, and then multiply the result by psi.

$ShearStrength_2 := 2 \cdot \phi \cdot \sqrt{\frac{f'_c}{psi}} \cdot psi$ $\quad ShearStrength_2 := 107.52 \, psi$ — Correct result.

Figure 4.23 Units in empirical formulas

If an empirical formula expects a number to be in a particular unit (in this case psi), then it is important to divide the number by psi, even if the variable was input in a different unit system. For example, if you are using a US formula, and if f'_c were input in MPa, you would still need to divide by psi, not MPa. The result can be converted to MPa. See below for an example.

$f'_c := 27.579MPa$ — Change the input to MPa. (This step isn't really necessary. Mathcad already knew that f'_c was 27.579 MPa. It is done only to emphasize that f'_c was input as SI.)

$f'_c = 4000 \, psi$ — Even though f'_c was input in metric units, Mathcad knows that it is the same as 4000 psi and the result will be the same.

$ShearStrength_3 := 2 \cdot \phi \cdot \sqrt{\frac{f'_c}{psi}} \, psi$ — Since the equation is meant for US units, divide by psi, not MPa.

$ShearStrength_3 = 107.52 \, psi$ — The result is the same, even though f'_c was input in metric units.

$ShearStrength_3 = 0.74 \, MPa$ — The result can be displayed as MPa.

$ShearStrength_4 := 2 \cdot \phi \cdot \sqrt{\frac{f'_c}{MPa}} \cdot MPa$ — Result is incorrect if you try to divide by MPa. (Because the formula was written for US units.)

$ShearStrength_4 = 1294.85 \, psi$

$ShearStrength_4 = 8.93 \, MPa$ — Incorrect result

Figure 4.24 Units in empirical formulas

The same shear strength equation in SI form is: $0.166 \cdot \phi \cdot \sqrt{f'_c}$ where f'_c is in MPa.

$f'_c = 4000$ psi $f'_c = 27.58$ MPa   These values were input in the previous example.

$$SI\_ShearStrength := 0.166 \cdot \phi \cdot \sqrt{\frac{f'_c}{MPa}} \cdot MPa$$

This equation was written for SI units; therefore, divide by MPa.

$SI\_ShearStrength = 107.47$ psi

$SI\_ShearStrength = 0.74$ MPa

The results are the same as using the US equation.

When using empirical formulas it is critical to know what units the equation was written for. It is then critical to divide by the units the equation was written for, not what the input units are.

The following illustrates why this is important. Notice how f'_c divided by psi = 4000 and f'_c divided by MPa = 27.58. These are the numeric numbers expected in the US and SI shear equations respectively.

$$\frac{f'_c}{psi} = 4000 \qquad \frac{f'_c}{MPa} = 27.58$$

Figure 4.25 Same example as in Figure 4.24, but in SI form

The Hazen-Williams equation for calculating the fluid velocity in a pipe system is Velocity=1.318*C_H*$R^{0.63}$*$S^{0.54}$. Where C_H is a coefficient related to the pipe material, L is length of pipe in feet, R is the hydraulic radius (which for a pipe flowing full is the Area in feet divided by the circumference in feet), and S is the slope of the energy line or the hydraulic gradient h_f/L.

Calculate the velocity given the following:

Hydraulic gradient: 90 feet over 1000 feet

$C_H := 110 \quad D := 24in \quad h_f := 90 \cdot ft \quad Length := 1000ft$

$$Velocity := 1.32 \cdot C_H \cdot \left(\frac{\frac{\pi \cdot D^2}{4}}{\pi \cdot D}\right)^{0.63} \cdot \left(\frac{h_f}{Length}\right)^{0.54}$$

Result is not accurate because of the empirical formula and units.

$Velocity = 25.56\ ft^{0.63}$

$$Velocity := 1.32 \cdot C_H \cdot \left[\frac{\frac{\pi \cdot \left(\frac{D}{ft}\right)^2}{4}}{\pi \cdot \frac{D}{ft}}\right]^{0.63} \cdot \left(\frac{h_f}{Length}\right)^{0.54} \cdot \frac{ft}{sec} \qquad Velocity = 25.56\ \frac{ft}{s}$$

$$1.32 \cdot 110 \cdot \left(\frac{2.0}{4}\right)^{0.63} \cdot \left(\frac{90}{1000}\right)^{0.54} = 25.56$$

With no units attached

Figure 4.26 Example of another empirical formula

SIUnitsOf()

Mathcad has a function called ***SIUnitsOf(x)***. The Mathcad definition of this function is, "Returns the units of x scaled to the default SI unit, regardless of your chosen unit system. If x has no units, returns 1." This means Mathcad takes the unit dimension and returns the SI base unit for that unit dimension, or a number equivalent to the SI base unit if another default unit system is used. Let's give an example. If you have chosen SI as your default unit system and you define "Length:=2 m," then ***SIUnitsOf(Length)*** returns 1 m, because meter is the default unit of length in the SI unit system. If you have chosen U.S. as your default unit system and you define "length:=2 ft," then ***SIUnitsOf(Length)*** returns 3.281 ft. This is 1 m converted to 3.281 ft. If you have a customized default unit system, and have changed the base unit of length to cm, the ***SIUnitsOf(x)*** function will still return 1 m. Remember, the ***SIUnitsOf(x)*** function returns the default SI unit, not customized units. The actual quantity of the dimension returned by the function is the same no matter what default unit system you have chosen; it is only the displayed unit value that changes. See Figure 4.27.

The function *SIUnitsOf(x)* returns the SI units of the variable "x".

When the SI is chosen as the default unit system, then the *SIUnitsOf(x)* function always returns the value of 1 times the default base unit. It does not matter what the magnitude of the unit is, the *SIUnitOf(x)* function always returns the value of one times the default base unit.

$\text{Length}_1 := 5\text{cm}$

$\text{SIUnitsOf}\ (\text{Length}_1) = 1\ \text{m}$

$\text{Length}_2 := 500\text{m}$

$\text{SIUnitsOf}(\text{Length}_2) = 1\ \text{m}$

$\text{Torque} := 25\text{N}\cdot\text{m}$

$\text{SIUnitsOf}(\text{Torquea}) = 1\ \text{J}$

$\text{Pressure} := 25\text{Pa}$

$\text{SIUnitsOf}(\text{Pressure}) = 1\ \text{Pa}$

$\text{Power} := 4\ \dfrac{\text{N}\cdot\text{m}}{\text{s}}$

$\text{SIUnitsOf}\ (\text{Power}) = 1\ \text{W}$

The quantity of the result of *SIUnitOf(x)* is always the same; however, it may be displayed differently depending on the chosen default unit system. The following values are displayed when the U.S. default unit system is chosen. When converted back to the SI default units, these quantities are exactly the same as shown above.

$\text{SIUnitsOf}(\text{Length}_1) = 3.281\ \text{ft}$

$\text{SIUnitsOf}(\text{Length}_2) = 3.281\ \text{ft}$

$\text{SIUnitsOf}(\text{Torque}) = 23.73\ \dfrac{1}{\text{s}^2}\ \text{ft}^2\text{lb}$

$\text{SIUnitsOf}(\text{Pressure}) = 1.45 \times 10^{-4}\text{psi}$

$\text{SIUnitsOf}(\text{Power}) = 1\text{W}$

Figure 4.27 SIUnitsOf(x)

WARNING: You may see examples of where the ***SIUnitsOf(x)*** function is used to create a unitless number. Do not use this function to create a unitless number. It is much better to divide the number by the unit you want the unitless number to represent. For example, if you have a number with length units attached, and you want to display a unitless number representing the length in meters, then divide the number by m. If you want to display a unitless number representing the length in feet, then divide by ft. This way you are sure to get the unitless number you want. Dividing by ***SIUnitsOf()*** can give you a different result than what you want. Let's relook at Figure 4.25, but using the ***SIUnitsOf()*** function. See Figure 4.28.

The *SIUnitsOf()* function is difficult to understand and use.
Use the same information as used in Figures 4.25.

$f'_c = 4000\ \text{psi}$ $\qquad \dfrac{f'_c}{\text{psi}} = 4000$ $\qquad \dfrac{f'_c}{\text{MPa}} = 27.58$

The following example shows the values Mathcad returns when using the *SIUnitsOf()* function.

$$SI\_ShearStrength := 0.166 \cdot \phi \cdot \sqrt{\frac{f'_c}{SIUnitsOf(f'_c)}} \cdot SIUnitsOf(f'_c)$$

$SI\_ShearStrength = 0.11\ \text{psi}$ Should be 107 psi

$SI\_ShearStrength = 7.41 \times 10^{-4}\ \text{MPa}$ Should be 0.74 MPa

Let's examine why unexpected results were returned.

$SIUnitsOf(f'_c) = 1.45 \times 10^{-4}\ \text{psi}$ Value Mathcad returns. (psi is shown because the default unit system is set to US. This is the US value of 1 Pa.)

$SIUnitsOf(f'_c) = 1\ \text{Pa}$ Pa is the SI default unit for the base dimension of pressure.

$\dfrac{f'_c}{SIUnitsOf(f'_c)} = 2.76 \times 10^7$ The unitless number is the result of f'_c being divided by Pa. For the metric shear equation f'_c must be divided by MPa, not Pa. Therefore, the result of using the *SIUnitsOf()* function is inaccurate.

$\dfrac{f'_c}{\text{Pa}} = 2.76 \times 10^7$ Do not use the *SIUnitsOf()* function. In order to avoid getting an unexpected result, divide by the appropriate unit and not *SIUnitsOf()*.

Figure 4.28 Example of using the SIUnitsOf() function

As seen in Figure 4.28, you receive unexpected results when you use the function ***SIUnitsOf()*** incorrectly. It is best to divide by the expected units instead of using the ***SIUnitsOf()*** function.

Custom scaling units

Beginning with Mathcad 13, functions can be used in the units placeholder. This allows the use of degree Fahrenheit and degree Celsius, which could not be used in previous versions of Mathcad.

Fahrenheit and celsius

The °F and °C are actually functions, which means that you type the function name and then list its arguments in parenthesis. In order to make it easier to input these functions, there is a Custom Character toolbar that contains the °F and °C function icons. If you type °F(32) = you get 273.15 K. If you click the units placeholder and insert °C, then you get the display °F(32)=0.00°C. Figure 4.29 shows some examples of using different units systems.

°F(32) = 273.15 K	°C(100) = 373.15 K	32°C = ▮
°F(32) = 491.67 R	°C(100) = 671.67 R	32°F = ▮
°F(32) = 0°C	°C(100) = 100°C	
°F(32) = 32°F	°C(100) = 212°F	
250K = 250 K	300R = 166.667 K	
250K = 450 R	300R = 300 R	
250K = −23.15°C	300R = −106.483°C	
250K = −9.67°F	300R = −159.67°F	

Notice how the degree Fahrenheit and degree Celsius must be entered as functions. You cannot type 32°C or 32°F.

Figure 4.29 Fahrenheit and Celsius

Even though °F and °C are functions, there is a way to make it appear as if these are unit assignments. This is by the use of a postfix operator. What this means is that the argument of the function is displayed first without parentheses, and the name of the function is displayed second. The postfix operator "xf" is on the Evaluation toolbar ("x" meaning argument, and "f" meaning function). When you click the "xf" icon, it inserts two blank placeholders. Place the argument in the first placeholder, and place the function in the second placeholder. See Figure 4.30.

Use the "xf" icon on the Evaluation toolbar to get the postfix operator.

32°F = 273.15 K °F(32) = 273.15 K	This is a function in postfix notation, but it appears as a unit. It functions the same as using a standard function notation. It is not the same as typing "32°F=" or typing "32*" and then inserting "°F" from the Insert Unit dialog box.
100°C = 373.15 K °C (100) = 373.15 K	This is the °C function in postfix notation.

Figure 4.30 Postfix notation

Change in temperature

You may add or subtract Kelvin and Rankine temperatures as you would any units. For example 400K−300K=100K, or 200R+50R=250R.

When adding or subtracting Fahrenheit and Celsius temperatures, you must remember that Mathcad converts to Kelvin before doing any addition or subtraction. Thus, 212°F−200°F, is the same thing as taking 373.15K−366.48K=6.67K (−447.67°F) or 671.67R−659.67R=12R (−447.67°F). The answer you wish to obtain is 12°F. Mathcad has solved this issue by creating a unit called Δ°F. This simply means the change in temperature in degree Fahrenheit. Therefore, if you do a Fahrenheit subtraction you must display the result in Δ°F, not °F. If you do a Celsius subtraction, you must display the result in Δ°C, not °C. The Δ°F and Δ°C are available on the Insert Unit dialog box. See Figure 4.31.

You may subtract Kelvin and Rankine temperatures as you would any units.The result is an absolute temperature. When subtracting Fahrenheit or Celsius temperatures you usually want to find the difference between the temperatures.

When you calculate the change in temperature you must display the result in terms of the change in degrees Δ°F and not the actual temperature. The Δ°F and Δ°C are available in the Insert Unit dialog box.

$TempChange_1 := F(212) - °(F200)$

$TempChange_1 = 6.667K$

$TempChange_1 = 12\Delta°F$

$TempChange_1 = 12R$

$TempChange_1 = 6.667\Delta°C$

If you do not display the result in Δ° For Δ°C, then Mathcad will display the temperature difference as the Rankine or Kelvin temperature.

$TempChange_1 = -447.67°F$ $\quad 12R = -447.67°F$

$TempChange_1 = -266.483°C$ $\quad 12R = -266.483°C$

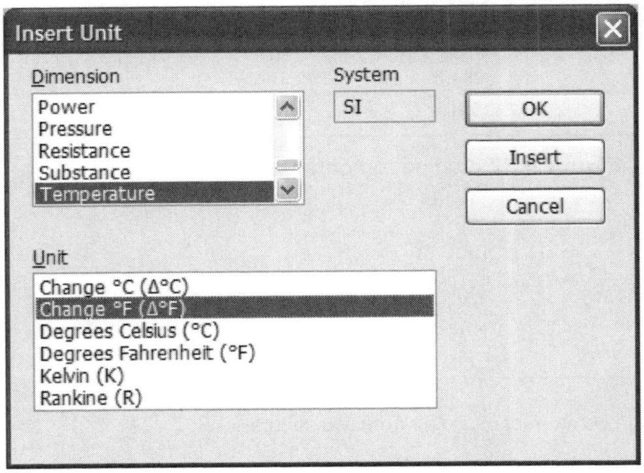

Figure 4.31 Change in temperature

When adding Fahrenheit and Celsius temperatures, you actually want to add a change in temperature, not add an absolute Kelvin or Rankine temperature. This is illustrated in Figure 4.32.

Figure 4.33 gives an example of using Fahrenheit in engineering calculation.

You may add Kelvin and Rankine temperatures just as you would any unit.

$200K + 50K = 250\ K$

$200R + 50R = 138.889\ K$ $\quad 200R + 50R = 250\ R$

If you are adding Fahrenheit or Celsius temperatures, you must use Δ°F and Δ°C

$TempChange_2 := °F(50) + °F(30)$

$TempChange_2 = 555.189\ K$

$TempChange_2 = 539.67\ °F$

Adding 50°F and 30°F is the same as adding 283.15K and 272.04K, which is equal to 555.19K or 539.67°F.

$°F(50) = 283.15\ K$

$°F(30) = 272.039\ K$

$283.15K + 272.039K = 555.189\ K$

$TempChange_3 := °F(50) + 30\Delta°F$

$TempChange_3 = 80\ °F$

$TempChange_3 = 299.817\ K$

$°F(80) = 299.817\ K$

To be accurate, you must take 50°F and add a change of 30°F.

This is the same thing as taking 50°F and adding 30R or 30*(5/9)K.

In other words 1Δ°F=1R and 1Δ°C=1K.

$283.15 + 30\cdot\frac{5}{9} = 299.817$

Figure 4.32 Adding temperatures

Calculate the temperature of mixed air from two sources.

$Temp_1 := °F(85)$ $\quad$ Remember that °F is a function, not a unit.

$Temp_2 := °F(20)$

$CFM_1 := 200cfm$

$CFM_2 := 50cfm$

$$Temp_{Final} := \frac{Temp_2\cdot CFM_1 + Temp_2\cdot CFM_1}{CFM_1 + CFM_2}$$

$Temp_{Final} = 295.372\ K$

$Temp_{Final} = 72°F$

Figure 4.33 Example of using fahrenheit in engineering calculations

Degrees minutes seconds (DMS)

Another custom scaling function added in Mathcad 13 is the ***DMS*** function. This function converts degrees, minutes and seconds to decimal degrees. It will also display radians or decimal degrees as degrees, minutes and seconds. In order to use this function, we need to introduce the concept of a Mathcad vector. Vectors will be discussed in detail in Chapter 9. We will just go through the mechanics here. A full discussion can be found in Chapter 9. To insert a vector type **CTRL+m**. This opens the Insert Matrix dialog box. For the ***DMS*** function, you need a 3 row, 1 column matrix.

The format and use of the ***DMS*** function is illustrated in Figure 4.34.

Using *DMS* as a function

DMS(▪) Insert the DMS function from the Insert Unit dialog box.

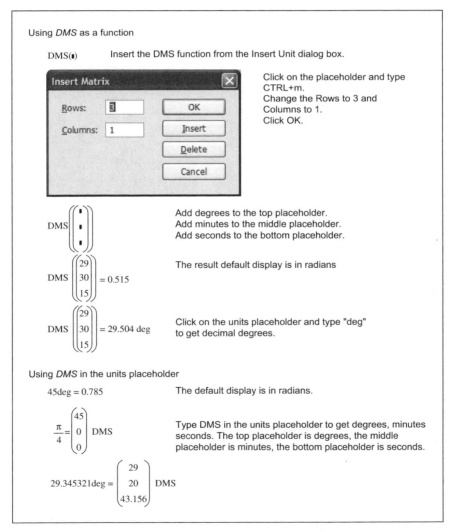

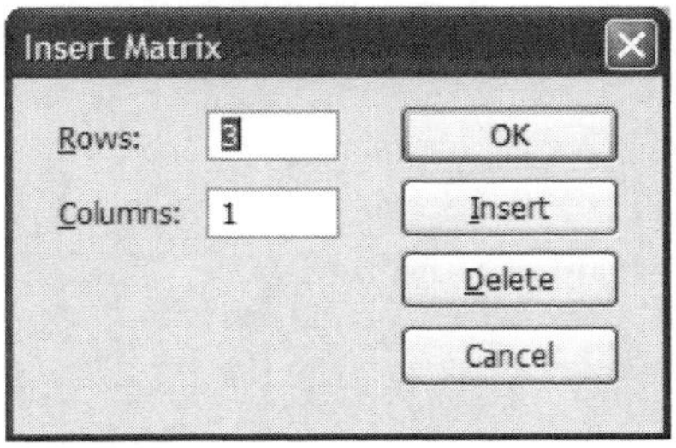

Click on the placeholder and type CTRL+m.
Change the Rows to 3 and Columns to 1.
Click OK.

$\mathrm{DMS}\left(\begin{pmatrix} \blacksquare \\ \blacksquare \\ \blacksquare \end{pmatrix}\right)$ Add degrees to the top placeholder. Add minutes to the middle placeholder. Add seconds to the bottom placeholder.

$\mathrm{DMS}\left(\begin{pmatrix} 29 \\ 30 \\ 15 \end{pmatrix}\right) = 0.515$ The result default display is in radians

$\mathrm{DMS}\left(\begin{pmatrix} 29 \\ 30 \\ 15 \end{pmatrix}\right) = 29.504\ \mathrm{deg}$ Click on the units placeholder and type "deg" to get decimal degrees.

Using *DMS* in the units placeholder

$45\mathrm{deg} = 0.785$ The default display is in radians.

$\frac{\pi}{4} = \begin{pmatrix} 45 \\ 0 \\ 0 \end{pmatrix} \mathrm{DMS}$ Type DMS in the units placeholder to get degrees, minutes seconds. The top placeholder is degrees, the middle placeholder is minutes, the bottom placeholder is seconds.

$29.345321\mathrm{deg} = \begin{pmatrix} 29 \\ 20 \\ 43.156 \end{pmatrix} \mathrm{DMS}$

Figure 4.34 Degree minutes seconds

Hours minutes seconds (hhmmss)

The ***hhmmss*** function converts hour, minutes and seconds into decimal time—either seconds, minutes, hours, days, etc. It can also be used in the units placeholder to convert decimal units of time into hours, minutes and seconds. The input and display of ***hhmmss*** differs from the ***DMS*** function. Instead of using a vector, the ***hhmmss*** function uses a text string with the hours, minutes and seconds separated by colons. Thus 2 hours, 32 minutes, and 14 seconds would be: "2:32:14." Figure 4.35 gives some examples.

Using *hhmmss* as a function

$Time_1 := hhmmss("12:32:15")$

The hours, minutes, and seconds must be input between quote marks " and must be separated by colons.

$Time_1 = 45135.00\ s$

$Time_1 = 752.25\ min$

$Time_1 = 12.537\ hr$

$Time_1 = 0.522\ day$

Using *hhmmss* in the units placeholder

$Time_2 := 40000s$

$Time_2 := "11:6:40"\ hhmmss$

$54.25min = "0:54:15"\ hhmmss$

$4.7525day = "114:3:36"\ hhmmss$

$0.125yr = "1095:43:35.747"\ hhmmss$

Figure 4.35 Hours minutes seconds *(hhmmss)*

Feet inch fraction (FIF)

The ***FIF*** function is very similar to the ***hhmmss*** function. It uses a string as the argument for the function, and when used in the units placeholder it also displays a string. The format uses a string with a single quote for feet and double quote for inches and fractions of inches. A dimension of 2 feet 3½ inches would be input as a text string like this: FIF("2' 3-1/2" "). Notice the (2) double quotes at the end. One to indicate inches, the other to close the text string. The keystrokes are: `FIF, left parenthesis, double quote, feet, single quote, space, inches, dash, numerator/denominator, double quote, double quote, right parenthesis. FIF ("feet' inches-numerator/denominator"").` Figure 4.36 provides examples of its use.

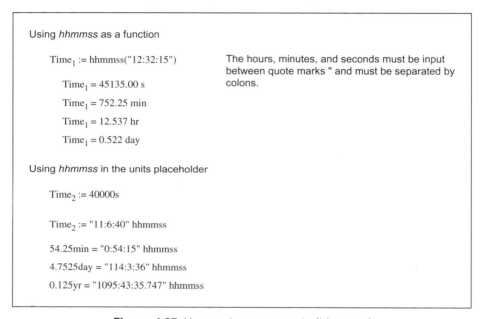

Using *FIF* as a function

FIF_Length := FIF("1'2-1/2"")

FIF_Length = 0.368m

FIF_Length = 14.5in

FIF_Length = 1.208ft

The feet, inches, and fractions must be input between quote marks. The format is: FIF, left parenthesis, double quote, feet, single quote, space, inches, dash, numerator/denominator, double quote, double quote, right parenthesis.
FIF("feet' inches-numerator/denominator"")

Using *FIF* in the units placeholder

18.5ft = "18' 6""FIF

25.5in = "2' 1-1/2""FIF

157.75in = "13' 1-3/4""FIF

2.575ft = "2' 6-9/10""FIF

1m = "3' 3-47/127""FIF

Figure 4.36 Feet inch fraction *(FIF)*

Creating your own custom scaling function

Beginning with Mathcad 13 you can create your own custom scaling function by defining the function and its inverse function. The inverse function is the same name as the function preceded by a forward slash. You must use the special text mode (**CTRL+SHIFT+K**) to input the forward slash.

The Figure 4.37 illustrates an example given in the Mathcad Help to create a custom scaling unit for decibels.

Create a unit function for decibels.

$dB_{20}(x) := 10^{\frac{x}{20}}$	Create a user-defined function. This is called the "forward function."
$/dB_{20}(x) := 20\log(x)$	Create its "inverse function." The name uses the forward slash in front of the "forward function." To do this, type a letter to begin a math region, then type **CTRL+SHIFT+K**, then arrow to the beginning of the the name and type / and the rest of the variable name. Type **CTRL+SHIFT+K** to close the special text mode.
$Test\_1 := dB_{20}(23)$	
$Test\_2 := 40dB_{20}$	You can also use the postfix operator to input the value.
$Test\_1 = 14.125$	Result with no units attached.
$Test\_1 = 23dB_{20}$	Result with dB_{20} attached to the units placeholder.
$Test\_2 = 100$	Result with no units attached.
$Test\_2 = 40dB_{20}$	Result with dB_{20} attached to the units placeholder.

Figure 4.37 Creating your own custom scaling function

Dimensionless units

There may be times when you want to attach a unit to a number that is not one of the seven basic dimensions of length, mass, time, temperature, luminous intensity, substance or current or charge. In order to do this, type the name of your unitless dimension and define it as the number 1. This means that you can attach the unit to a number and not affect its value. Once you have defined a dimensionless unit, you can create other units that have relationships with this unit. Figure 4.38 gives two examples of dimensionless units. You can create dimensionless units for just about anything.

Here are a few things to consider when using dimensionless units.

- Mathcad will not automatically attach these types of units.
- You must attach them yourself.
- Mathcad does not do a consistency check and warn you if you have attached inconsistent units.
- Inconsistent units can create wrong results. See Figure 4.38.

Currency Conversion

$USD := 1$

$EUR := 1.2USD$

$GBP := 1.8USD$

Create a dimensionless unit of USD. This unit is set to one. Attaching it to a number is the same as multiplying by one. Create additional units and set their relationship to USD.

$500USD = 500$	$10EUR = 12$	$100GBP = 180$
$500USD = 500\,USD$	$10EUR = 12\,USD$	$100GBP = 180\,USD$
$500USD = 416.667\,EUR$	$10EUR = 10\,EUR$	$100GBP = 150\,EUR$
$500USD = 277.778\,GBP$	$10EUR = 6.667\,GBP$	$100GBP = 100\,GBP$

Mathcad cannot automatically attach custom dimensionless units. You must place a unit in the unit placeholder.

Nonsense Example

$Widget := 1$	$Product_1 := 4Widget$
$Gadget := 2Widget$	$Product_2 := 4Gadget$
$Watzut := 3Gadget$	$Product_3 := 4Watzut$

$Product_1 = 4$	$Product_2 = 8$	$Product_3 = 24$
$Product_1 = 4\,Widget$	$Product_2 = 8\,Widget$	$Product_3 = 24\,Widget$
$Product_1 = 2\,Gadget$	$Product_2 = 4\,Gadget$	$Product_3 = 12\,Gadget$
$Product_1 = 0.667\,Watzut$	$Product_2 = 1.333\,Watzut$	$Product_3 = 4\,Watzut$

Be careful to not attach the wrong unit.

$Product_1 = 4\,USD$

$Product_1 = 3.333\,EUR$

Mathcad will not warn you if you use inconsistent units. The unit USD has a value of 1. So attaching USD to "$Product_1$" did not change its value, but attaching EUR did change its value.

Figure 4.38 Dimensionless units

Limitation of units

There are some limitations when using units with the exponent function. When using the exponent function x^y, when x is a dimensioned variable, y must be a constant—it cannot be a variable. In addition y must be a multiple of 1/60000. (Previous versions of Mathcad limited this to 1/60). See Figure 4.39.

$x := 2m \quad y := 3$

$Sample_1 := x^y$ — A dimensioned value cannot be raised to a variable power

$Sample_1 := \blacksquare$

$Sample_2 := x^3$ — However, you may raise a dimensioned value to a constant power.

$Sample_2 = 8 \times 10^3 \ L$ — Liter is the Mathcad default unit display for volume.

$Sample_2 = 8 \ m^3$

The constant power must be a multiple of 1/60000

$Sample_3 := x^{\frac{1}{70}}$

$Sample_3 = \blacksquare$

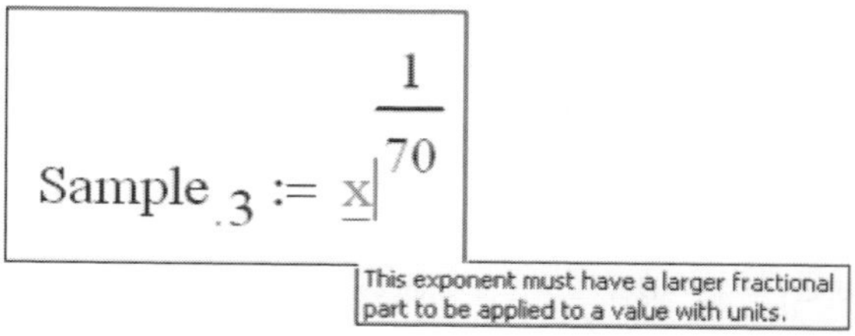
Sample .3 := x^(1/70)
This exponent must have a larger fractional part to be applied to a value with units.

$Sample_4 := x^{\frac{1}{60}}$

$Sample_4 = 1.012 \ m^{0.016666666666666666}$

Note: If you want the exponent to be displayed as a fraction, rather than as a decimal, then check "Show unit exponent as a fraction" from the Result Format dialog box.

Figure 4.39 Limitation of units

Summary

Units are essential when engineering with Mathcad. If you don't use them, you will be missing out on one of the greatest features of Mathcad.

In Chapter 4 we:

- Showed how to attach units to numbers by multiplying the number by the unit.
- Explained that Mathcad will keep track of all units in variables with similar unit dimensions no matter what units are used.
- Illustrated that units in equations do not need to match as long as the unit dimensions are the same. Mathcad does the conversion for you!
- Discussed how results can be displayed in any unit system.
- Demonstrated how to use the tab key to move you to the unit placeholder after pressing [=].
- Showed that if Mathcad does not have the unit you want, you can define it yourself.
- Encouraged the creation of a custom unit "lbm" to avoid confusion of mass and force in the U.S. unit system.
- Emphasized that user-defined functions need to have units attached to the numbers in the arguments, not the function.
- Learned that when using empirical formulas, divide the numbers (which need to be unitless) by the units expected in the equation, then multiply by the expected units.
- Suggested to avoid using the function ***SIUnitOf()*** when creating a unitless number.
- Explained the use of custom scaling functions such as °F, °C, DMS, hhmmss, and FIF.
- Introduced the concept of dimensionless units.

Practice

1. On a Mathcad worksheet type `Unit.1=10*m`. Now display this variable with the following units: Meters, centimeters, millimeters, kilometers, feet, inches, yards, miles, furlongs, and bohr.
2. In order to get a feel for the Mathcad default unit systems, open four different Mathcad worksheets and set each worksheet to a different default unit system (SI, MKS, CGS, and U.S.). Type the following in each of the four different default unit systems.

[a]	g=	(Gravity)
[b]	25ft^2=	(Area)
[c]	25F=	(Capacitance)
[d]	25C=	(Charge)

[e]	25A=	(Current)
[f]	25BTU=	(Energy)
[g]	25N*m=	(Energy)
[h]	25ft^3[spacebar]/min=	(Flowrate)
[i]	25lbf=	(Force)
[j]	25N/m=	(Force per length)
[k]	25km=	(Length)
[l]	25cd=	(Luminous intensity)
[m]	25G=	(Magnetic flux density)
[n]	25kg=	(Mass)
[o]	25lbf/ft^3=	(Mass density)
[p]	25V=	(Potential)
[q]	25kW=	(Power)
[r]	25N/m^2=	(Pressure)
[s]	25K=	(Temperature)
[t]	25m/s=	(Velocity)
[u]	25ft^3=	(Volume)

3. Go through each of the displayed units from the previous exercise and change the displayed units to a different unit (if it is available).
4. Create a custom unit system and display the results from exercise 2 in the custom unit system.
5. From your field of study (or another field), try to find five (or more) units (or combination of units) that are not defined by Mathcad. Create custom units for these units. Use the global definition. (For example, **cfm~ft^3[spacebar]/min**.)
6. Go back to the user-defined functions you created in Chapter 3 practice exercises. Attach units to the input arguments. Make sure that the result is in the unit dimension you expect to see. Then change the displayed units to other units of your choice.
7. Create four dimensionless units that are all related, and perform arithmetic operations using them.

5
Mathcad settings

This chapter will give you detailed information about many of the Mathcad settings. With this information you can make informed decisions as to what each different setting should be for your situation. The chapter will focus on the Preferences dialog box the Worksheet Options dialog box and the Result Format dialog box.

Some of the settings discussed in this chapter will not mean anything to you until you become familiar with many of the Mathcad features. Other settings will be useful to you now. This chapter will discuss each setting in the order they appear in the dialog boxes. Skip over any item that is not clear or not useful. After you become more familiar with Mathcad you can refer back to this chapter to review the different settings. Much of the information in this chapter is taken directly from the Mathcad Help with additional comments added.

Chapter 5 will:

- Discuss the Preferences dialog box and show how different settings affect the way Mathcad starts-up and how it operates.
- Discuss the Worksheet Options dialog box and show how different settings can affect a specific worksheet.
- Recommend specific settings for various features.
- Discuss the Result Format dialog box and show how to control the way results are displayed.

Preferences dialog box

The Preferences dialog box sets global features. The changes made in this dialog box are effective for all Mathcad worksheets. They will be effective every time you open Mathcad.

To open the Preferences dialog box, click **Preferences** from the **Tools** menu. Figure 5.1 shows the Preferences dialog box.

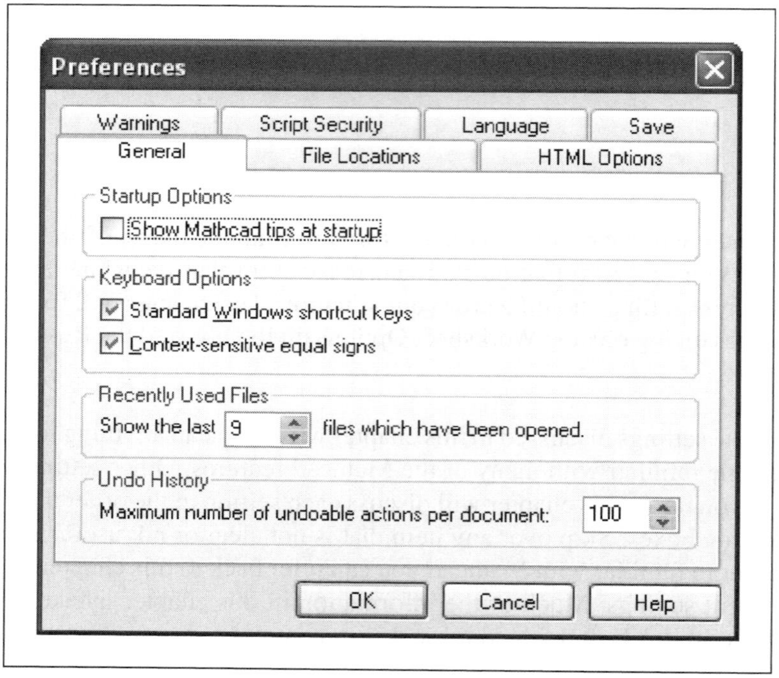

Figure 5.1 Preferences dialog box

General tab

Startup options

Checking the "Show Mathcad tips at startup" box will cause a Mathcad Tips dialog box to appear every time you start Mathcad. If you would like to see a Mathcad tip every time you open Mathcad, then make sure that this box is checked. If you are tired of seeing the tips, then uncheck this box. The tips are very useful if you are just starting to learn Mathcad. They give you many worthwhile suggestions concerning the many features of Mathcad.

Keyboard options

Leave the "Standard Windows shortcut keys" box checked, unless you are a long-time Mathcad user and want to use some of the keystrokes from the early versions of Mathcad.

The "Context-sensitive equal signs" box is checked by default. Leave this box checked. This feature allows the [=] key to insert the evaluation equal sign (=) when it is to the right of a defined variable and the definition symbol (:=) when it is to the right of an undefined variable.

Recently used files

This feature sets the number of recently opened worksheets displayed in the File menu. The default is 4 worksheets.

> **Tip!** I like to see as many files as possible, so I increase this number to 9.

Undo history

This feature controls how many actions can be undone when using the undo command. The undo history can be set to any number between 20 and 200. The default value is 100. The higher the number, the more system memory is used. Unless you are doing some very critical calculations where you would need more than 100 undos, the 100 default should be adequate.

File Locations tab

See Figure 5.2 for an example of the File Locations tab.

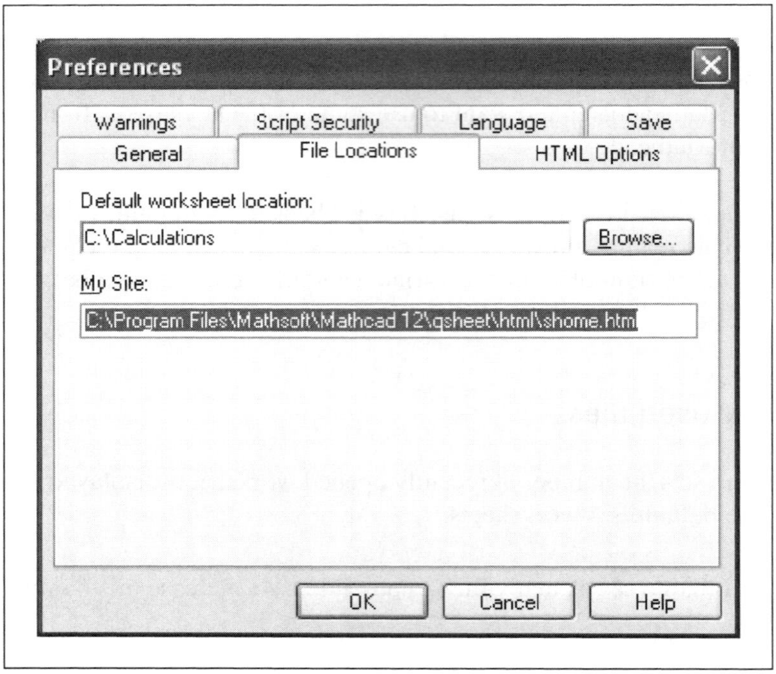

Figure 5.2 File Locations tab

Default worksheet location

This is where you set the directory location where Mathcad looks for files after starting. It is also the default location where files will be stored. Note that this location is the location just after Mathcad starts. Once Mathcad has started and you open a file from another location, or save a file to another location, Mathcad will go to the last used folder location, not this default location. The next time you start Mathcad this default location is used.

Tip! I suggest changing the default worksheet location to a corporate calculation location or to a specific calculation folder.

My Site

This is the path that Mathcad uses to map to the My Site on the Resources toolbar. The default path opens a Mathcad window that has several useful links. Unless you have a reason to map to a specific site, stay with the Mathcad default.

HTML Options tab

See Figure 5.3 for an example of the HTML Options tab.

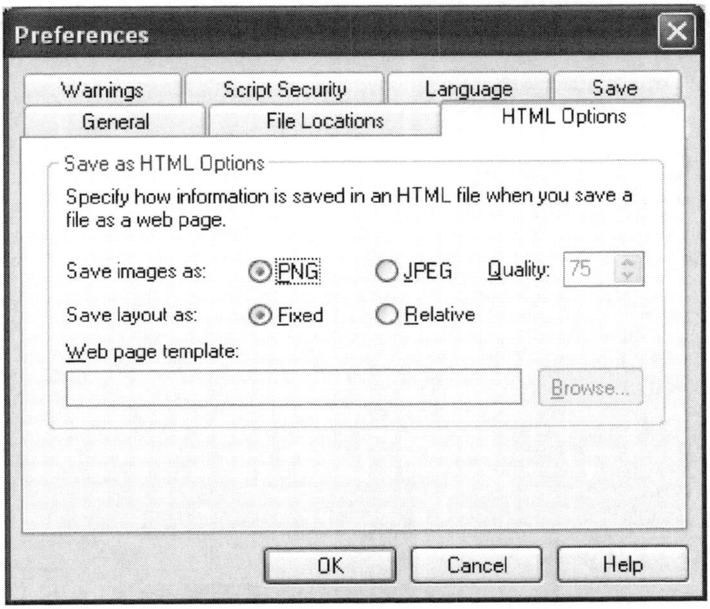

Figure 5.3 HTML Options tab

With Mathcad you can export your files to HTML format to create web pages. This tab sets the way Mathcad saves the files as a web page. In a nutshell: if you want your web page to look like your Mathcad worksheet then select **PNG** and **Fixed**. Your HTML worksheet will be an exact image of your Mathcad worksheet. The math cells, however, will not be interactive. They will be graphic images. After saving your Mathcad worksheet into HTML format, you can still use Mathcad to open the HTML file, and all of the math cells become interactive again.

If you want the file size of your HTML document to be smaller than **PNG** provides, then you can select **JPEG** and set **Quality** to about 50. This not only reduces the quality of the image in the HTML format, but it also reduces the file size. The lower the quality number, the lower the quality of the image in the HTML file. It becomes a trade-off between file size and image quality. If you can afford the larger file size, then always use **PNG**.

The options on this tab set the defaults for when you select **Save As** from the **File** menu, and then choose **HTML File (*.htm)** from the "Save as type" list. See Figure 5.4.

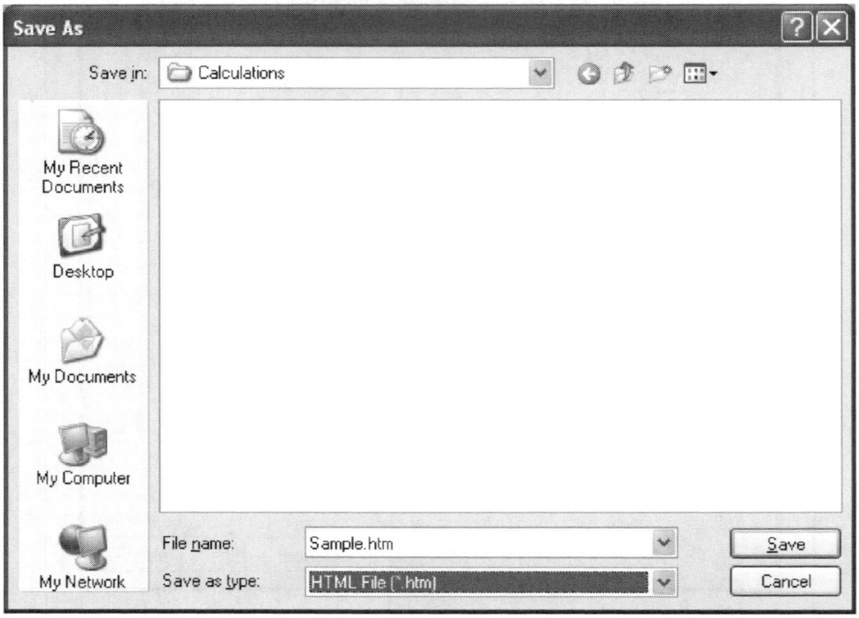

Figure 5.4 Using Save As HTML file

If you choose **Save As Web Page** from the **File** menu, then you will be given a dialog box similar to Figure 5.5 after you type a file name and click **Save**.

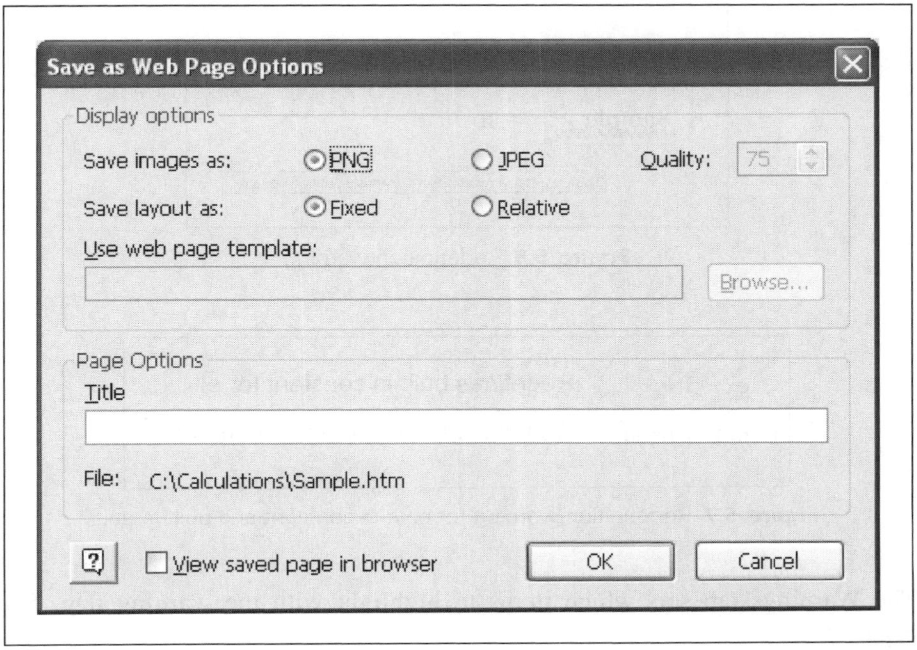

Figure 5.5 Dialog box after using Save as Web Page

From this dialog box, you can select the same features as from the HTML tab in the Preferences dialog box.

If you are familiar with web page design, then the additional options on this tab will mean something to you. This is sufficient for our discussion.

Warnings tab

Mathcad can warn you if you redefine a previously defined variable. Mathcad can also warn you if you define a variable with the same name as a built-in Mathcad variable name such as a built-in constant or a built-in unit. The warning flag Mathcad uses is a green wavy underscore beneath the variable name being redefined. See Figures 5.6 and 5.7.

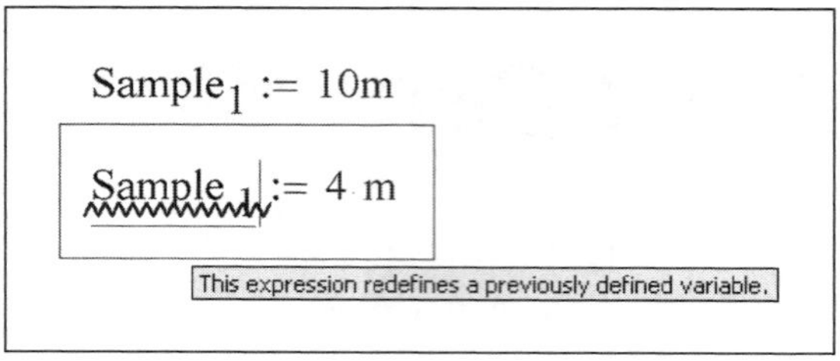

Figure 5.6 Redefinition warning

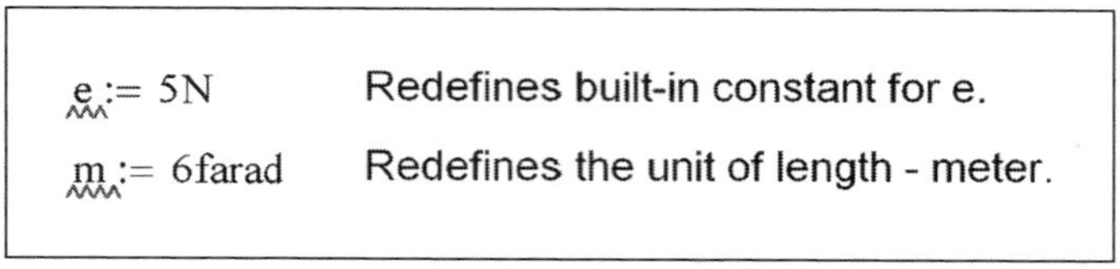

Figure 5.7 Redefinition warning for built-in constant and built-in unit

The Warnings tab sets which items to highlight with the warning flag. See Figure 5.8.

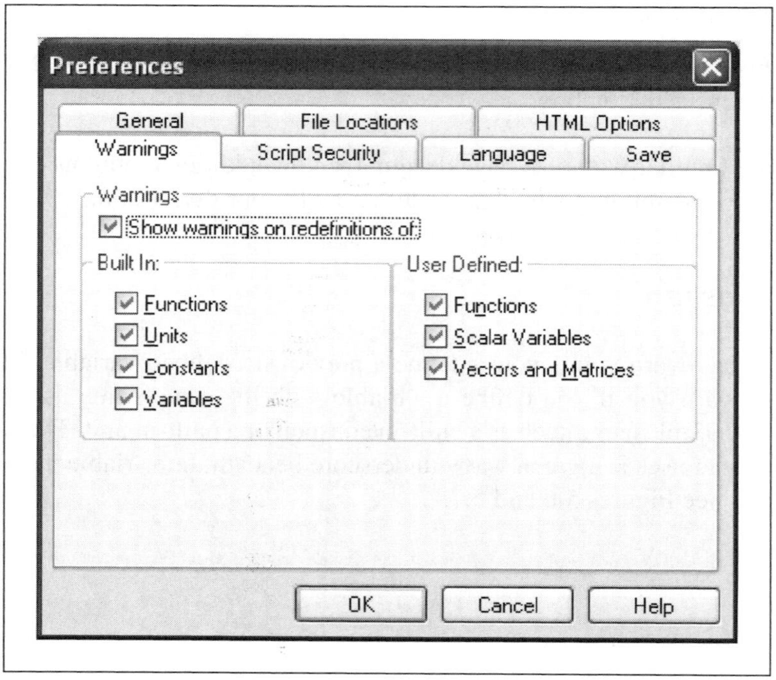

Figure 5.8 Warnings tab to set redefinition warnings

Make sure that there is a check in the box next to "Show warnings on redefinitions of:."

When using Mathcad for engineering calculations, it is important to know when any variable or function is redefined. It is suggested that all the checkboxes be checked. This will better protect you as you prepare your engineering or technical calculations.

Built-in functions include things such as: sin, cos, ln, etc. It includes any function that is included in the Insert Function dialog box. Figure 5.9 shows what happens when a built-in function is redefined.

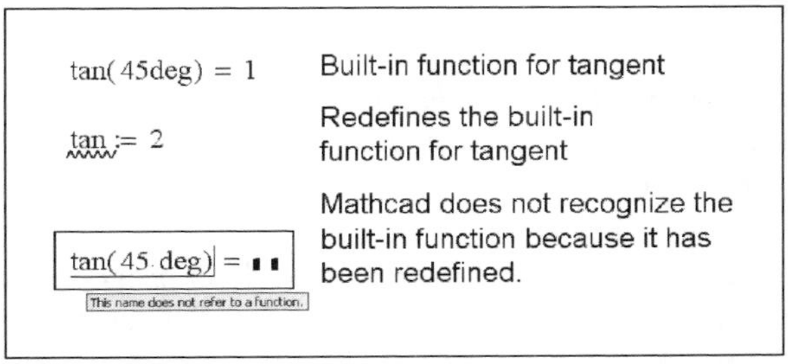

Figure 5.9 Redefinition warning for built-in function

Mathcad has many built-in units. These units are listed in the Insert Unit dialog box. If any of these units are redefined, Mathcad will warn you.

Built-in constants include such things as e, pi, i, and j. For a complete list of all the Mathcad built-in constants see Mathcad Help and search for "constants."

To see a list of the Mathcad built-in variables, click **Worksheet Options** from the **Tools** menu, and then select the Built-in Variables tab. This tab will be discussed later in this chapter.

The "User-Defined" Functions, Scalar Variables, and Vectors and Matrices are all definitions that you create in your worksheets. If you overwrite any of these definitions, Mathcad will give you a warning.

Script security tab

Mathcad worksheets may contain scriptable components. These components, such as VBScript, Jscript or macros, may contain harmful code. This tab allows you to control how Mathcad deals with scriptable code when it opens a worksheet that contains scriptable code.

There are three security options to select from:

High Security is the most secure. When this option is checked, Mathcad automatically disables all scripts when a worksheet is opened. In order to enable a scripted component when this option is selected, right-click on the component and choose Enable Evaluation.

Medium Security has Mathcad prompt you whenever a worksheet containing scriptable components is opened. You are given an option of whether or not to disable the script.

Low security allows Mathcad to enable all scripts whenever a Mathcad worksheet is open. This option is the least secure, because harmful scripts could be opened without your knowledge.

It is recommended selecting Medium Security. This prevents unknown scripts from being opened, and it allows you to select which scripts to enable. If you have full confidence in the worksheet being opened, you can then choose to enable the scripts when opening the file.

If you have Microsoft Excel worksheets embedded into your Mathcad worksheets, and if your Excel worksheets have macros, then you may also get an Excel warning. This warning is controlled by the Excel macro security level. The use of Microsoft Excel will be discussed in Chapter 22.

Tip! If you have the Mathcad script security set to medium and you open a file containing several scripts, Mathcad stops at each instance of a script and asks whether or not to enable the script. This can be annoying as you try to scroll through the document after opening the file. If you are clicking the right scroll bar to move down through the document and Mathcad comes to a script, a dialog box is opened asking if you want to enable the script. The default is to disable the script. If you have a smart mouse enabled, your mouse may jump to the default button of the dialog box. If you are quickly clicking the scroll bar, you may accidentally disable a script you wanted to enable. In order to avoid this, type **CTRL+END**. This causes Mathcad to go to the end of the document. You may also type **CTRL+F9**. This Forces Mathcad to calculate the entire worksheet. Now all instances of scripts will be encountered, and all of the dialog boxes will appear one after another, so that you can answer all the questions at one time.

Language

Language options

The options on the Language tab control how Mathcad displays languages. If additional languages are installed, you can select which language to display for menus and Mathcad's math language. English versions of Mathcad will only have the English option. Other language versions of Mathcad will have additional language options available.

Spell check options

These options set the language to use when spell checking the Mathcad worksheet. Some languages have different spellings of words. Mathcad allows you to select which dialect to use for spell checking. For example, in English you can choose from American English, British English (ise) or British English (ize). In German, you can choose from (ss) or (ß).

Save tab

Default format

Starting with version 12, Mathcad allows you to save in two new formats: **Mathcad XML Document** and **Mathcad Compressed XML Document**.

The default is **Mathcad XML Document**. This should be adequate for most applications. This option saves the worksheet as an XML document and allows the Mathcad worksheet to be read by any text editor or XML editor. There are many benefits to saving your documents in XML format, which will be discussed later in Chapter 19. When files are saved in this format, the file extension is .XMCD.

If you select **Mathcad Compressed XML Document**, then the Mathcad file will be zipped. It will not be available as an XML file until it is unzipped. If you desire to send a smaller size Mathcad file to someone else, you can select this option. Files saved in this format have the file extension .XMCDZ.

Mathcad version 12 allows another save option: **Mathcad Worksheet**. Files saved in this format are similar to the files saved in previous versions of Mathcad. This option saves the file in Mathcad's native binary file format. The files are only able to be read by Mathcad.

Autosave

Starting with version 13 Mathcad finally introduced an autosave function. Yea! Be sure to check this box and set a time interval for the autosave.

Summary of the Preference tab

The items in the Preferences dialog box are Mathcad defaults. They are applicable to all new documents and all existing documents. Once the settings in the Preferences dialog box are set, they remain the same for all documents until the settings are changed. These settings affect the way Mathcad operates and affect how it saves documents.

Worksheet Options dialog box

We just discussed the Preferences dialog box, which sets global settings affecting all documents. We will now discuss the Worksheet Options dialog box. The settings in this dialog box only affect the current worksheet. Every worksheet can have different settings. This dialog box is opened by clicking **Worksheet Options** from the **Tools** menu.

Built-in variables tab

We discussed built-in variables in Chapter 1. This is the location where these built-in variables are defined. This tab is shown in Figure 5.10.

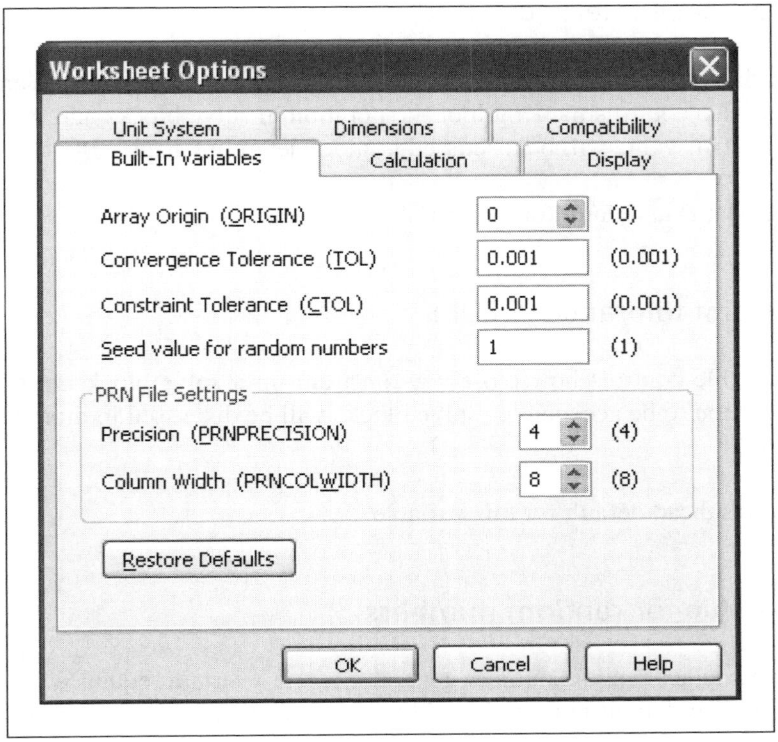

Figure 5.10 Built-in variables tab

The Mathcad default values for these built-in variables are noted in parenthesis to the right of the boxes.

Array origin

This option controls the value for the built-in variable "ORIGIN." This variable represents the starting index of all arrays in your worksheet. Arrays are simply vectors and matrices. They will be discussed in detail in Chapter 9. The Mathcad default for this variable is 0.

> **Tip!** There may be times when you want to begin vector numbering with 0, but I find it awkward. I like the first variable in an array to be labeled 1 rather than 0. I like to set the built-in variable ORIGIN to the value 1.

Convergence tolerance (TOL)

This variable controls the precision to which integrals and derivatives are evaluated. It also controls the length of the iteration in solve blocks and in the root function. This value will be discussed in more detail in Chapter 14.

Use the Mathcad default for this variable.

Constraint tolerance (CTOL)

This variable controls how closely a constraint in a solve block must be met for a solution to be acceptable. Solve blocks will be discussed in more detail in Chapter 14.

Use the Mathcad default for this variable.

Seed value for random numbers

This value tells a random number generator to use a certain sequence of random numbers. The default number is 1, but any number may be used to generate a different sequence of random numbers.

Precision (PRNPRECISION)

This variable controls the number of significant digits that are used when writing to an ASCII data file. Writing to data files is beyond the scope of this book.

Use the Mathcad default for this variable.

Column width (PRNCOLWIDTH)

This variable controls the width of columns when writing to an ASCII data file.

Use the Mathcad default for this variable.

Calculation tab

See Figure 5.11 for an example of the Calculation tab.

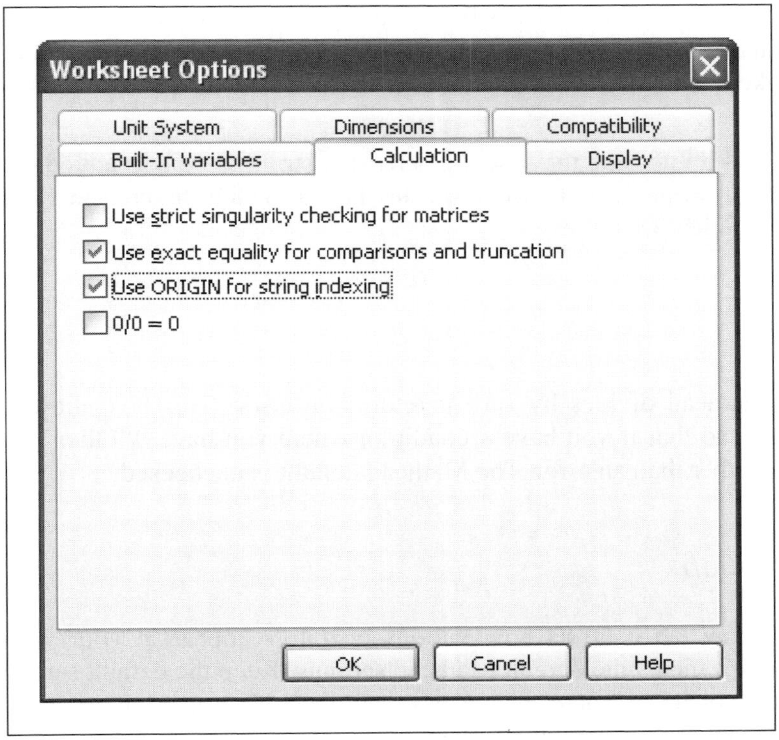

Figure 5.11 Calculation tab

Use strict singularity checking for matrices

This topic is beyond the scope of this book. The Mathcad default is unchecked. See Mathcad Help for more information.

Use exact equality for Boolean comparisons

This check box controls the standard use of Boolean comparisons. The Mathcad default is checked. This is adequate for most applications.

Use ORIGIN for string indexing

Mathcad defaults to considering the first character in a string as 0, similar to the default value for the built-in variable ORIGIN. Checking this box will consider the first character to be the value for ORIGIN. If you changed the default value of the built-in variable ORIGIN from 0 to 1, you can check this box and Mathcad will consider the first character in the string to be 1 instead of 0. If this box is unchecked, the integer associated with the first character in a string will be 0.

Tip! I do not like the first character in a string to be considered 0; I like it to be 1. I like to check this box to match the change I make to ORIGIN.

0/0=0

When Mathcad divides by zero it returns an error. This box allows you to tell Mathcad that if you have a condition where you have 0/0 then return the value 0 rather than an error. The Mathcad default is unchecked.

Display tab

The Display tab controls how various operators appear in your worksheet. Figure 5.12 shows the default Mathcad settings. Keep the default settings. The only time you may want to change these settings is if you are doing a presentation and do want to show different operators. For example you may want to show = instead of := for the definition of an expression. Refer to Mathcad Help for specific information on the different options.

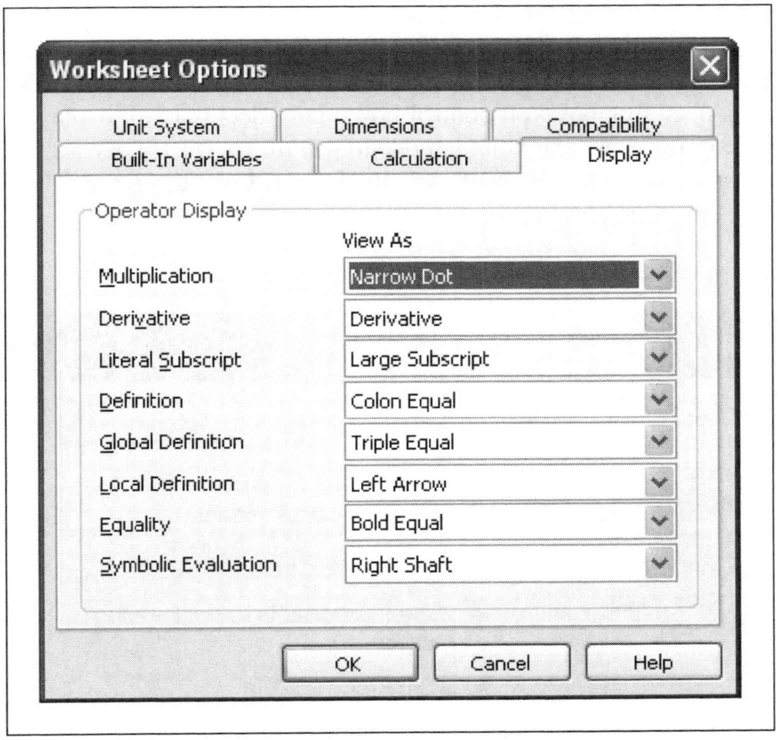

Figure 5.12 Display tab showing default Mathcad settings

Unit system tab

This tab sets the default system of units used by Mathcad. The default system used by Mathcad is SI (International). Units were discussed at length in Chapter 4. Refer to Chapter 4 for additional information about this tab.

Dimensions tab

This tab is not useful for engineering calculations. Do not use the options on this tab.

Compatibility tab

These settings help in the transition between different versions of Mathcad. If you have worksheets from previous versions of Mathcad, the Mathcad Help associated with this tab will be very important.

Result Format dialog box

The Result Format dialog box controls how results are displayed in Mathcad. For example, in this box you control how many decimals to display, whether to show trailing zeros, and what exponential form to use. To access the Result Format dialog box, select **Result** from the **Format** menu. Figure 5.13 shows a sample Result Format dialog box.

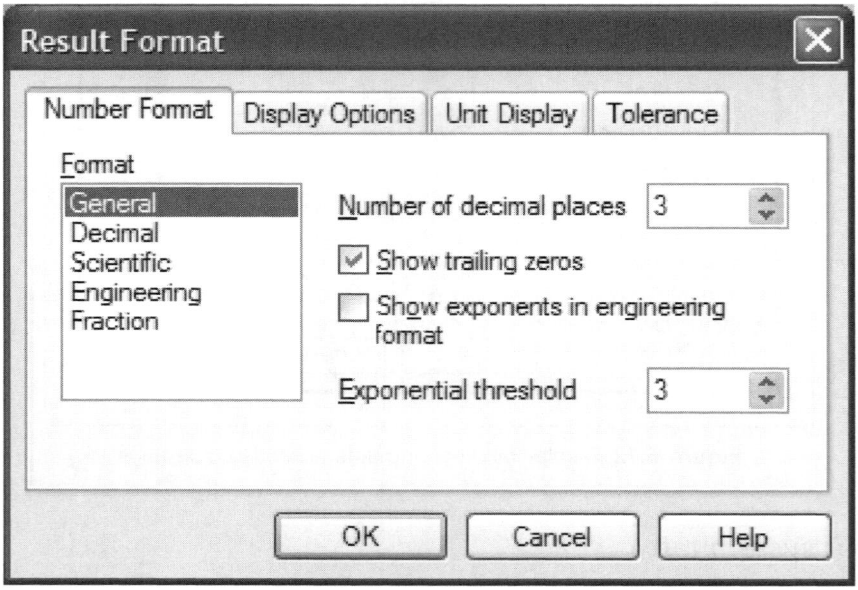

Figure 5.13 Result Format dialog box with General format selected

The changes you make in this dialog box change the display for all results in your worksheet. They are also only applicable to this worksheet. Chapter 7 will discuss how to make the changes affect all your worksheets. You can also override the settings made in this dialog box for any specific result. To change the results of a specific result, double-click on the result and a similar dialog box will appear. The changes that you then make will only affect the specific result.

Number Format tab

This tab controls the decimal places and exponents used in displaying your results. There are five different types of formats to select.

When General is selected, the results are displayed in decimal format until the exponential threshold is reached. After the exponential threshold is reached, then results are displayed in exponential notation.

When Decimal is selected, the results are never in exponential notation. The number of decimal places displayed can be set.

When Scientific is selected, the results are always in exponential notation.

When Engineering is selected, the results are always in exponential notation and the exponents are in multiple of three.

When Fraction is selected, the results are displayed as fractions. This also brings up a "Level of accuracy" box. This box controls how close the fraction approximation is to the decimal value. The higher the accuracy, the closer the approximation. The Mathcad default is 12. The "Use mixed numbers" checkbox allows the use of integers and fractions. It keeps the fraction part of the number less than one. See Figure 5.14.

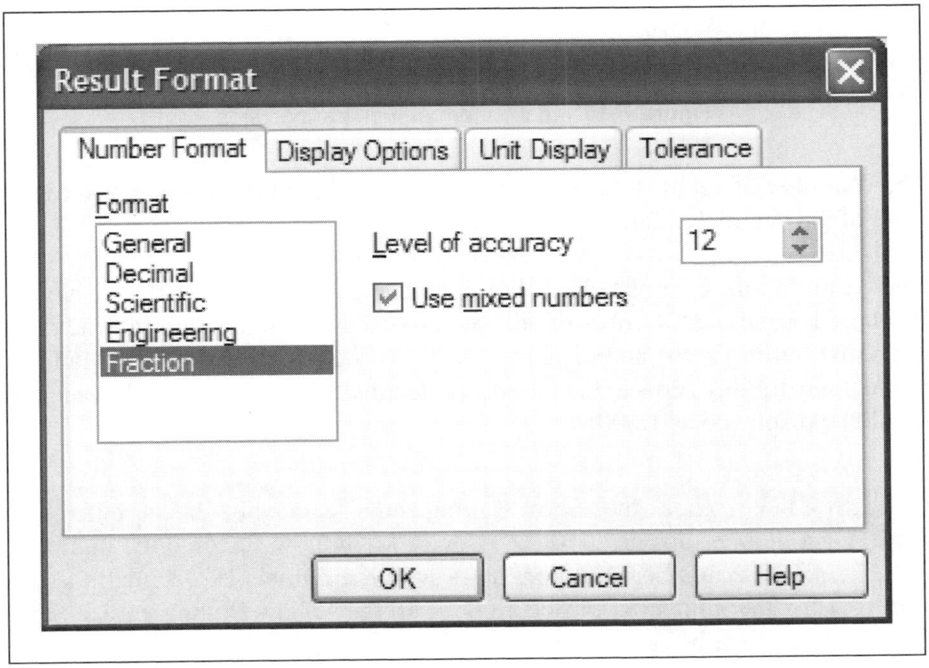

Figure 5.14 Result Format dialog box with Fraction format selected

Tip! When you use the Fraction format, if you have the "Show trailing zeros" box checked in the General format, then Mathcad displays the zeros in the fractions to the number of decimal places selected from the General format. For example, if you have the number of decimal places set to 3 in the General format and you have the "Show trailing zeros" box checked, then your fractions will display with three decimal places. When using fractions, first set the decimal places to zero in the General format, then select the Fraction format. You can also uncheck the "Show trailing zeros" box to solve the problem. See Figure 5.15.

$\frac{5}{3}+\frac{1}{7}=1.000\frac{17.000}{21.000}$	Result with decimal places set to three and "Show trailing zeros" box checked.
$\frac{5}{3}+\frac{1}{7}=1\frac{17}{21}$	Result with decimal places set to zero and "Show trailing zeros" box checked.
$\frac{5}{3}+\frac{1}{7}=1\frac{17}{21}$	Result with decimal places set to three and "Show trailing zeros" box unchecked.

Figure 5.15 Fraction format with trailing zeros

The "Number of decimal places" controls the number of digits displayed to the right of the decimal point.

The "Show trailing zeros" check box is unchecked by default. If you have a result of 1.6 and the "Number of decimal places" is set to 3, then Mathcad does not show trailing zeros and displays 1.6. If this box is checked, then Mathcad will display trailing zeros to the number of decimal places. Thus the above result would be displayed as 1.600.

Tip! I like to check the "Show trailing zeros" check box, because if I see a number displayed as 1.6, I don't know how accurate the number is. The number could be 1.649 with the number of decimal places set to 1, or the number could be 1.600 with the "Show trailing zeros" box unchecked. If the box is checked, then there is not a question.

The "Show exponents in engineering format" is similar to the Engineering format discussed above.

The "Exponential threshold" tells Mathcad when to start displaying results in exponential notation. This is only used when the General format is selected. The other formats either always display exponential notation or they never display exponential notation.

> **Tip!** I like numbers less than one million to be displayed in decimal format so I usually set the "Exponential threshold" to 6.

More information about the Number Format tab can be found in Mathcad Help.

Display Options tab

See Figure 5.16 for a sample of the Display Options tab.

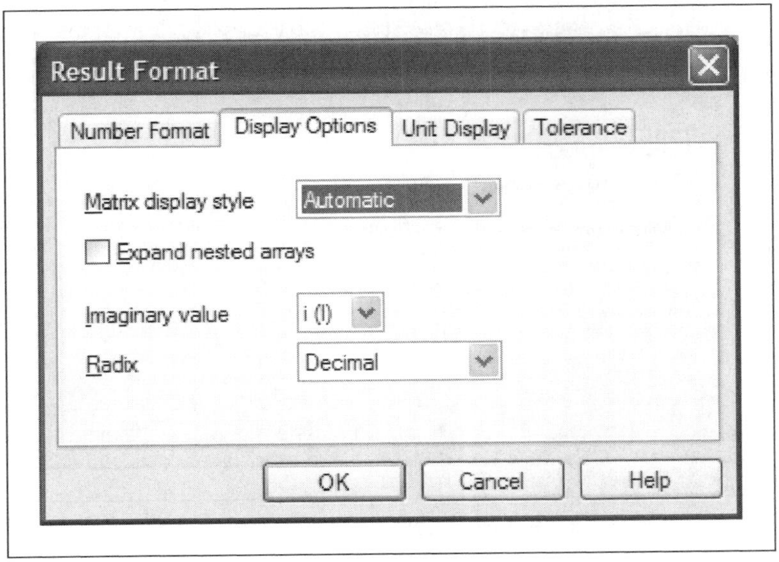

Figure 5.16 Display Options tab

Matrix display style

This box selects whether an array is displayed in an output table or in matrix form. These options will be discussed in more detail in Chapter 9.

Imaginary value

This box tells Mathcad whether to use an i or j when displaying the imaginary part of an imaginary number.

Radix

This box allows you to display results as decimal, binary, octal or hexadecimal numbers.

Unit Display tab

See Figure 5.17 for a sample of the Unit Display tab.

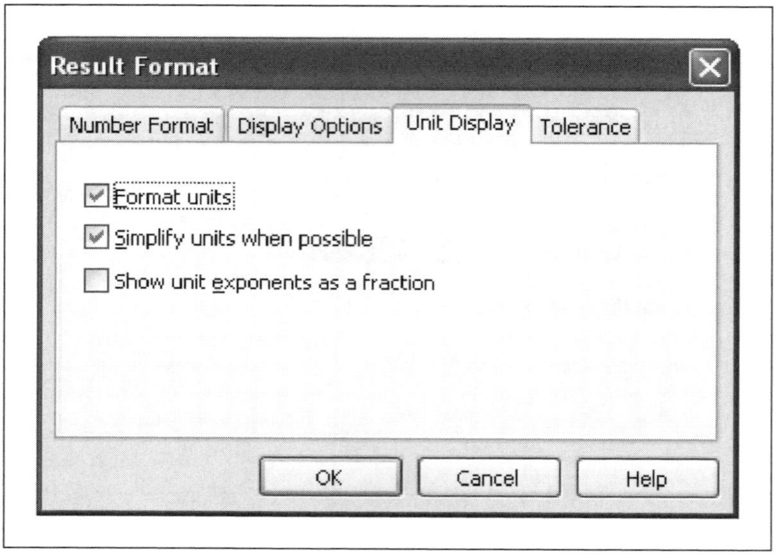

Figure 5.17 Unit Display tab

There are three check boxes on this tab. The first two are checked by default. The "Format units" box reformats the units displayed to a more common notation. For example, sec^{-1} displays as 1/sec. The "Simplify units when possible" box displays the simplest unit possible. For example, kg*m/sec^2 is simplified to N (Newton). The "Show unit exponents as a fraction" box displays unit exponents as rational fractions when checked. Otherwise, they are displayed as decimals.

This option can only be set for the whole worksheet, not on a region-by-region basis.

Tolerance tab

See Figure 5.18 for a sample of the Tolerance tab.

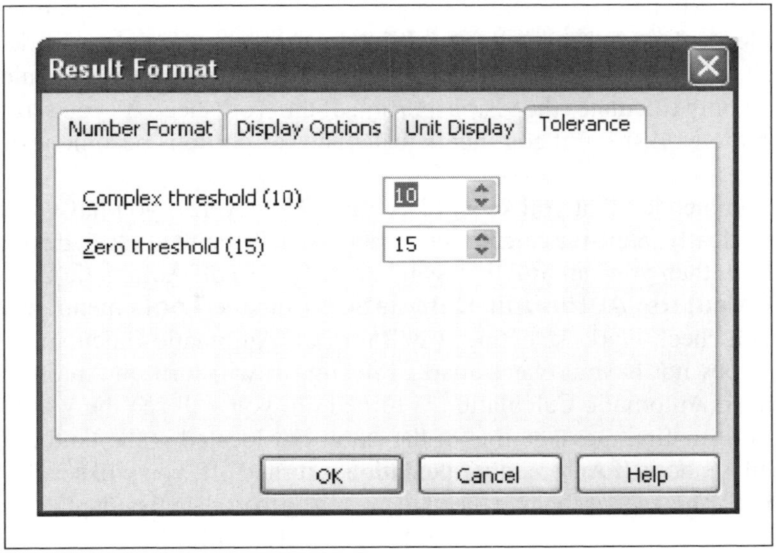

Figure 5.18 Tolerance tab

The "Complex threshold" box controls how much larger the real or imaginary part of a number must be before the display of the smaller part is suppressed. See Mathcad Help for examples.

The "Zero threshold" box controls how close a result must be to zero before it is displayed as zero. This means that if the zero threshold is set to 10, numbers smaller than 10^{-10} will be displayed as zero, even if the decimal places is set to a number greater than 15. The Mathcad default is 15. See Mathcad Help for additional examples.

Individual result formatting

The above discussion relates to default formatting for the entire worksheet. You can overwrite the worksheet default settings and format the display of a single result by double clicking on the result. This opens the Result Format dialog box. You may then set any of the formats discussed above. When you click **OK** the dialog box closes and only the selected result will change. All other results follow the worksheet default. If you later change the worksheet defaults, this result will not change because you have overwritten the worksheet defaults.

Automatic Calculation

By default, Mathcad automatically updates results any time an equation or plot is visible in the Mathcad window. Prior to printing, the entire worksheet is recalculated. After opening a document, the entire document is not calculated. Mathcad only calculates the visible portion of the worksheet. As you scroll down through the worksheet, it continually calculates as portions become visible.

It is recommended that you keep this setting. However, if you have some very mathematically intensive calculations that make it difficult to scroll through your worksheet, then you can turn this feature off. To turn off Automatic Calculation select **Calculate>Automatic Calculation** from the **Tools** menu. If this feature has a check mark, selecting it will turn off Automatic Calculation. If this feature does not have a check mark, selecting it will turn on Automatic Calculation. If Automatic Calculation is turned on, you will see the word "Auto" on the information message line of the status bar located at the bottom of your Mathcad window. If Automatic Calculation is turned off, you will see the words, "Calc F9." This means that you need to press `F9` to update the displayed results. See Figure 5.19. You can also select **Calculate>Calculate Now** from the **Tools** menu. Remember that the results you see displayed may not be accurate until you recalculate.

If you are scrolling through a worksheet, it is easier to calculate the entire worksheet first. That way you do not have to wait for each visible page to calculate as you scroll. To calculate the entire worksheet you can:

- Type `CTRL+F9`.
- Select **Calculate>Calculate Worksheet** from the **Tools** menu.
- Type `CTRL+END` to take you to the end of the worksheet. If Auto Calculate is turned on this will calculate the entire worksheet.

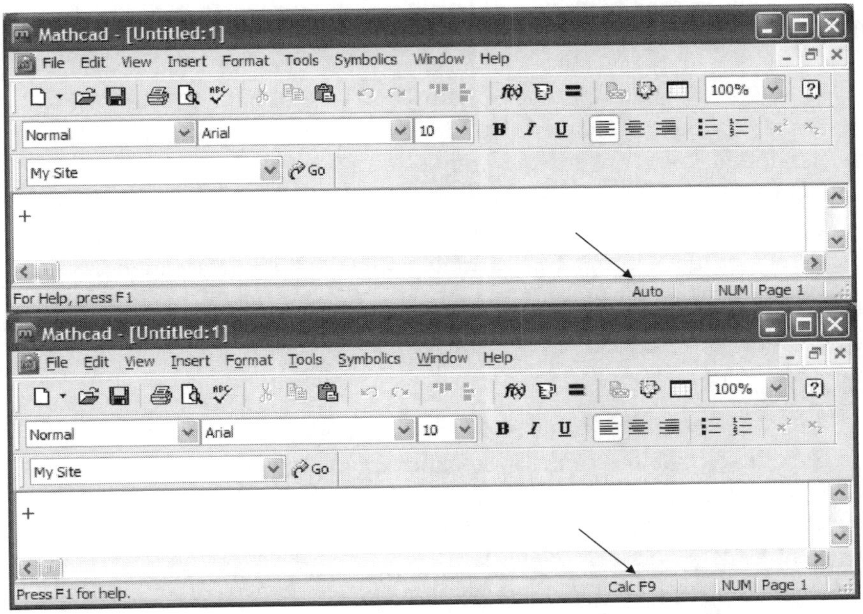

Figure 5.19 Automatic Calculation turned on (top) and turned off (bottom)

Tip! A warning about manual calculation mode: If you print a worksheet prior to calculating the worksheet when in manual calculation mode, the results on the printout are not necessarily up-to-date. This can be a serious problem if someone is relying on your printed calculations. There is no warning on the printed page alerting you to the fact that the worksheet has not been calculated. For this reason, it is suggested that you stay in automatic calculation mode.

Summary

Chapter 5—Mathcad Settings discussed the dialog boxes that affect how Mathcad functions and how Mathcad displays results. Some of the settings affect Mathcad globally and others affect only the specific worksheet. The Preferences dialog box settings affect all Mathcad worksheets. The Worksheet Settings dialog box and the Result Format dialog box only affect the current worksheet.

In Chapter 5 we:

- Showed how the Preferences dialog box sets global Mathcad settings.
- Discussed the general settings for how Mathcad operates and starts-up.
- Set default locations for saving files.
- Explained the ways of saving files as web pages.
- Discussed how Mathcad can warn you when redefining variables.
- Discussed the settings for warning against scriptable components.
- Compared the settings for the default file format for saving Mathcad worksheets.
- Showed how the Worksheet Options dialog box can set worksheet specific settings.
- Recommended the settings for values of built-in variables.
- Encouraged the use of the automatic calculation setting.
- Told how to set the default unit system.
- Showed how the Result Format dialog box controls how results are displayed.
- Showed how to set the number of displayed decimal places, and how to set the exponent threshold.

Practice

1. Open the Preferences dialog box. Change the "Default worksheet location" to a location other than the Mathcad default. Next, set the Warnings tab so that all check boxes are checked.
2. Open the Worksheet Options dialog box. Set the Array Origin (ORIGIN) to 1. Next, on the Calculation tab place a check in the "Use ORIGIN for string indexing" box.
3. Type the following on a blank Mathcad worksheet:

   [a] 1.6=
   [b] 1.06=
   [c] 1.006=
   [d] 1.0006=
   [e] 1.00006=
   [f] 1.000006=
   [g] 1.0000006=
   [h] 1.00000006=
   [i] 1=
   [j] 11=

[k] 111=
[l] 1111=
[m] 11111=
[n] 111111=
[o] 1111111=
[p] 11111111=

4. Now open the Result Format dialog box. Choose "General" format, and uncheck "Show trailing zeros." Change the "Number of decimal places" to zero and look at the displayed results. Now change the "Number of decimal places" to 1 and look at the displayed results. Next change the "Number of decimal places" incrementally from 2 to 9 and see how each number affects the displayed results.
5. Open the Result Format dialog box. Place a check in the "Show trailing zeros" box, and look at the displayed results. Now change the "Number of decimal places" incrementally from 0 to 9 and see how each number affects the displayed results.
6. Open the Result Format dialog box. Change the "Exponential threshold" to zero and look at the displayed results. Next change the "Exponential threshold" incrementally from 1 to 7 and see how each number affects the displayed results.
7. Open the Result Format dialog box. Uncheck the "Show trailing zeros" check box and change the format to "Fraction." Set the "Level of accuracy" to 12. Uncheck the "Use mixed numbers" and look at the displayed results. Next, place a check in the "Use mixed numbers" box and look at the displayed results. Next, change the "Level of accuracy" incrementally from 0 to 9 and see how each number affects the displayed results.
8. Open the Result Format dialog box. Experiment with the "Decimal," "Scientific," and "Engineering" formats. See how the different settings affect the displayed results.

6 Customizing Mathcad

With some customizing, Mathcad calculations can look as nice as a published textbook. This chapter will teach you how to set-up customizations to improve the appearance of your worksheets.

One way to achieve a consistent professional look to your calculation is by the use of styles. Styles allow numbers and constants to have different appearances. They also allow you to change the look of text for titles, headings, explanations, and conclusions. You can even change the look of specific variables, such as vectors, so that they stand-out from other variables.

Headers and footers are critical to engineering calculations. They identify you, your company, the project information, and the date the calculations were performed.

Chapter 6 will:

- Discuss Mathcad styles for variables, constants and text.
- Tell about the advantages of using styles.
- Show how to create and modify math and text styles.
- Explain how to create headers and footers.
- Describe how to create a standard header that includes a graphic logo.
- Give suggestions about information that should be included in headers and footers.
- Discuss margins, including how to use information located to the right of the right margin.
- Discuss how to customize the icons on the toolbar.

Default Mathcad styles

Whether you know it or not, you always use styles when you use Mathcad. Every time you type a definition or enter text, Mathcad assigns a default style to the typed information. A style is a specific set of formatting characteristics associated with the items displayed on your Mathcad worksheet. The formatting characteristics of a style include such things as: font type, font size, font color, bold, underline, italics, margins, indents, etc. There are two types of Mathcad styles: math styles and text styles. The style being used by Mathcad is shown on the left side of the Formatting Toolbar. See Figure 6.1.

Default math styles

Math styles are associated with variable definitions and expressions. There are many possible math styles, but there are two default math styles: "Variables" and "Constants." Variables are letters. Constants are numbers. Whenever you type letters for an expression (or a combination of letters and numbers), Mathcad assigns the "Variables" style. When you type numbers, or Mathcad displays numerical results, Mathcad assigns the "Constants" style. The Mathcad default for both of these styles is Times New Roman, 12 point, black font, with no bold, italics, or underline. Let's look at some examples of the default Mathcad math styles. See Figures 6.1–6.4.

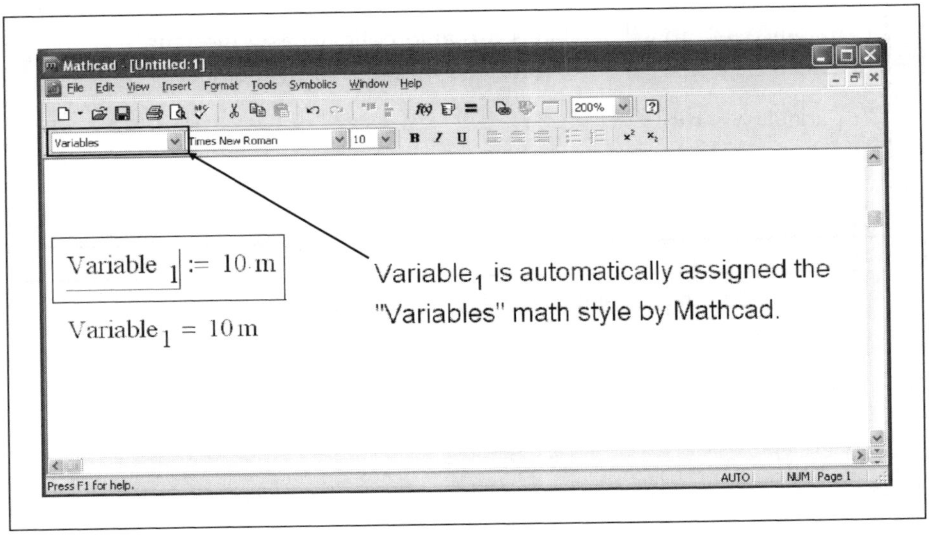

Figure 6.1 "Variables" math style for variable name

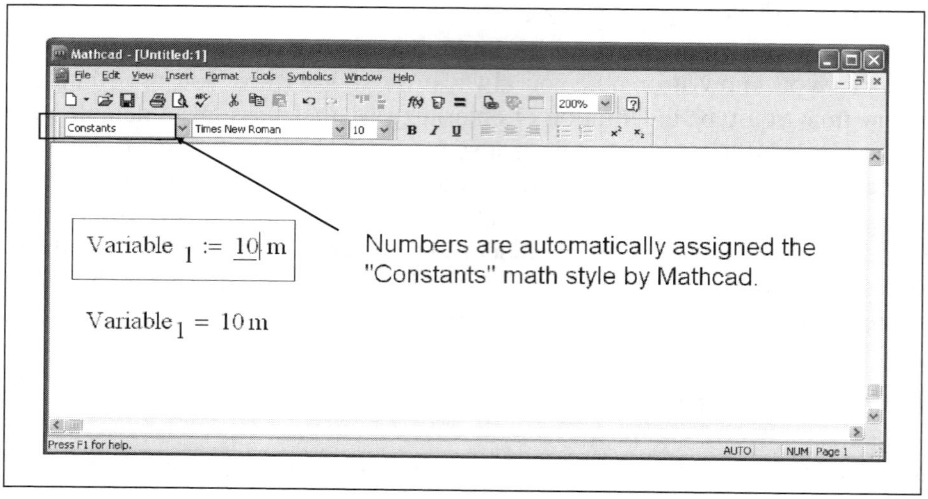

Figure 6.2 "Constants" math style for number

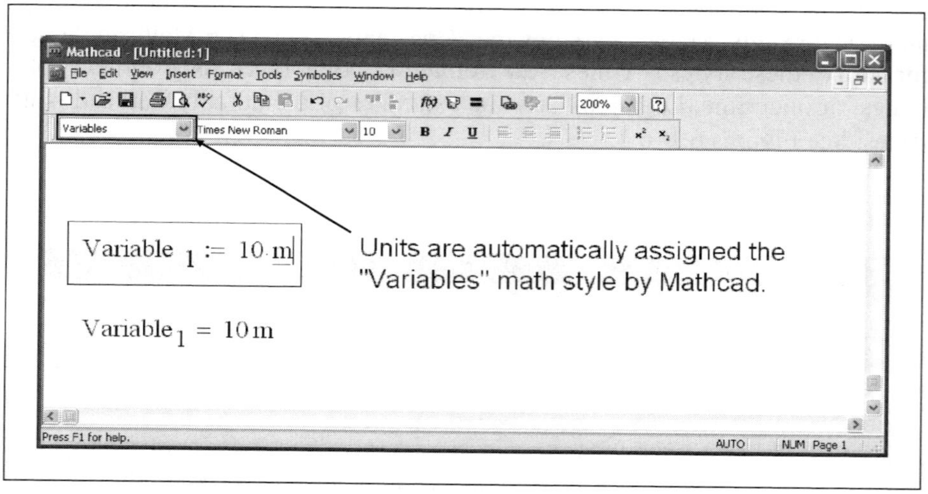

Figure 6.3 "Variables" math style for unit

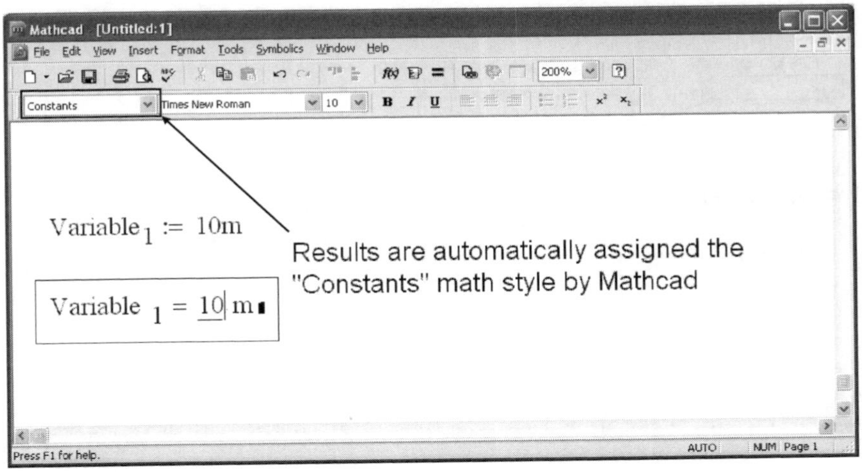

Figure 6.4 "Constants" math style for result

Additional math styles will be discussed later in this chapter.

Default text styles

Text styles control the appearance of text within Text Regions. Text styles control font and paragraph characteristics. When you create a text region, Mathcad assigns the "Normal" text style to the text in the region. The Mathcad default "Normal" text style is Arial, 10 point, black font, with no bold, italics, or underline. It has left justification, and there are no paragraph indents. See Figure 6.5.

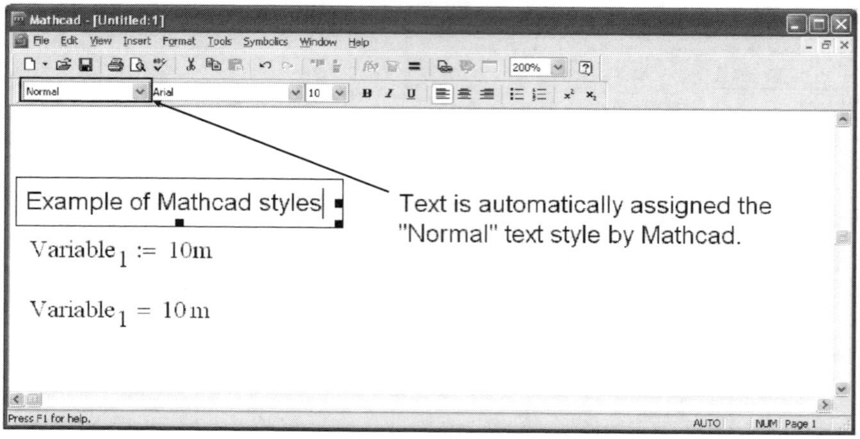

Figure 6.5 "Normal" text style

Mathcad comes with several other styles that can be assigned to the text regions. These additional text styles will be discussed shortly.

Additional Mathcad styles

Mathcad comes with many styles in addition to the two default math styles and the one default text style mentioned above.

Additional math styles

If you click within any expression, the styles drop-down box on the formatting toolbar changes to show the available math styles. Clicking on the drop-down arrow will reveal additional available math styles. From Figure 6.6, you will see that Mathcad comes with 10 math styles: Constants, Math Text Font, User 1 through User 7 and Variables.

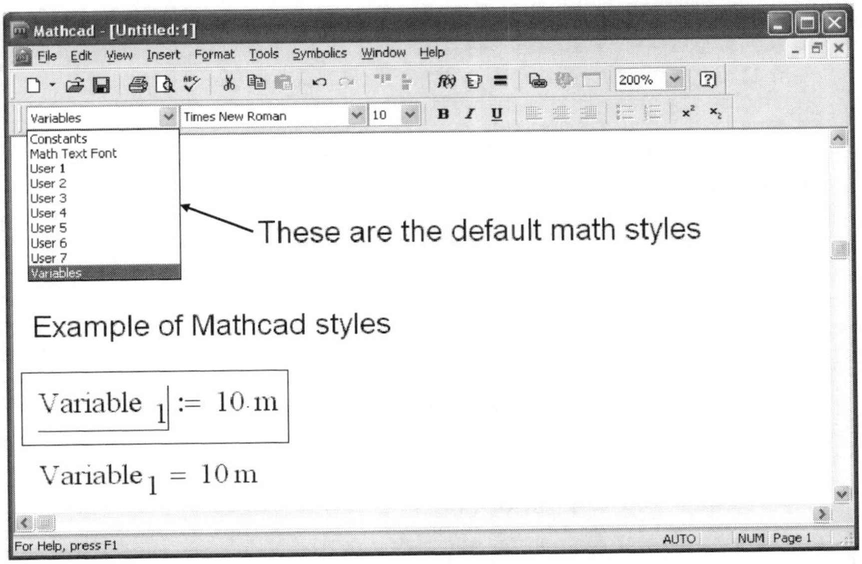

Figure 6.6 Default math styles

We have already discussed the "Variables" style and the "Constants" style. The "Math Text Font" style is the default style used in plot labels. The use of this style will be discussed in Chapter 11.

The other math styles allow you to change the way specific variables look. For example, suppose you had a very important variable in your document that you wanted to stand-out and look different from all the other variables. You can assign this variable to use "User 3" style. This assigns the font characteristics associated with the style "User 3" to the variable. To assign this style to a variable, click within the variable name, and then click the arrow in the drop-down styles box and then click, "User 3." The variable name now takes on the font characteristics associated with the style "User 3." See Figure 6.7 to see how the appearance of variables change with the different styles.

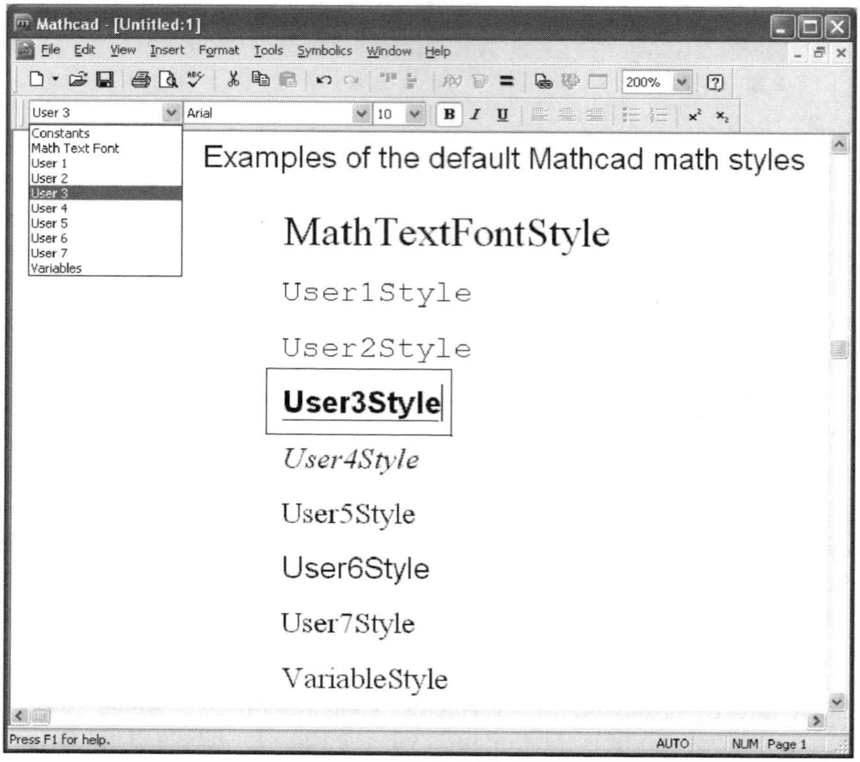

Figure 6.7 Default math styles

It is important to remember that variable names are style sensitive. When you assign a style different than the default "Variables" style to a specific variable name, you must also assign the same style to all occurrences of that variable throughout your worksheet. Every time you type the variable you must stop and assign that same style to the variable name. Mathcad will not recognize the variable until you assign the correct style to all occurrences of the variable. This can be very cumbersome if you have many variables with specific styles associated with them. You will constantly be stopping your typing and clicking on the styles drop-down box. This can greatly reduce your efficiency in creating a worksheet. For this reason, use judgment in assigning styles to variable names. Use variable styles only when the added benefit of having a unique looking variable outweighs the extra effort of making it look unique. See Figure 6.8 to see how variables are style sensitive.

$\mathbf{Variable_2} := 20N$	$Variable_2$ is assigned to "User3" style.
$Variable_2 = ▮$	Mathcad doesn't recognize this variable because it is "Variables" style.
$\mathbf{Variable_2} = 20N$	Mathcad recognizes the variable after changing to "User3" style.

Figure 6.8 Variable names are style sensitive

It is possible to have the same variable name use nine different math styles and have nine unique variables. See Figure 6.9 for an example of this.

Because variable names are style sensitive, it is possible to use the same variable name yet have several different variables. The following example illustrates this. Notice how there are nine different variables with the same name.
WARNING: THIS IS FOR ILLUSTRATION PURPOSES ONLY. IT IS STRONGLY ADVISED NOT TO USE THE SAME VARIABLE NAME WITH DIFFERENT STYLES.

Variable Names		Style Used
$Variable_3 := 1$	$Variable_3 = 1$	Variables (Mathcad default)
$Variable_3 := 2$	$Variable_3 = 2$	User 1
$Variable_3 := 3$	$Variable_3 = 3$	User 2
$\mathbf{Variable_3} := 4$	$\mathbf{Variable_3} = 4$	User 3
$Variable_3 := 5$	$Variable_3 = 5$	User 4
$Variable_3 := 6$	$Variable_3 = 6$	User 5
$Variable_3 := 7$	$Variable_3 = 7$	User 6
$Variable_3 := 8$	$Variable_3 = 8$	User 7
$Variable_3 := 9$	$Variable_3 = 9$	Math Text Font

Figure 6.9 Variable names are style sensitive

If you have the same variable name with two or more styles, it can (and most likely will) introduce errors into your worksheet. You may forget to assign the correct style to the variable in your expression, and the result will be different from what you expect. This can be very dangerous in engineering calculations. For this reason, it is suggested to use only one style with each variable name.

Additional text styles

Clicking the drop-down arrow in the styles drop-down box on the formatting menu will list the text styles that come with Mathcad. The beauty of using different styles with your text is that you can create a different look for different parts of your calculations. You can create titles, subtitles, headings and subheadings. You can also emphasize different parts of your calculations by creating special styles for input and output information.

Figure 6.10 lists the text styles that come with Mathcad. It shows the different font and paragraph characteristics associated with each style.

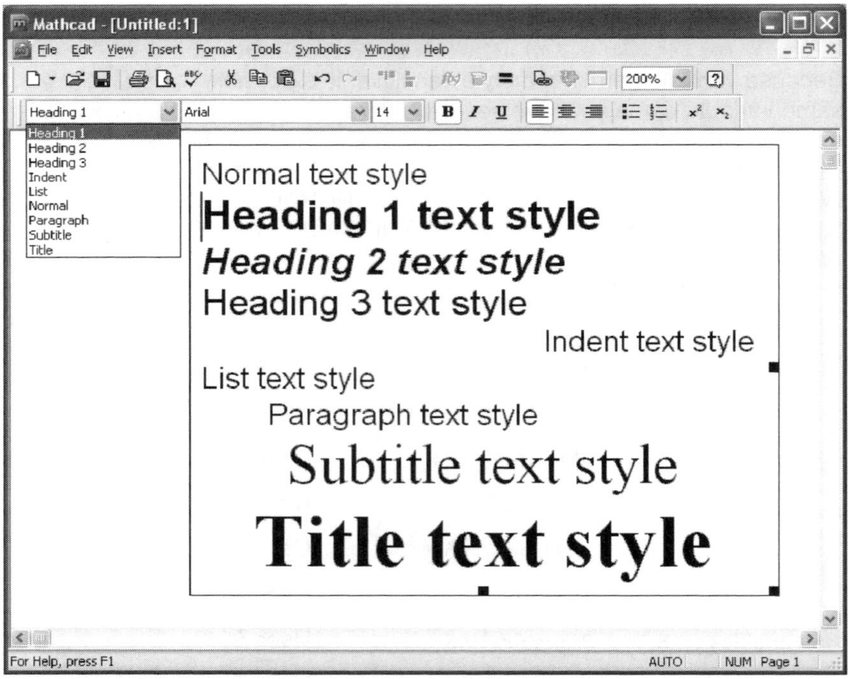

Figure 6.10 Default text styles

The default font styles are only a beginning. The next section discusses how to change existing styles and how to create new styles.

Changing and creating new math styles

We have just discussed the styles that come with the standard Mathcad installation. The real power in using styles is in creating styles that fit your own specific needs. For example, if you are part of a corporation, you can have customized styles so that all worksheets have a consistent look. If you are a student, you may want to create specific styles to give your worksheets a unique look. Professors may want homework to be submitted with consistent styles. This section focuses on teaching how to create and change styles. The changes you make to styles will only be effective in the current worksheet, not to all worksheets. Chapter 7 will discuss the ways to save your styles to be used over and over again as a template.

Changing the "Variables" style

The "Variables" style controls how the text in your equation looks. A very quick way to change the "Variables" style is to select any portion of the text in any expression, and then from the formatting toolbar change the font style or add bold, italics or underline. This will change the "Variables" style. Because most variables in your worksheet have been attached to the "Variables" style, they will all be changed to match the revised style. Another way to change the "Variables" style is to select **Equation** from the **Format** menu. You will see a dialog box similar to Figure 6.11. Select **Variables** from the "Style Name," and click on **Modify**.

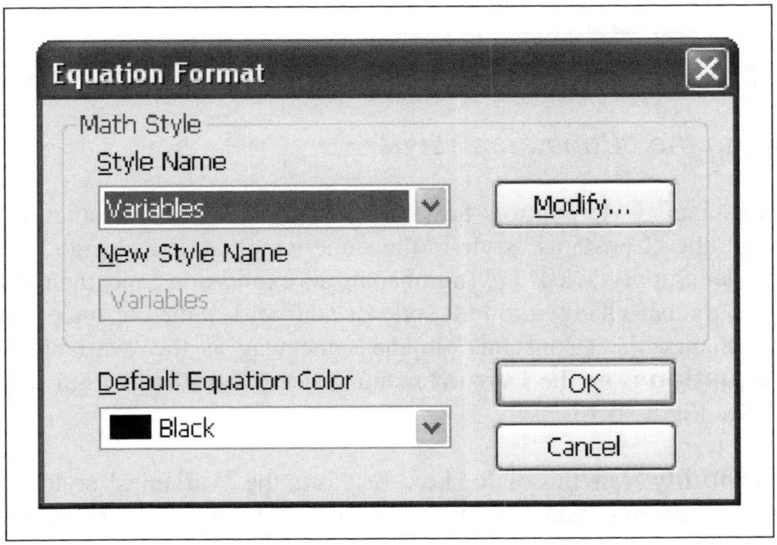

Figure 6.11 Changing the "Variables" style

From here you will be able to change font type, font style, font size, and font color. See Figure 6.12 for an example of the Variables font dialog box.

The changes you make here will occur for all variables in your worksheet.

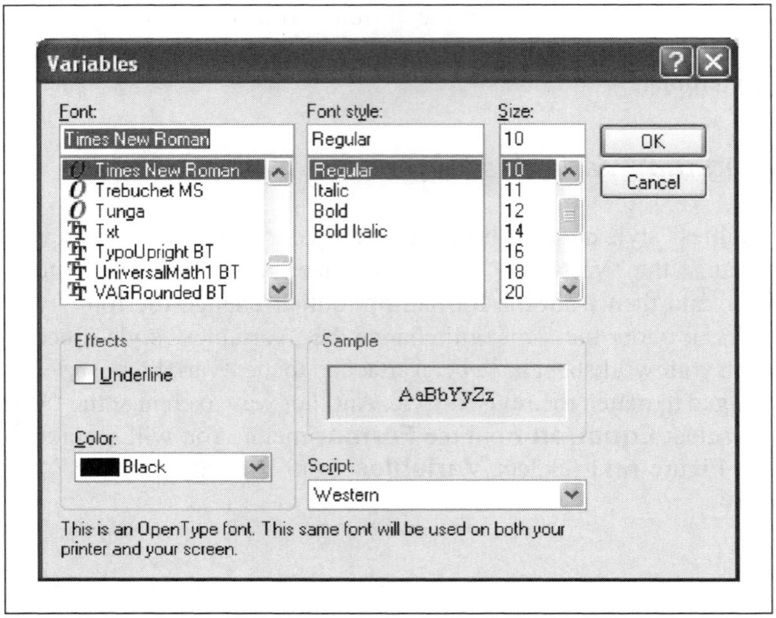

Figure 6.12 Variables dialog box

Changing the "Constants" style

The "Constants" style controls how numbers look in your calculations. You can change the "Constants" style in the same way that you change the "Variables" style. Simply select any number in an expression, and then from the formatting toolbar change the font style or add bold, italics or underline. You can also change the "Constants" in the same way as the "Variables" style. Select **Equation** from the **Format** menu. Select **Constants** from the "Style Name." See Figure 6.13.

Click on **Modify** from this dialog box. Just as in the "Variables" style, you will be able to change font type, font style, font size, and font color. The changes you make will affect all constants in your worksheet.

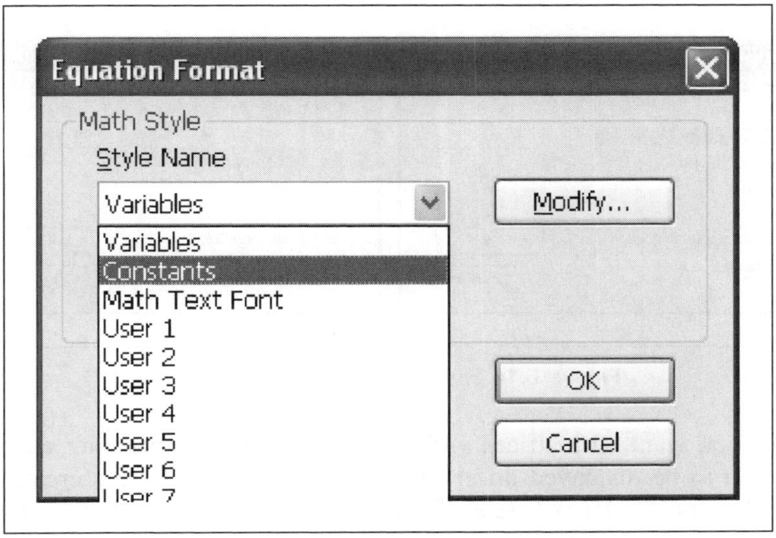

Figure 6.13 Changing the "Constants" style

Creating new math styles

You may want to have certain types of variables stand-out from the other variables in your worksheet. You can either change the User 1 through User 7 styles mentioned earlier, or you can rename any one of the User styles to a new name and change the characteristics of the style. To change the name of a User style, select **Equation** from the **Format** menu. Click on the drop-down arrow and select one of the User styles. The "New Style Name" box should now contain the name of the User style. You can now overwrite the name of the User style with your new style name. You can also now customize your new style by clicking on the **Modify** button. In Figure 6.14, the "User 1" style is renamed "Sample Style 1."

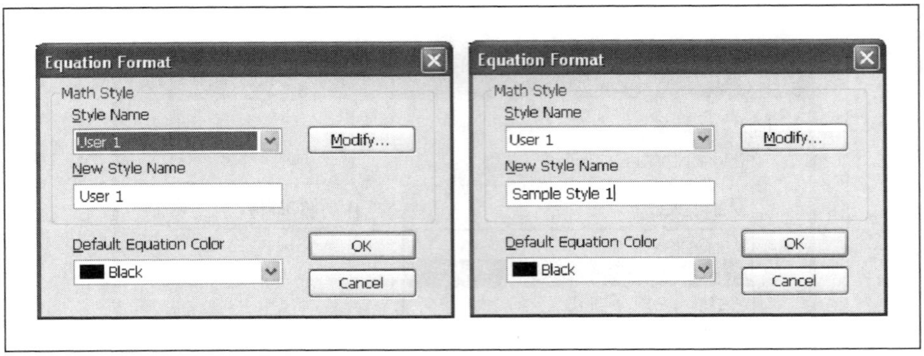

Figure 6.14 Renaming default style names

Suppose you want all matrices variables (arrays and matrices are covered in Chapter 9) to be displayed different than other variables. Let's create a new math style named "Matrix." Select **Equation** from the **Format** menu. Click on the drop-down arrow to the right of the box and select **User 7** from the list. Click in the "New Style Name" box and type **`Matrix`**. Now click on the **Modify** button. For this example we select Roman font, 12 point font size, with bold. Now when we attach this new math style to a variable it stands-out from other variable names. See Figure 6.15.

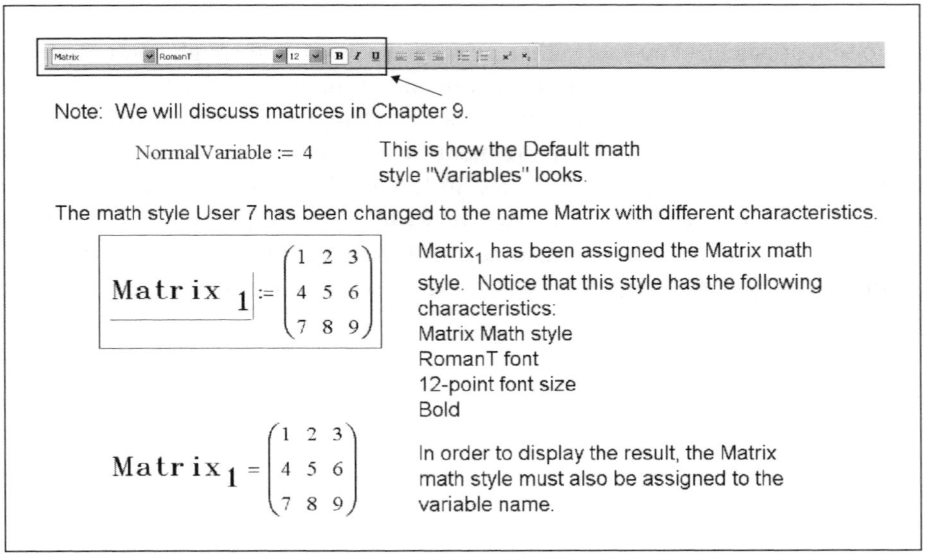

Figure 6.15 Variable with the new "Matrix" math style

Each time you have a matrix in your worksheet you can now attach this "Matrix" math style to the variable names.

Changing and creating new text styles

Having and using a good selection of text styles will add variety to your engineering calculations, and make them easier to follow. This section discusses how to change the default Mathcad text styles, and how to create new text styles.

There are more characteristics that can be changed with text styles than with math styles. Math styles allow you to change the font, bold, italic, underline, font size and color. With text styles, you can change all of the above, but you are also allowed the following font characteristics: strikeout, subscript, and superscript. Text styles also allow you to set many paragraph features such as: left indent, right indent, first line indent, hanging indent, bullets, numbering, paragraph alignment, and tab settings. Another feature about text styles is the ability to base one text style on another text style. See Figure 6.16. In this figure the style "Subtitle" is based on the style "Title." This means that the "Subtitle" style will have all the font and paragraph characteristics of the "Title" style, except it will have 18 point font and the bold is turned off. (See Figure 6.18 for a description of the "Title" style.)

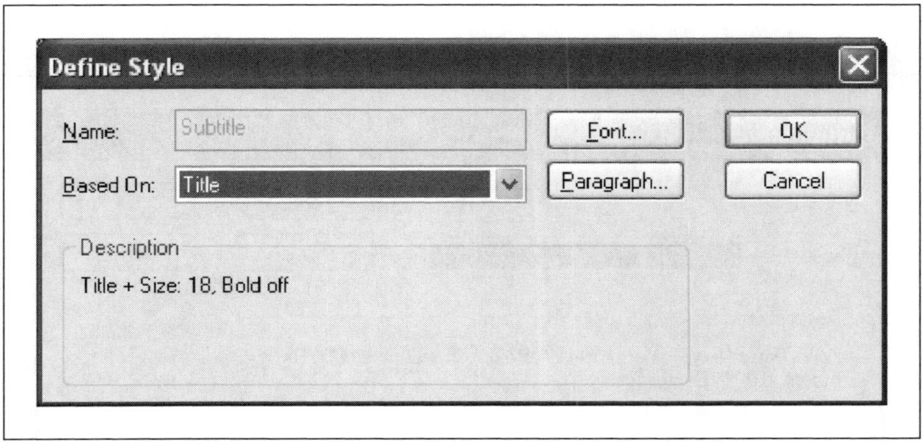

Figure 6.16 Basing one text style on another text style

Changing text styles

To change a text style you must use a dialog box. It is not possible to change text styles just by selecting text and changing the font properties, as we did with math styles. To change a text style, select **Style** from the **Format** menu. This brings up a dialog box similar to Figure 6.17.

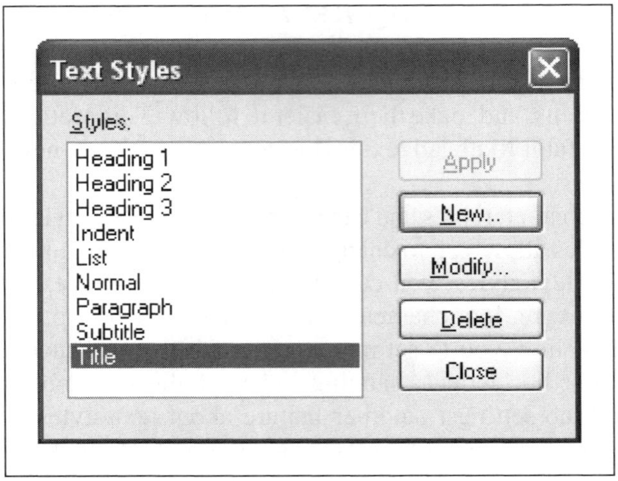

Figure 6.17 Text Styles dialog box

Click on the style you want to edit, and then click on **Modify**. This brings up the Define Style dialog box. See Figure 6.18.

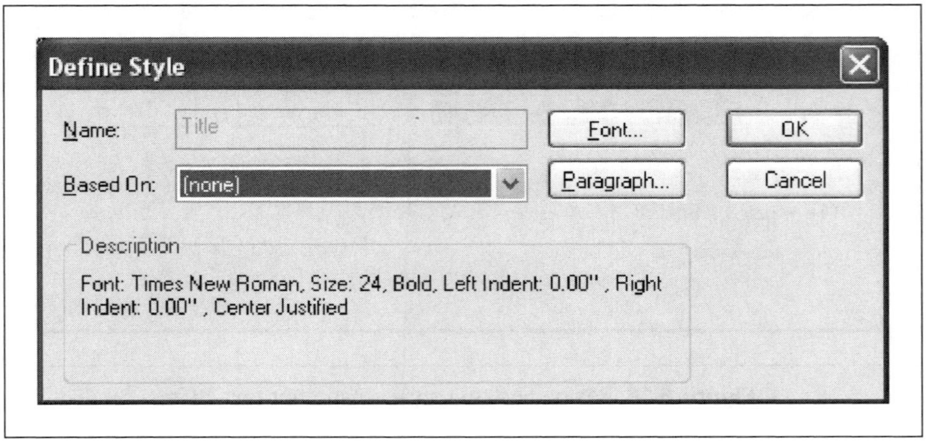

Figure 6.18 Define Style dialog box for text styles

In Figure 6.18, we have chosen to edit the "Title" text style. Notice the description of the font and paragraph characteristics of the style. The font characteristics are: Times New Roman font, Size 24, and Bold. The paragraph characteristics are: No Indent and Center Justified. If you click on the **Font** button, Mathcad brings up the Text Format dialog box. See Figure 6.19.

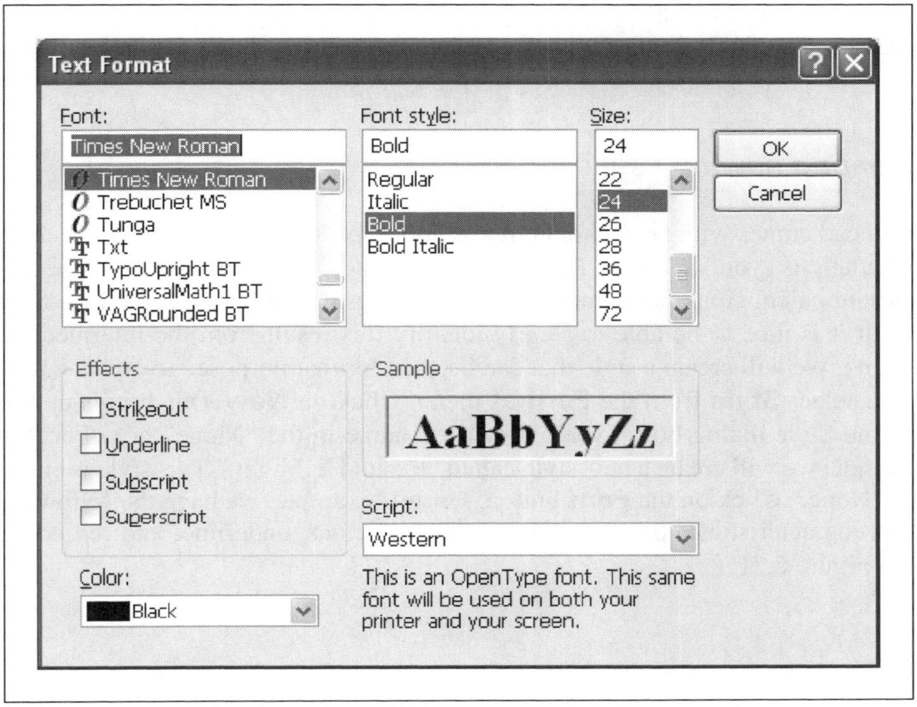

Figure 6.19 Text Format dialog box for text styles

From this box you can change many different font characteristics for the selected style. If you click on the **Paragraph** button from the Define Style dialog box, Mathcad brings up the Paragraph Format dialog box. See Figure 6.20.

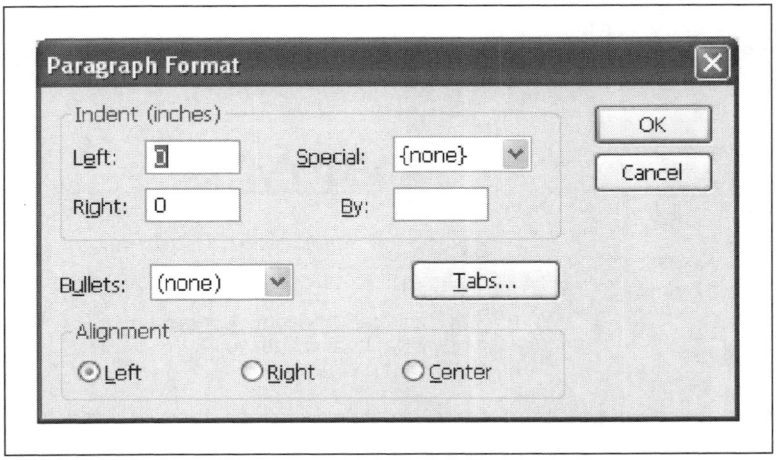

Figure 6.20 Paragraph Format dialog box for text styles

From this box you can change many different paragraph characteristics for the selected style. We will discuss more about the specifics of this dialog box later.

Creating new text styles

Mathcad comes with several default text styles. As you create your engineering calculations, you will most likely want to create some new text styles. Engineering calculations have many intermediate results. When you get to a final result it is nice to be able to clearly identify this result from the intermediate results. We will create a style that can be used for that purpose. To create a new style select **Style** from the **Format** menu. Click on **New** This brings up the Define Style dialog box. Type a new style name in the "Name" box. For this example we will create a new style called "Result Highlight." This style is based on "None." Click on the **Font** button. For this example, we have the following font characteristics: Arial Font, Bold, 20 point font, underline, and red color. See Figure 6.21.

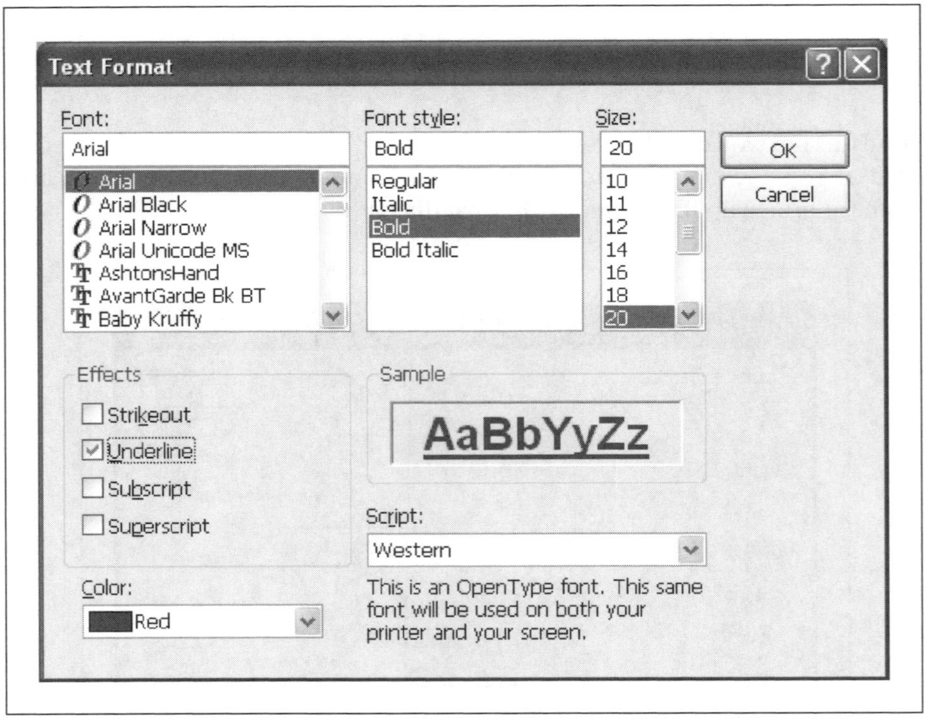

Figure 6.21 Font characteristics of the new text style "Result Highlight"

When you are done changing the font characteristics, click on **OK**. Next, click on the **Paragraph** button from the Define Style dialog box. For this example **Bullets** is selected. All other paragraph characteristics are left at the default. See Figure 6.22.

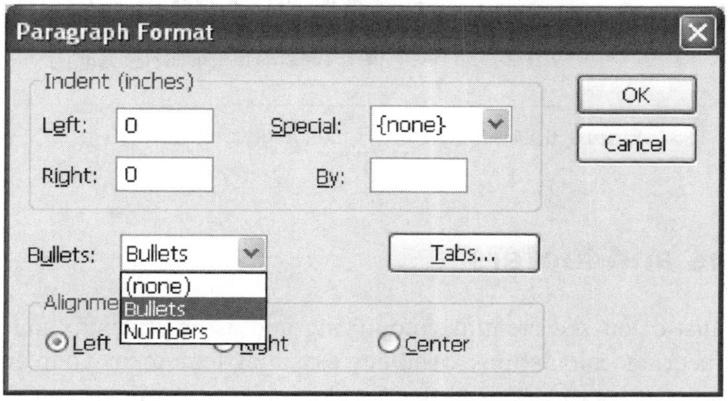

Figure 6.22 Paragraph characteristics for the new text style "Result Highlight"

Figure 6.23 shows the characteristics now assigned to the new style. Click **OK** in the Define Style dialog box, and click **Close** in the Text Styles dialog box.

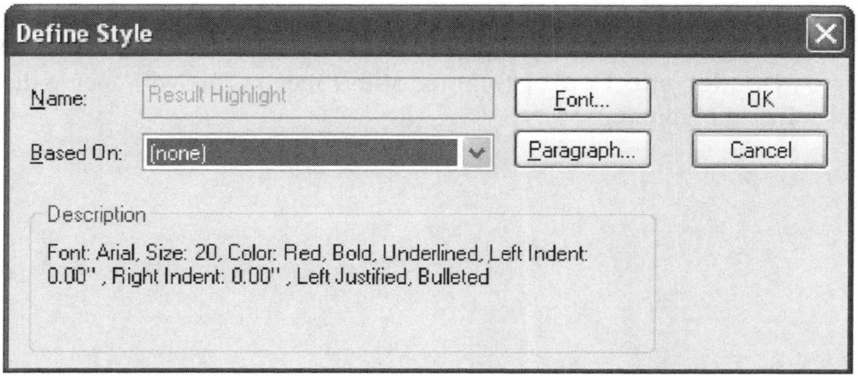

Figure 6.23 Style characteristics of the new text style "Result Highlight"

Figure 6.24 shows how this new text style can be used to emphasize the final results of a series of equations.

$Sample_A := 5m + 3.25m + 0.50m$

$Sample_A = 8.75\,m$ Intermediate result

$Sample_B := 5N + 4.5N + 3.25N$

$Sample_B = 12.75\,N$ Intermediate result

$Sample_C := Sample_A \cdot Sample_B$

$Sample_C = 111.563\,J$ • **This is a final result**

(The above style is "Result Highlight")

Figure 6.24 Using the "Result Highlight" text style

Headers and footers

We have just discussed creating, modifying and using styles to add variety to your engineering calculations. Another essential feature in your engineering calculations is a way to clearly identify the project information at the top and bottom of each page. This can easily be done by using headers and footers. When using Mathcad to create and organize engineering calculations, headers and footers are absolutely critical. This section will discuss how to create headers and footers, and it will discuss what information is important to include in the headers and footers.

Creating headers and footers

To create a header or footer, open the Header and Footer dialog box. To do this, click on **Header and Footer** from the **View** menu. This will open a dialog box similar to Figure 6.25.

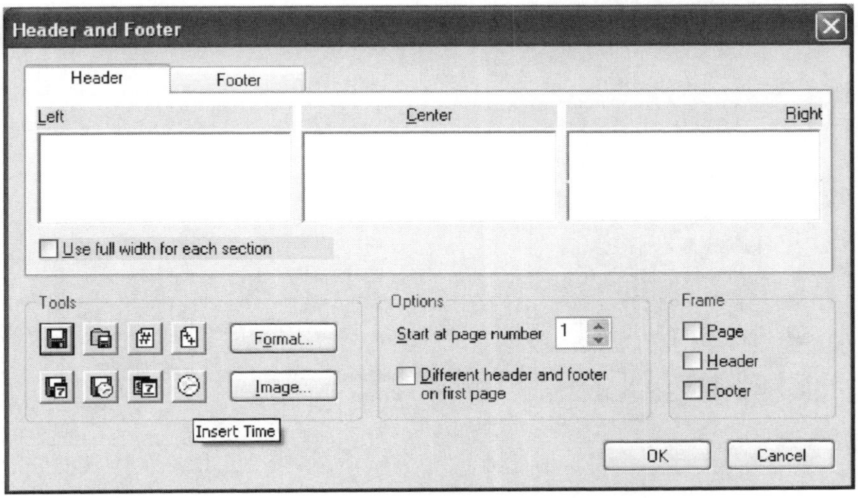

Figure 6.25 Header and Footer dialog box

You will notice that there are left, center, and right sections. There are icons in the "Tools" section that can be used to insert codes that will automatically update as information changes. These icons allow you to insert the following: file name, file path, page number, total number of pages, save date, save time, current date, and current time. The **Format** button allows you to change font characteristics for your header or footer. The **Image** button allows you to insert a bit map (.bmp) image. The "Start at page number" allows you to tell Mathcad on which page to begin your header or footer. If there is a check in front of "Different header and footer on first page," then two additional tabs are added at the top for Page 1. See Figure 6.26. The check boxes in the "Frame" section put borders around the page, the header or the footer. The tab for Footer looks identical to the tab for Header.

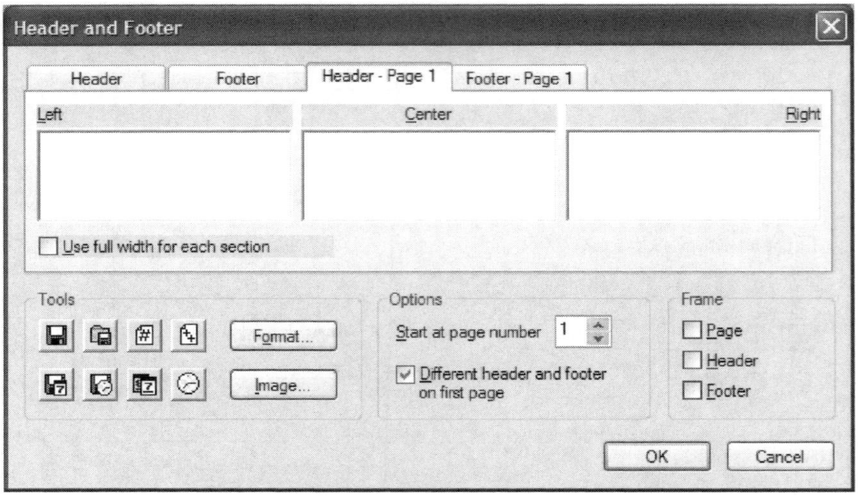

Figure 6.26 If there is a check by "Different header and footer on first page," then two additional tabs appear at the top

Information to include in headers and footers

Headers and footers are essential for printing engineering calculations created using Mathcad. Since Mathcad calculations can be changed very easily, it is important to know how current the printed calculations are. Thus, it is important to know when the file was saved, and when the file was printed. The save date is important because you will want to be able to make sure that the printed calculations are the same as the saved calculations. You can do this by comparing the save date on the printed calculations to the save date of the worksheet file.

The file name is also useful to have on printed calculations. This will help in locating the file for future reference. It will also help in making sure that the calculations are from the correct file. If you use several network drives, then the path is also a useful item to include on the printed calculations.

Page numbers are essential. It is suggested that you use this format: Page n of nn. This helps ensure that all printed calculations are kept together. If your calculations are contained in several files, it is also helpful to list a subject in front of the page number.

Company logos, photos, or scanned images can be included in a header or footer. The image must be a bit map (.bmp) file. The size of the bit map image must be adjusted to fit properly within the header or footer. You may need to use a separate imaging program to resize the image to best fit within your header or footer.

Other things you may want to consider adding to your header or footer include: project title, project number, the part of the project the calculations are for, who created the calculations, who checked the calculations, etc.

Examples

Figures 6.27 and 6.28 show a sample header and footer.

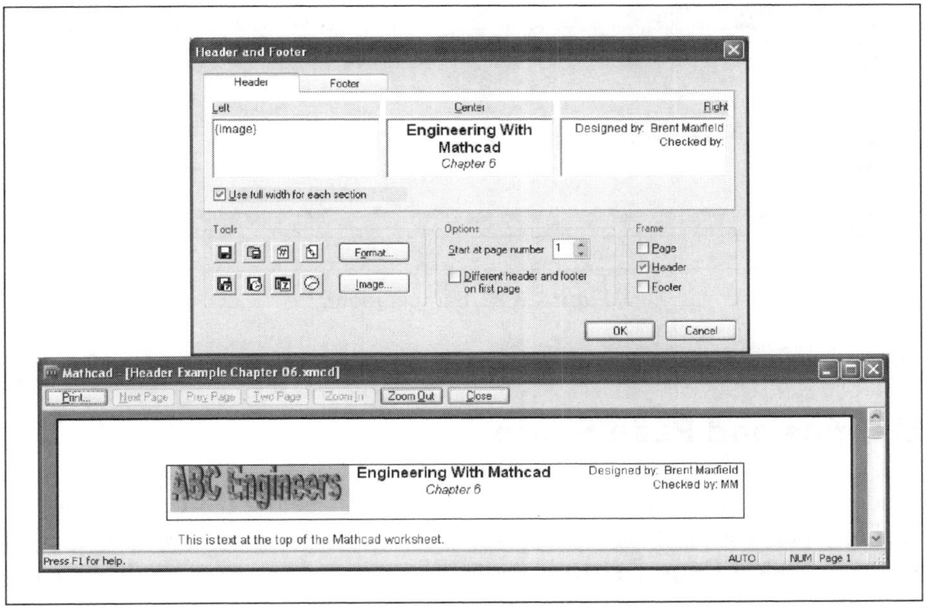

Figure 6.27 Sample header

Note that the center portion of the header in Figure 6.27 has some font format characteristics that were added. The image of the company logo was inserted from a bit map file. The image size needed to be adjusted to fit the header space. In order to fit the title in the center section on a single line, the check box, "Use full width for each section," needed to be checked.

The footer in Figure 6.28 has both the save date and the print date. The save date is more important that the print date, because you will want to be able to compare how current the printed calculations are, compared to the saved worksheet file.

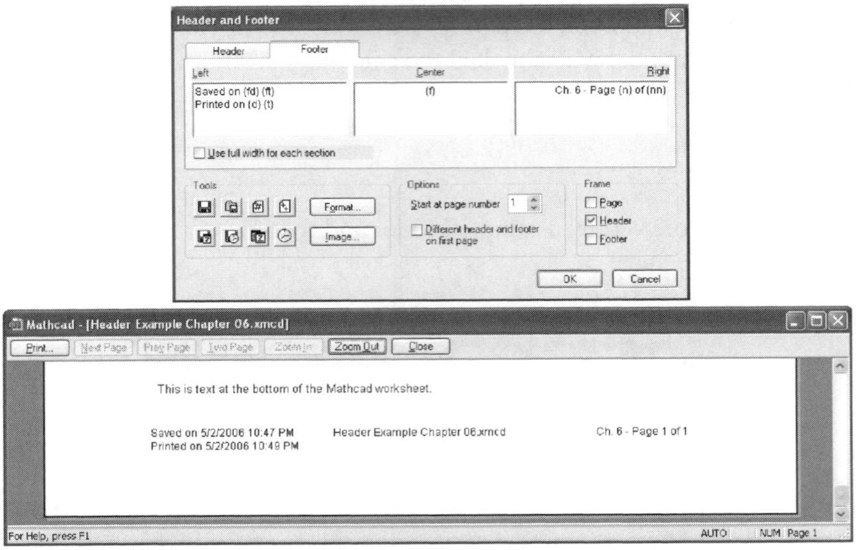

Figure 6.28 Sample footer

Margins and Page Setup

Headers fit above the top margin, and footers fit below the bottom margin. Setting paper size and margins is very similar to other Microsoft Windows based programs. Select **Page Setup** from the **File** menu. This opens the Page Setup dialog box. From this dialog box, you can select your paper size and your page margins. See Figure 6.29.

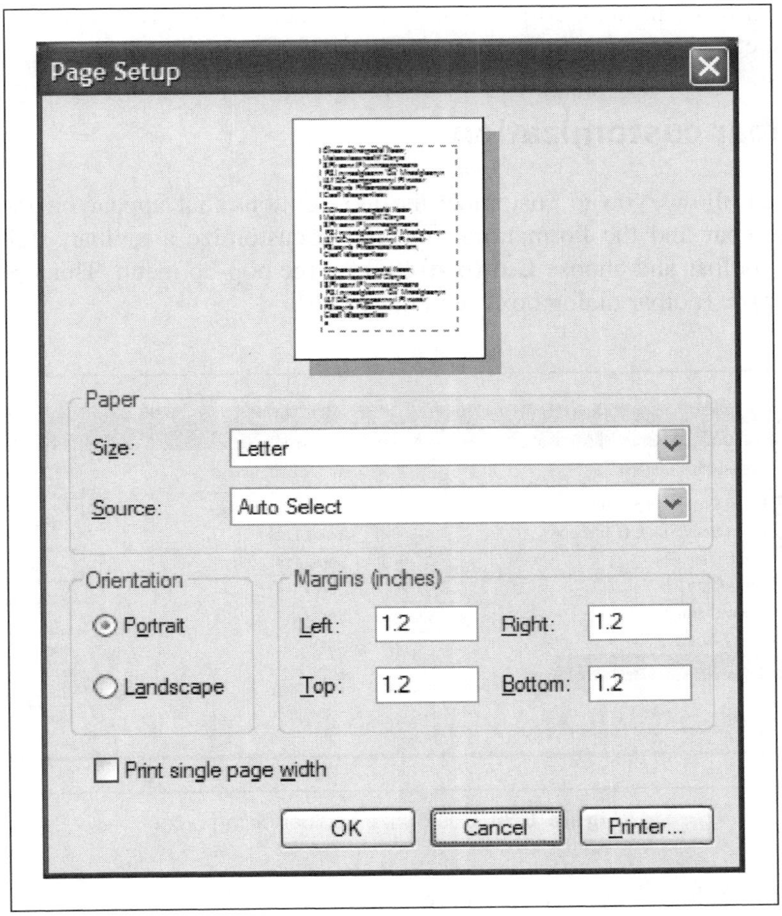

Figure 6.29 Page Setup dialog box

The top and bottom margins you select will need to be large enough so that the amount of information contained in your header or footer will display. If you print a worksheet and you are missing information from the header or footer, you will need to increase the top or bottom margin settings. You could also try to use a smaller font in your header or footer in order to fit the information within the specified margin.

If you put a check in the "Print single page width" box, then Mathcad will not print anything to the right of the right margin. This is a very useful feature for engineering calculations. It allows you to use the area to the right of the right margin for non-printing information. You can include things such as notes to yourself, instructions for others using the worksheet, custom units that do not need to be printed, formulas that do not need to be printed, intermediate results, etc.

Toolbar customization

Mathcad allows you to customize the icon buttons that appear on the Standard toolbar and the Formatting toolbar. To customize a toolbar, right-click on the toolbar and choose **Customize** from the pop-up menu. This opens the Customize Toolbar dialog box. See Figure 6.30.

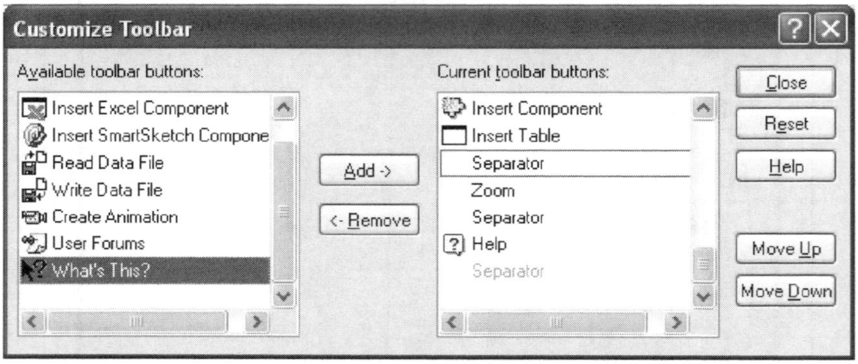

Figure 6.30 Customize Toolbar dialog box

From this dialog box you can select from the available buttons on the left and add them to the current toolbar buttons on the right. You may also remove buttons from the current toolbar.

A quick way to remove a button from a toolbar is to press the **Alt** key and drag the button off the toolbar.

Summary

In Chapter 6 we:

- Learned that Mathcad uses styles for all variables, constants, and text.
- Showed how you can customize the default Mathcad styles.
- Discussed how to create your own custom styles.

- Explained when you may want to use a different math style.
- Explained how to use different text styles to highlight different parts of your calculations.
- Discussed headers and footers.
- Recommended what information is critical to have in your headers and footers.
- Showed how to customize the Mathcad toolbars.

Practice

1. On a blank worksheet type `Sample.1:25kg`. Then type `Sample.1=`. Next type `Sample.2:50kg` and `Sample.2=`. Notice how all variable names have the "Variables" math style. Change the math style of $Sample_1$ to "User 1." Notice the difference between the variable names "$Sample_1$" and "$Sample_2$." Now change the style of $Sample_1$ to "User 2" through "User 7." Compare the different look for each different style.
2. In a text box, type a heading and two paragraphs. Choose any topic to write about. Now experiment with the different text styles. Change the heading to each of the different heading and title styles. Change the two paragraphs to different text styles. Experiment with different combinations of styles.
3. Change all the variable names in exercise 1, back to the "Variables" style. Now experiment with changing the characteristics of the "Variables" style. Change the font style, the font size, the color, bold, etc. If the regions overlap, use the **Separate Regions** command from the **Format** menu.
4. Create a new math style. Select a name for your style. Make it unique from the "Variables" style. List the characteristics of your new math style. Change the math style of $Sample_1$ to your new style.
5. Change the characteristics of the text style to "Normal." Change the font style, font color, and other characteristics. Type several text boxes to see how the new style looks.
6. Create five new text styles and assign these styles to several different text boxes.
7. Create a header and footer for use with your company or school. Include the items mentioned in this chapter. Also include a graphic with your header or footer. Place some text at the top and bottom of your worksheet. Print your worksheet page. Did the entire header and footer print? Did the text at the top and bottom of your page print?

8. Open the Page Setup dialog box. Adjust the top and bottom margins so that your full header and footer will print. Adjust the left and right margins to see how it affects your worksheet. Place a check in the "Print single page width" checkbox. Now print your worksheet page. Did the entire header and footer print? Did the text at the top and bottom of your page print?
9. Add new icons to your Standard toolbar and to your Formatting toolbar.

7 Templates

In the previous chapters we have discussed many different things that will make your worksheets unique. We have discussed how to change many of the default Mathcad features. We have discussed styles and how they give a consistent look to your calculations. We have also discussed how to set specific formats for your numerical results. In most cases, the changes made only affected the specific Mathcad worksheet you were working in. In Chapter 7—Templates, we discuss how to save all of these customizations so that they can be used over and over again. We do this through the use of templates.

Templates are essential in order to have a consistent look for all your engineering calculations, especially if you are working with other engineers. Templates allow Mathcad customizations to be applied consistently to all calculations. This chapter discusses the benefits of templates.

Chapter 7 will:

- Tell what a template is.
- Discuss the type of information that is stored in a template.
- Show the templates that are shipped with Mathcad.
- Explain when to make use of these templates.
- Review the items discussed in Chapters 4, 5, and 6, and show how to include these items in a customized template.
- Suggest items to include in a customized template.
- Create a sample template.
- Show how to modify an existing template.
- Discuss the normal.xmct file, and show how to store a customized template in this file, so that it opens whenever Mathcad is opened.

Information saved in a template

A template is essentially a collection of information that Mathcad uses to set various settings when opening a new document. This information is stored in a template file. Every time Mathcad opens a file based on that template every thing is formatted the same way. A template can save the following information:

- Worksheet settings.
- Headers and footers.
- On-screen information.
- Margins.
- Unit settings.
- Result display settings.
- Styles.
- Fonts.
- Unit system.

Mathcad templates

Mathcad comes with many templates. Every time you open Mathcad you are actually opening a new worksheet based on the normal.xmct template. You can also open worksheets based on the other templates that are shipped with Mathcad. To do this, select **New** from the **File** menu. This opens the New dialog box. See Figure 7.1.

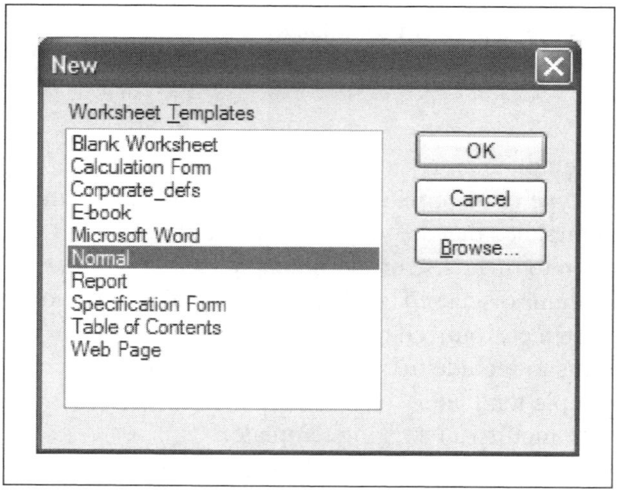

Figure 7.1 New dialog box for opening a new worksheet based on a template

From this dialog box, you can open a new file based on one of the many different templates. Take time to open worksheets based on each different template, and explore the differences between the worksheets opened with each template. Compare the math styles, text styles, headers, footers, margins, and tab settings. These templates are more useful as examples than as templates that you would use on a regular basis.

Another way to open a worksheet based on these templates is to click on the down-arrow adjacent to the new worksheet icon. A list of available templates will be shown. Click on one of these templates and a new worksheet based on that template will open. See Figure 7.2.

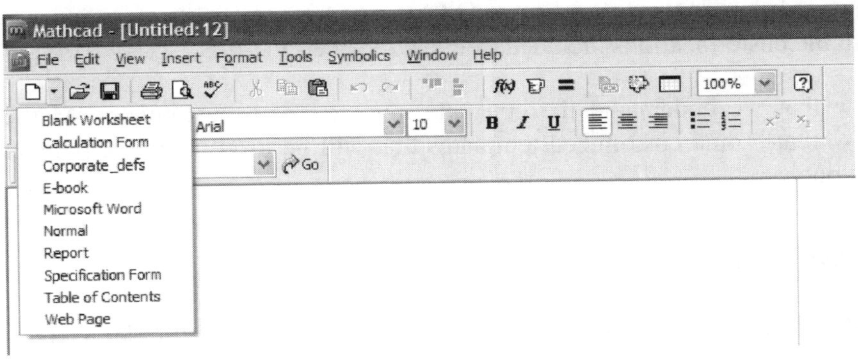

Figure 7.2 Using the new worksheet icon for opening a worksheet based on a template

Review of chapters 4, 5, and 6

Let's make a quick review of some of the things discussed in the previous three chapters. These are things that can be saved in a template.

In Chapter 4 we discussed units. The default unit system can be saved in a template. Custom units that you create can also be saved in a template.

In Chapter 5 we discussed Mathcad settings. Any of the settings changed in the Worksheet Options dialog box can be saved in a template. Also, any of the result formats set from the Results Format dialog box can also be saved in a template.

In Chapter 6 we discussed various math and text styles. We also discussed headers and footers. Each of these items can be saved in a template.

Creating your own customized template

Saving your own custom template is easy. Simply open a new worksheet based on the normal.xmct or on another existing template. Next, make the changes to the worksheet as discussed above. Once you have the worksheet to a point where you want to use it for a template, simply select **Save As** from the **File** menu. From the "Save as type" box choose **Mathcad XML Template (*.xmct)**. Mathcad versions earlier than version 12 will need to choose **Mathcad Template**. Type in a file name for the template, and store the template file in the "template" folder in the Mathcad program folder.

To open a new worksheet based on this new template, choose **New** from the **File** menu. The name of the template should appear in the list of templates. Click on the template, and then click on the **OK** button. A new worksheet should appear with the customizations included in the worksheet.

Now that we understand the concepts involved with customizing Mathcad, let's create some customized templates that will be used for the remainder of this book.

EWM Metric

Let's call the first template EWM Metric (for Engineering With Mathcad). Start by opening a new worksheet based on the Blank Worksheet template. This is easy to do, just click on the down-arrow next to the new worksheet icon and select **Blank Worksheet**. Since this will be a metric template, let's change the ruler to centimeters. Open the ruler by clicking **Ruler** from the **View** menu. Now right click on the ruler and select **Centimeters**.

Worksheet options

Let's start by setting worksheet options. Select **Worksheet Options** from the **Tools** menu. On the Built-In Variables tab, change the "Array Origin" from 0 to 1. Next, click on the Calculation tab. Add a check in the "Use ORIGIN for string indexing" box. This tab should look like Figure 7.3.

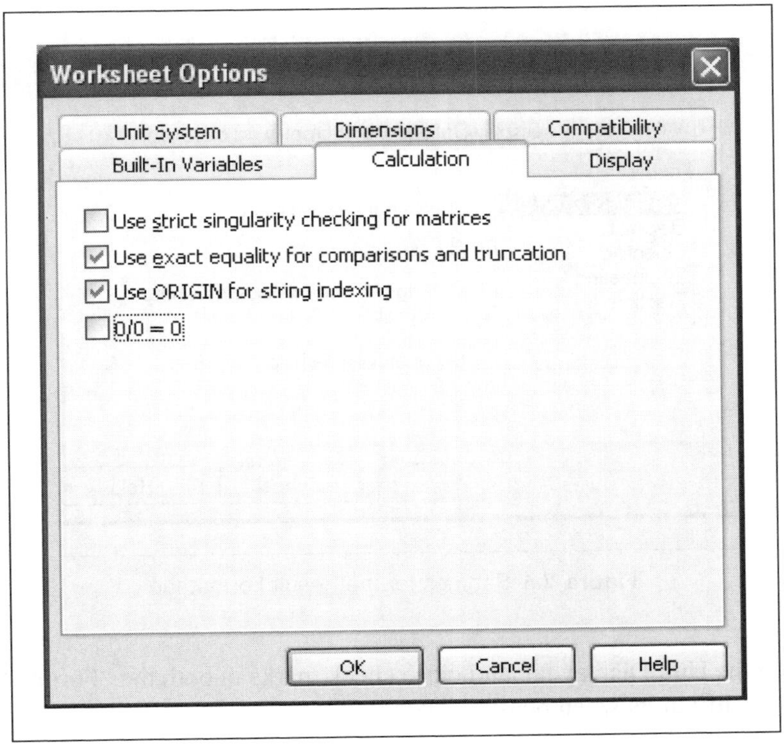

Figure 7.3 Settings for the Calculation tab

Now let's create a Custom Unit System based on SI units. On the Unit System tab select the "Custom" button and choose "Based on" **SI**. Keep the same "Base Dimensions." In the "Derived units," highlight **Volume (Liter)** from the list and click **Remove**. Click **OK** to close the Worksheet Options dialog box.

Next, let's set the Result Format. Select **Result** from the **Format** menu. On the Number Format tab select **General**. Set the "Number of decimal places" to **2** and place a check in the "Show trailing zeros" box. Change the "Exponential threshold" to **6**. This tab should look like Figure 7.4.

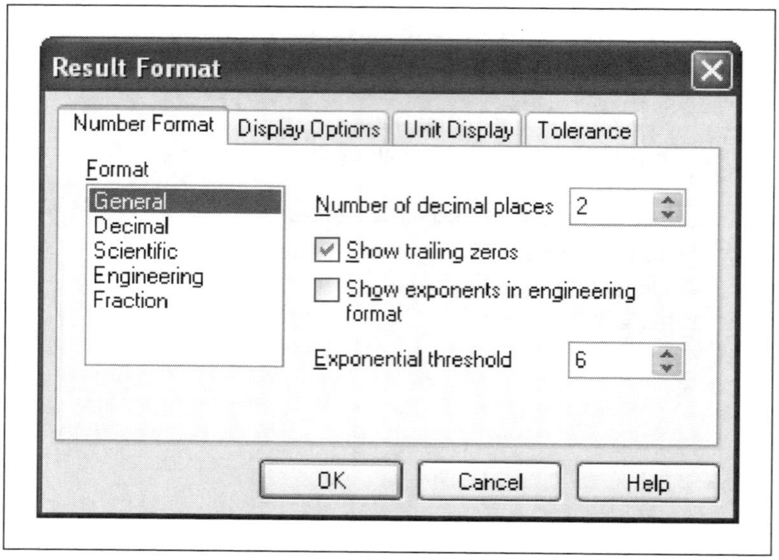

Figure 7.4 Settings for the Result Format tab

Click on the Unit Display tab and place check marks in both the "Format Units" and "Simplify units when possible" boxes.

Math styles

We will now modify the math styles for our template.

Select **Equation** from the **Format** menu. Select **Variables** from the "Style Name" and then click **Modify**. Choose **Times New Roman** "font," **bold** "Font Style," and **9** "Size". Leave the "Color" **Black**. The bold will make variables more prominent, and the 9-point font will allow more text to fit on the page. This tab should look like Figure 7.5. Click on **OK**.

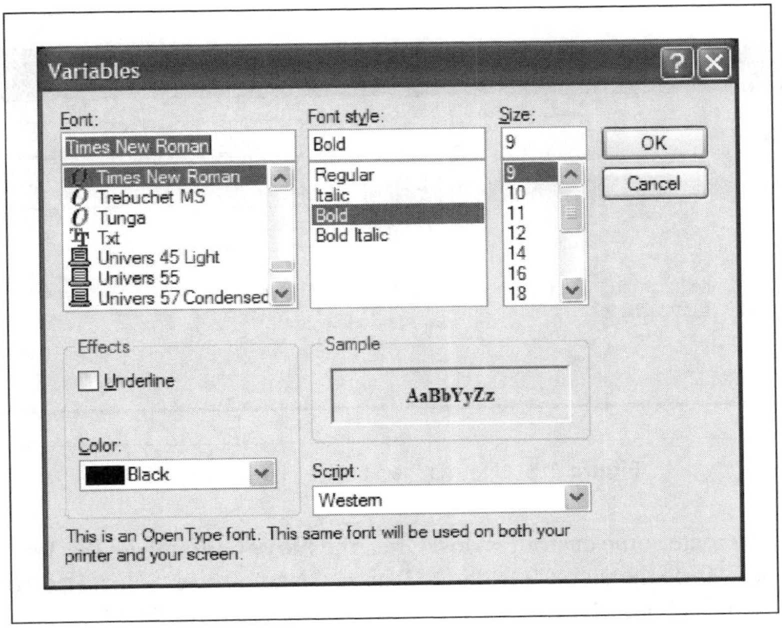

Figure 7.5 "Variables" math style characteristics

Now select **Constants** from the "Style Name," and then click **Modify**. Choose **Times New Roman** "font," **Regular** "Font Style," and **9** "size." Do not use bold. We want the variables to be more prominent than the constants. Click on **OK** to close the dialog box.

Text styles

Let's create some text styles for our template.

First, let's modify the normal text style. Select **Style** from the **Format** menu. Select **Normal** from the "Styles" list, and then click **Modify**. This style will be based on **(none)**. Click on the **Font** button. Choose **Arial** "Font," **Regular** "Font style," and **10** "Size." Next click on the down-arrow in the "Color" box and select **Blue**. This will make the text stand-out from the variables and constants. Click on **OK**. This style should now look like Figure 7.6. Click on **OK** again to close the Define Style dialog box.

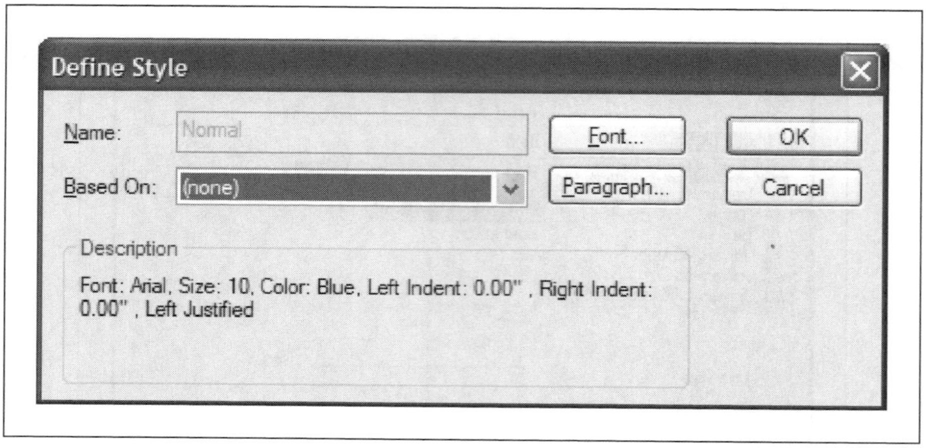

Figure 7.6 "Normal" text style characteristics

Now let's create some custom styles. Click the **New** button from the Text Styles dialog box. Type the name "Heading 1" in the "Name" box. Select **(none)** from the "Based On" box. Now click the **Font** button. Choose **Arial** "Font," **Bold** "Font style," and **18** "Size." Place a check mark next to "Underline" under the "Effects." Next, click on the down-arrow in "Color," and select **Green**. This will make the heading stand-out from other text. Click on **OK**. Your Heading 1 style should now look like Figure 7.7. Click on **OK** again to close the Define Style dialog box.

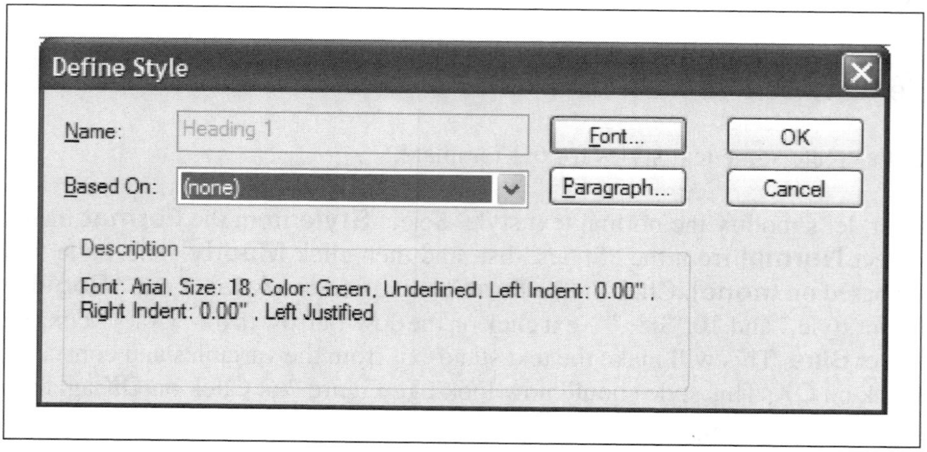

Figure 7.7 "Heading 1" text style characteristics

We will now create a "Heading 2" style. Select **New** from the Text Styles dialog box and type "Heading 2" in the "Name" box. Select **Heading 1** from the "Based On" box. Click the **Font** button and change the font size from 18 to 16.

Click on **OK**. Now click on the **Paragraph** button. Because we changed the ruler to centimeters, the indent is measured in centimeters. Change the left indent from 0 to 1.25. Click on **OK**. Your "Heading 2" style should now look like Figure 7.8.

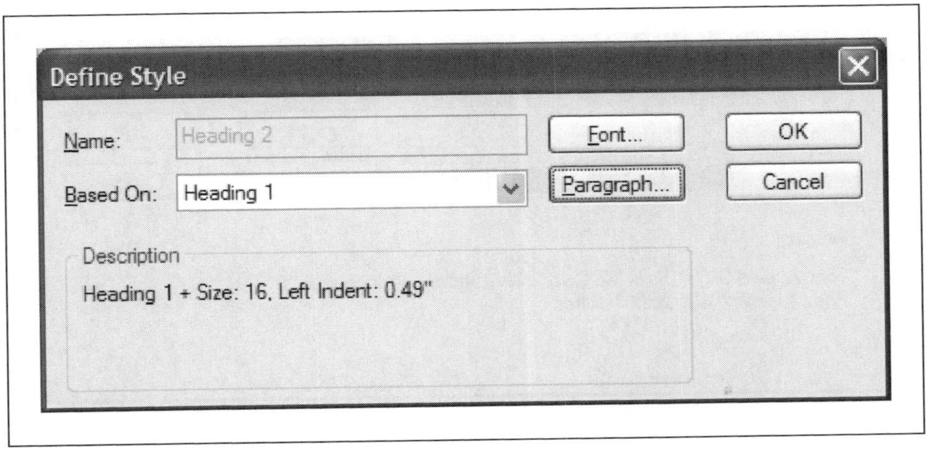

Figure 7.8 "Heading 2" text style characteristics

Because "Heading 2" is based on "Heading 1" if we change font characteristics of "Heading 1" such as adding italics or changing the font color, then the changes will also be reflected in "Heading 2."

For this example, let's add a couple more text styles. Add a new text style called "Results." This should have the following characteristics: Based on (none), Arial Black font, Bold, 12 point size, and Red color. The "Results" style should look like Figure 7.9.

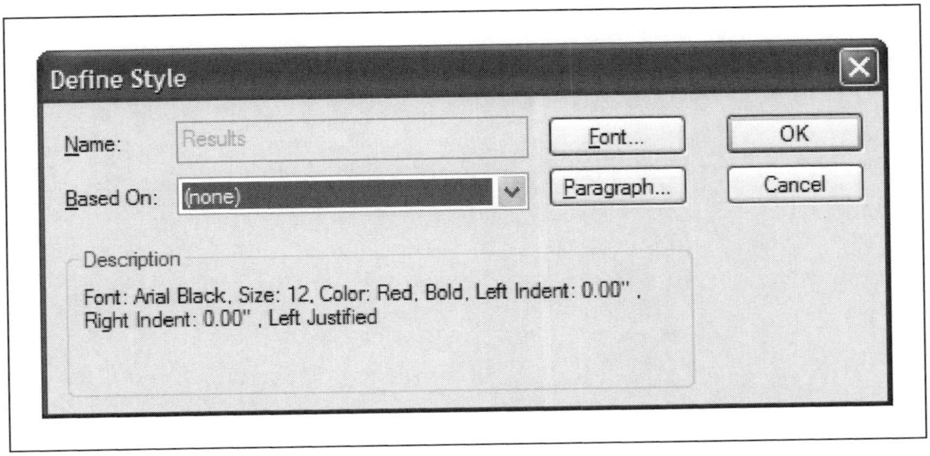

Figure 7.9 "Results" text style characteristics

The next text style will be called "Notes." This should have the following characteristics: Based on (none), Lucida Bright font, Italic, 12 point size, and Navy color. The "Notes" style should look like Figure 7.10.

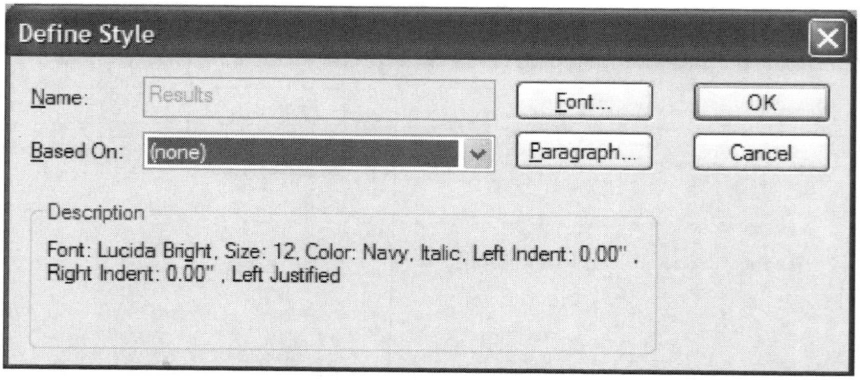

Figure 7.10 "Notes" text style characteristics

Margins

Select **Page Setup** from the **File** menu. Change both the left and right margin to 0.5. Change the top margin to 1.0 and the bottom margin to 0.75. Place a checkmark in the "Print single page width box." This will prevent Mathcad from printing information to the right of the right margin. You can now use the area on the right of the margin for notes and non-printing information. Your Page Setup dialog box should look like Figure 7.11.

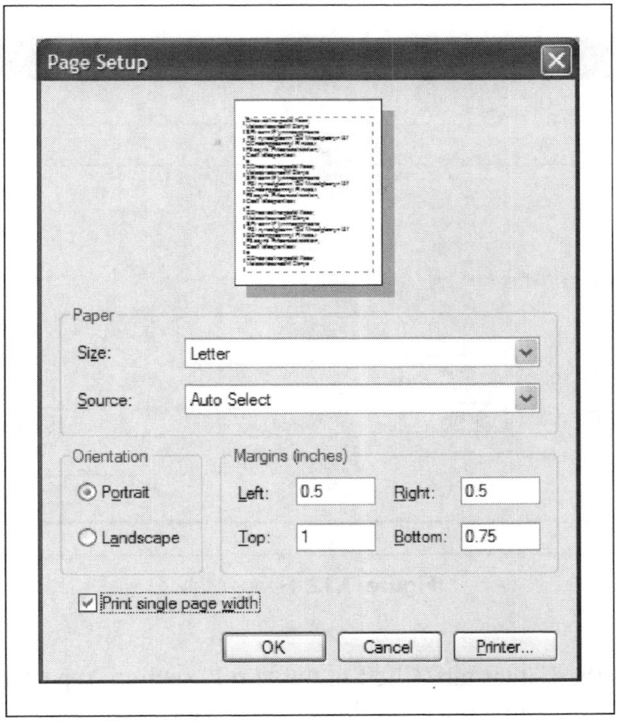

Figure 7.11 Page Setup settings

Headers and footers

Finally, let's add a header and footer. Select **Header and Footer** from the **View** menu. On the Header tab, type `Engineering with Mathcad` in the Center section. Your Header tab should look like Figure 7.12.

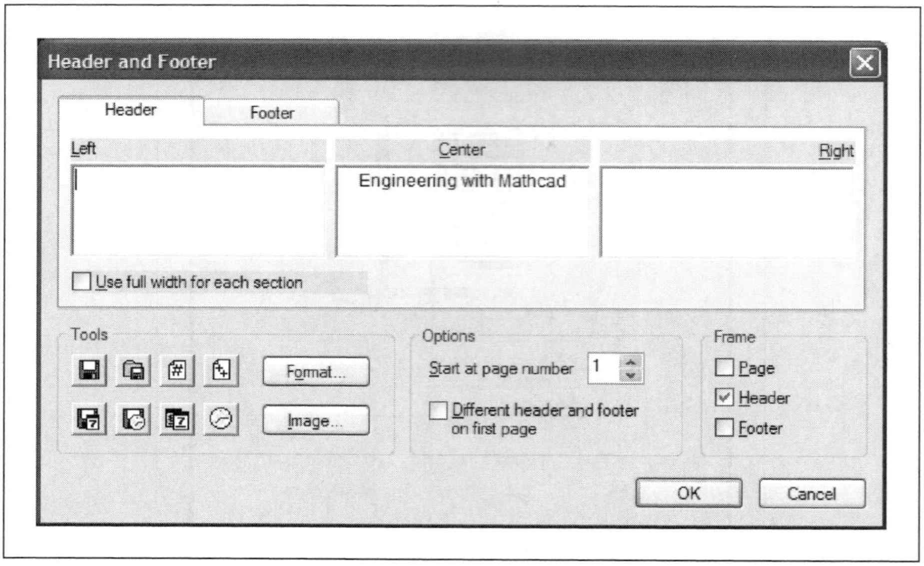

Figure 7.12 Header tab

Now click on the Footer tab. Click in the "Left" section, type **Printed**, and then in the "Tools" area click on the icon showing a calendar, and then click on the icon showing a clock. This inserts a {d} and {t} in the left section. This will print current date and time, thus printing the date and time the worksheet was printed. Press **Enter** to get a new line in the left section. Type **Saved**, and then in the "Tools" area click on the icon showing floppy disk with a number, and then click on the icon showing a floppy disk with a clock. This inserts a {fd} and {ft} in the left section. This will print the date and time that the file was saved.

Now click in the "Center" section. Type **Page** , and then click on the icon showing a sheet with the # symbol on it. Next type **of** and then click on the icon showing a sheet with two + symbols. This will print the current page number and the total page numbers.

Now click in the "Right" section. Type **File:**, and then click on the icon with a floppy disk. This inserts {f}. Mathcad will now print the name of the current file.

Now we will change the font size in the left and right sections. Select all the text in the "Left" section, and click on **Format**. Place your cursor in the "Size" box and type **6**. You cannot use the arrow keys to select 6. Click on **OK**. Now select the text in the "Right" section and click on **Format**. Change the font size to 6 just as you did in the "Left" section.

Your footer tab should now look like Figure 7.13.

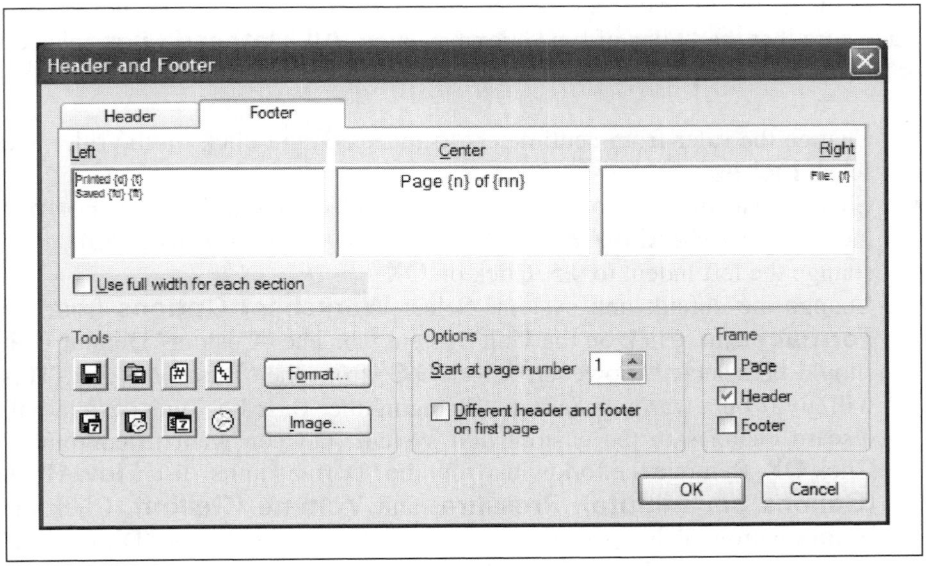

Figure 7.13 Footer tab

Saving the template

Now that we have made all these customizations, we are ready to save the template. Select **Save As** from the **File** menu. In the "Save as type" box, select **Mathcad XML Template (*.xmct)**. For versions prior to Version 12, select **Mathcad Template (*.mct)**. In the File name box type `EWM Metric`. In order to have the template available with all other Mathcad templates, you need to save the template in the template folder of the Mathcad program folder. This folder may possibly have the following path: C:\Program Files\Mathsoft\Mathcad 13\Template.

Once the template is saved, you can now open a new document based on this template. Simply select EWM Metric from the list of available templates as discussed earlier.

EWM US

Now let's create a similar template for US units. This will be much simpler. We will open the EWM Metric template, make a few changes, and then save it as EWM US.

To open the EWM Metric template, select **Open** from the **File** menu. Move to the Template folder of the Mathcad directory. The path may be as noted above. Select the EWM Metric.xmct file and click **Open**. If the file does not appear, make sure that the "Files of type" box has either **All Mathcad Files** selected or **Mathcad Templates** selected. We will make the following changes:

- Change the ruler from centimeters to inches. Right click on the ruler and select **Inches**.
- Change "Heading 2" to 0.5 inch indent. Select **Style** from the **Format** menu. Select **Heading 2** and click **Modify**. Click on **Paragraph**, then change the left indent to 0.5. Click on **OK**.
- Change the default unit system. Select **Worksheet Options** from the **Format** menu. Click on the Unit System tab. The "Custom" Default Unit should be selected; however, select **U.S.** from the "Based on" list. This will give you a warning stating, "Changing the 'Based on' unit system will discard changes to the custom unit system. Do you want to continue?" Click **OK**. Remove the following from the "Derived units" list: **Flow Rate (Gallons per minute)**, **Pressure**, and **Volume (Gallon)**. Click the **Insert** button. This opens the Insert Unit dialog box. In the "Dimension" list, scroll down to **Force Density**. Select **Pounds per cubic foot (pcf)** and click **OK**. Click **Insert** again. In the "Dimension" list scroll down to the **Force per Length**. Select **Pounds Force per linear foot (plf)**, and click **OK**. Click **Insert** again. Scroll down to **Pressure**, select **Pounds per square foot (psf)**, and click **OK**. Click on **OK** to close the Worksheet Options dialog box.

That is all we have to do. We want the rest of the template to remain the same. Now select **Save As** from the **File** menu. If you have not saved anything since opening EWM Metric, the directory should already be in the Template folder. Make sure the "Save as type" box reads **Mathcad XML Template (*.xmct)**. Change the file name to EWM US and click **Save**.

You should now have two customized templates that you can use anytime you want.

Normal.xmct file

To create a new file based on one of the two customized templates, you must manually select the template as explained earlier. However, there is a way to have the EWM Metric template be used every time you start Mathcad.

To have Mathcad always open with your customized template, open Windows File Manager or My Computer then go to the Mathcad Templates directory.

Rename the "Normal.xmct" template to "Original Normal.xmct." Then copy "EWM Metric.xmct" and paste it in the same folder. This will create a copy of the file "EWM Metric.xmct." Rename the duplicate file "Normal.xmct." When Mathcad starts, it uses "Normal.xmct" as the default template. Now the template "EWM Metric" is your default template.

> **Tip!** We could have just renamed "EWM Metric.xmct" to "Normal.xmct," but I think it is best to keep the original file name in the directory. If you want to change the template, make a change to "EWM Metric.xmct" and then make a copy again and overwrite the "Normal.xmct" file.

There is one drawback to having the "Normal.xmct" template stored on your local hard drive. If you want to have consistent templates in your organization, everyone must have the same "Normal.xmct" file stored on their local hard drive. If you change the default template very often, this makes it difficult to keep everyone using exactly the same default template. You can have a corporate template stored on a network drive, and everyone could browse to the corporate template when they open a new worksheet, but that is difficult to enforce. Hopefully in future releases, Mathcad will allow you to point to the location of "Normal.xmct." This will allow you to have it stored on a network drive.

Summary

Templates are essential to engineering with Mathcad. They will give your calculations a consistent look. They allow a consistent appearance for all corporate calculations. Instructors can create a template and have all students use the same template so that all assignments have a consistent appearance. You may want to create different templates for different clients. There are many uses for templates, but whatever the use, you must create your own template and begin using it.

In Chapter 7 we:

- Discussed what information is stored in a template.
- Told how a template works.
- Reviewed Chapters 4, 5, and 6.
- Showed how to save a template.
- Created a customized template with many unique styles and settings.

- Demonstrated how easy it is to create a new template when based on an existing template.
- Encouraged you to create your own customized template and store it as Normal.xmct.

Practice

1. Create the EWM Metric template discussed in this Chapter.
2. Create the EWM US template discussed in this Chapter.
3. Create and save your own custom template. Start your template from a worksheet based on the "Blank Worksheet" template. Select a name for your template. Include the following in your template: Header, footer, new math styles, new text styles, custom worksheet settings, and custom number format settings. List the characteristics of your new template.
4. Open the template file you created in exercise 3. Refer to the custom units you created in Chapter 4 practice exercises. Place these custom units on the right side of the right margin, and save your template. Open a new worksheet based on this template. Did the custom units come into the new worksheet?
5. Rename the "Normal.xmct" template file to "Original Normal.xmct." Make a copy of your own custom template file. Rename this copied file "Normal.xmct." Now open a new worksheet to see if the new worksheet is based on your custom template.

8
Useful information—Part I

This chapter is a collection of useful information and tips related to the chapters in Part 1. There will not be a summary or practice exercises.

Variables

Custom Character toolbar

The Custom Character toolbar has characters that are difficult to type into Mathcad.

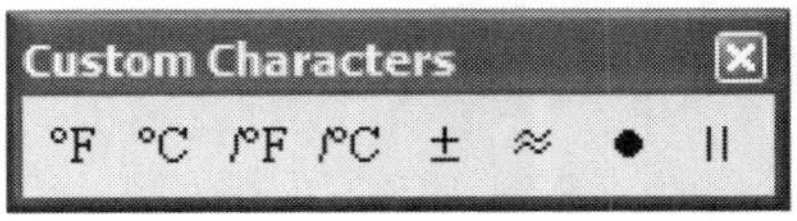

The temperature characters are used with the temperature functions. The remaining symbols are not predefined and can be used with any variable name.

QuickSheet

Open QuickSheets by clicking **QuickSheets** from the **Help** menu. Select **Extra Math Symbols** from the list of topics. This opens a list of symbols that can be used for variable names. See Figure 8.1. To use any of these characters copy and paste them into Mathcad, or drag and drop them into Mathcad.

This QuickSheet provides characters from the Symbol and extended Times New Roman fonts which can be reused for operators or user-defined functions.

°C	°F	ℵ	Å	ϒ
Σ	Π	∀	∃	∫
′	∝	∅	ƒ	•
∂	∩	∪	⊃	⊇
∈	∉	⊆	⊂	⊄
≈	÷	±	×	√
∠	⊗	°	∇	²
£	¥	$	□	¢
↔	↓	↑	〈	—
⇔	⇐	⇒	⇑	⇓
È	Æ	℘	ℑ	ℜ
§	®	ö	ð	‡
⎨	⎬	◊	∣	↵
♣	♥	♠	♦	Œ
…	¹	³	Þ	Đ

Figure 8.1 QuickSheet characters

Greek letters

Figure 8.2 is a list of the Greek letters and their Roman equivalents. To insert a Greek letter, use the Greek symbol toolbar, or type the Roman equivalent and then type **CTRL+G**. To insert π type **CTRL+SHIFT+P**.

To type a Greek letter into an equation or into text, press the Roman equivalent from the table below, followed by [Ctrl]G. Alternatively, use the Greek toolbar.

Name	Uppercase	Lowercase	Roman equivalent
alpha	A	α	A
beta	B	β	B
chi	X	χ	C
delta	Δ	δ	D
epsilon	E	ε	E
eta	H	η	H
gamma	Γ	γ	G
iota	I	ι	I
kappa	K	κ	K
lambda	Λ	λ	L
mu	M	μ	M
nu	N	ν	N
omega	Ω	ω	W
omicron	O	ο	O
phi	Φ	ϕ	F
phi (alternate)		φ	J
pi	Π	π	P
psi	Ψ	ψ	Y
rho	P	ρ	R
sigma	Σ	σ	S
tau	T	τ	T
theta	Θ	θ	Q
theta (alternate)	ϑ		J
upsilon	Y	υ	U
xi	Ξ	ξ	X
zeta	Z	ζ	Z

Figure 8.2 Greek Letters (*Source*: Greek Letters from Mathcad 11 User's Guide, Appendix H, Page 444. © 1986–2002 Mathsoft Engineering and Education, Inc)

Other characters

You can insert ASCII characters in variable names by holding the ALT key and then typing the ASCII code using the numeric keypad. Other characters are available by using the Windows Character Map, which can be opened by clicking: Start>Programs>Accessories>System Tools>Character Map.

List of predefined variables

Figure 8.3 gives a listing of Mathcad's predefined or built-in constants. Do not redefine these names.

Math Constants

Name	Keystroke	Default Value
∞	**[Ctrl][Shift]z**	10^{307}
e	**e**	Value of e to 17 digits, 2.7182818284590451
π	**[Ctrl][Shift]p** or **p[Ctrl]g**	Value of π to 17 digits, 3.1415926535897931
i or j	**1i** or **1j**	The imaginary unit.
%	**%**	0.01; multiplying by % gives you the appropriate conversion. You can type [expression]% for inferred multiplication or use it in the unit placeholder.
NaN	NaN	Not a number.

These constants retain their exact values in symbolic calculations. To redefine the value of any of these constants, use the equal sign for definition (:=) as you would to define any variable.

System Constants

Name	Default Value	Use
TOL	.001	Controls iterations on some numerical methods.
CTOL	.001	Controls convergence tolerance in Solve Blocks.
ORIGIN	0	Controls array indexing.
PRNPRECISION PRNCOLWIDTH	4 8	Controls data file writing preferences.
CWD	current working directory in the form of a string variable	Can be used as an argument to file handling functions.
FRAME	0	Controls animations.
ERR	NA	Size of the sum of squares error for the approximate solution to a Solve Block.
1L, 1M, 1T, 1Q, 1K, 1C, and 1S	assigned by the selected unit system	Define custom unit definitions for the base dimensions.

To redefine system constants, use the equal sign for definition (:=) in your worksheet, as you would to define any variable, or choose **Worksheet Options** from the **Tools** menu, and go to the Built-in Variables tab.

Figure 8.3 Mathcad's predefined variables

CWD

Current Working Directory (CWD) is a predefined Mathcad variable, and it can be used as an argument for a function. It is also handy to use when you are not sure what path your current worksheet is saved in. Just type **CWD:** and Mathcad will display the path name.

Information about Math regions

In-line division

In-line division is a way to save space when you have several divisions in your expression. It displays division similar to a textbook. To add an in-line division operator to your expression, type **CTRL+/** rather than just the **/**. You may also use the division ÷ icon on the Calculator toolbar. See Figure 8.4.

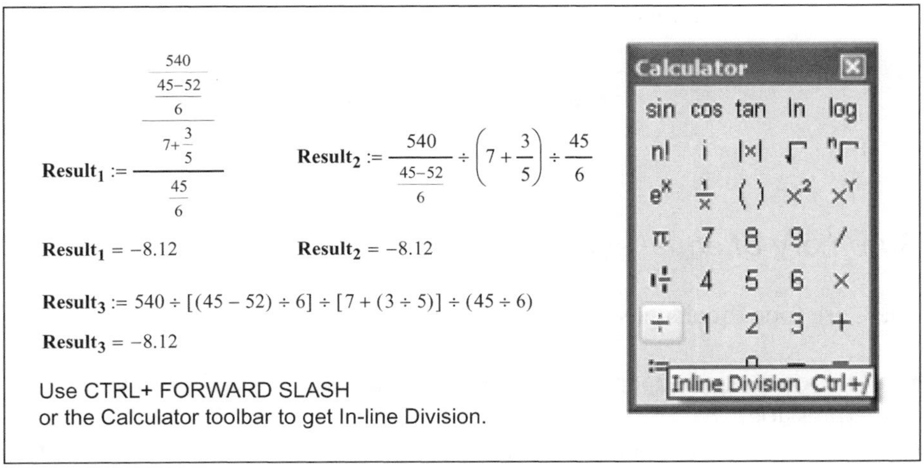

Figure 8.4 In-line division

Mixed numbers

We briefly discussed mixed numbers as a display result in Chapter 5—Mathcad settings. Here we will show you how to input a number in mixed number format. To enter a mixed number type **CTRL+SHIFT+PLUS** or use the icon on the Calculator toolbar. See Figure 8.5.

Conventional input. Note that first number was not input as 5.667.

$\mathbf{MixedNumber_1} := 5.666 + 3.333 + 5.625$

$\mathbf{MixedNumber_1} = 14.62$ Normal Result format

$\mathbf{MixedNumber_1} = \frac{1828}{125}$ Fraction Result format

$\mathbf{MixedNumber_1} = 14\frac{78}{125}$ Fraction Result format with "Use mixed Numbers" checked.

Same equation using mixed number format - CTRL+SHIFT+PLUS.

$\mathbf{MixedNumber_2} := 5\frac{2}{3} + 3\frac{1}{3} + 5\frac{5}{8}$

$\mathbf{MixedNumber_2} = 14.63$ Normal Result format

$\mathbf{MixedNumber_2} = \frac{117}{8}$ Fraction Result format

$\mathbf{MixedNumber_2} = 14\frac{5}{8}$ Fraction Result format with "Use mixed Numbers" checked.

Figure 8.5 Mixed numbers

Summary of equal signs

There are four equal signs used in Mathcad.

- The evaluation operator (=) `EQUAL SIGN` is used to evaluate an expression.
- The assignment operator (:=) `COLON` is used in variable or function definitions.
- The Boolean equality operator (**=**) `CTRL+EQUAL SIGN` is used to evaluate the equality condition in a Boolean statement. It is also used in symbolic equations.
- The global assignment operator (≡) `Tilda ~ or SHIFT+ACCENT` is used to assign a global variable. Global variables are evaluated before other variables, and thus can be assigned anywhere in the worksheet.

The evaluation operator, assignment operator and global assignment operator are found on the Evaluation toolbar. The Boolean equality operator is found on the Boolean toolbar.

Math Regions in text regions

To insert a math region in a text region click **Math Region** from the **Insert** menu. This places a blank placeholder in the text region, where you can type a math expression. After you are finished with the expression, use the right arrow to move you back into the text box, where you can continue typing text. See Figure 8.6.

You can include a math region $\mathbf{Example} := \frac{600\mathbf{N}}{3\mathbf{m} \cdot 2\mathbf{m}}$ in a text region by using the Math Region command from the Insert Menu.

$\mathbf{Example} = 100.00\ \mathbf{Pa}$

Figure 8.6 Including math regions in text regions

Tip! Before inserting a math region, I like to add one or two spaces after the insertion point. This makes it easier to continue typing after I insert the math region. It is not necessary, but it makes it easier to see the cursor after leaving the math region.

Math Regions that do not calculate

There will be many times when you want to display an equation prior to the point where the variables are defined. When you try to do this, Mathcad will give you an error message.

There are several ways to work around this.

- Type in the expression using variable names that have not yet been defined. Before clicking out of the Mathcad region, right-click and choose **Properties**. Click on the Calculation tab and place a check mark in the "Disable Evaluation" box. This will place a black box in the upper-right corner of your math region and will prevent Mathcad from evaluating the expression.
- Use the Boolean equality operator `CTRL+EQUAL SIGN` instead of the assignment operator `COLON`. This does not make a variable assignment, but it allows you to display what you want without an error.
- After you define your variables and expression, copy the math region that has the expression you want to display. Then move up to the location where you want to display the expression. Click **Paste Special** from the **Edit**

menu and select **Picture (Metafile)**. This displays a graphic image of the math region definition. This method is not recommended. You can imagine the confusion it can cause when trying to check a calculation. If you choose to use this method, then make it very clear that the region is a graphic image and not a Mathcad math region.

See Figure 8.7 for an example of using these different methods of displaying expressions.

$\mathbf{NoVariables_1} := \mathbf{a} \cdot \mathbf{x}^2 + \mathbf{b} \cdot \mathbf{x} + \mathbf{c}$ ∎	Using "Disable Evaluation" from the Properties dialog box.
$\mathbf{NoVariables_1} = ∎$	This value does not exist because the above expression was disabled.
$\mathbf{NoVariables_2} = \mathbf{a} \cdot \mathbf{x}^2 + \mathbf{b} \cdot \mathbf{x} + \mathbf{c}$	Using a Boolean equal
$\mathbf{NoVariables_2} = ∎$	This value does not exist because the Boolean equal sign does not define a variable.
$\mathbf{NoVariables_3} := \mathbf{a} \cdot \mathbf{x}^2 + \mathbf{b} \cdot \mathbf{x} + \mathbf{c}$	The image to the left is a pasted graphic image. It is not a math region. This method is not recommended, because it can be very confusing when trying to check the calculations. If you use this method, be sure to make it clear that it is a graphic and not a math region.
$a := 1 \quad b := 5 \quad c := 6 \quad x := 4$	
$\mathbf{NoVariables_3} := \mathbf{a} \cdot \mathbf{x}^2 + \mathbf{b} \cdot \mathbf{x} + \mathbf{c}$	This math region was copied and pasted as a graphic above.
$\mathbf{NoVariables_3} = 42.00$	

Figure 8.7 Displaying math regions without having variables defined

Customizing operator display

If you are publishing or presenting your calculations, you may want to display them using different operator symbols than are used in Mathcad. For example, you may want to see the equal sign $=$ instead of the $:=$ for variable definitions or the $\equiv$ for the global equal sign. You may want to see an x for a multiplication sign instead of a dot.

Mathcad has provided a means for doing this. It was discussed briefly in Chapter 5—Mathcad settings. To change the operator display, open the **Worksheet Options** from the **Tools** menu and choose the Display tab. See Figure 8.8. This dialog box allows you to change several operators, a few of which we have not yet discussed. The changes made here will affect the entire worksheet.

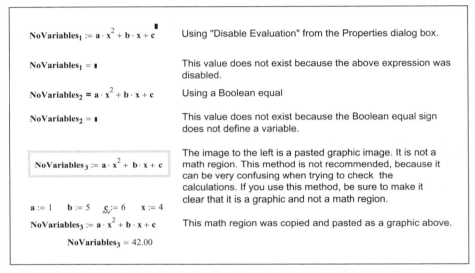

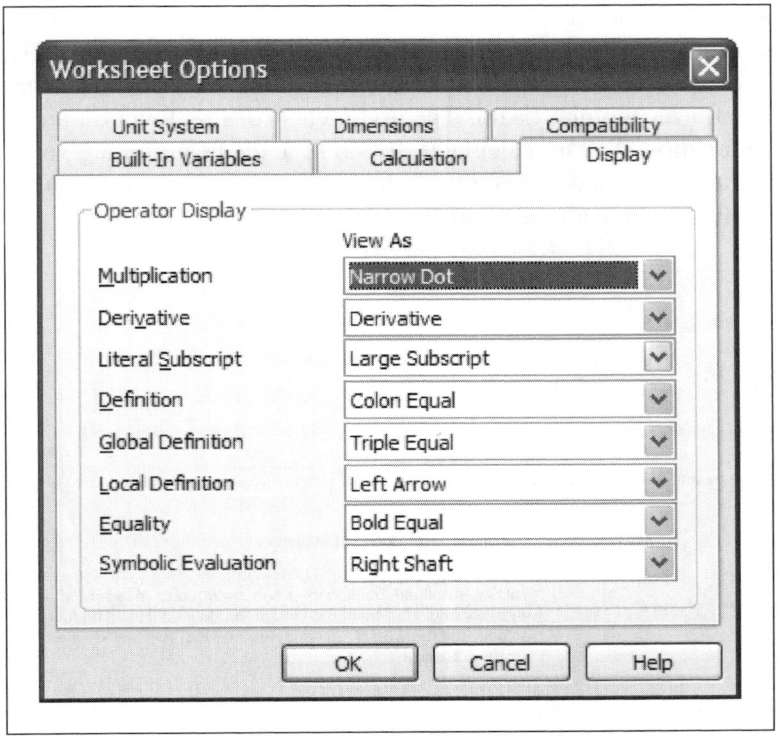

Figure 8.8 Changing Operator Display

When you change the display, you have not affected the way Mathcad operates. The default operator appears, when you select the math region.

You can also change the operator on specific regions instead of for the entire worksheet. To do this, right-click on the operator and select **View Definition As**." This will allow you to select from a menu of choices for the specific operator. If you want to switch back to the Mathcad default after making a change, choose the default option.

Functions

Passing a function to a function

Our discussion of functions in Chapter 3 was intentionally very basic.

Let's now add to our discussion by demonstrating that it is possible to use a previous user-defined function in a new user-defined function. Figure 8.9 gives an example of a user-defined function "SectionModulus" that calculates the section modulus of a rectangular beam. The arguments for this function are b and d. The new function "Stress" calculates the stress in the beam for a given moment M. This function uses the "SectionModulus" function. Thus, the arguments for "Stress" must include all the arguments needed for both functions.

$$\mathbf{SectionModulus(b,d)} := \frac{1}{6}\cdot \mathbf{b}\cdot \mathbf{d}^2$$

$$\mathbf{Stress(b,d,M)} := \frac{\mathbf{M}}{\mathbf{SectionModulus(b,d)}}$$

$$\mathbf{Stress(.5m, 1.3m, 800000N\cdot m)} = 823.88\,\mathbf{psi}$$

The above example used a fixed function. The next example shows the included function to be variable.

$\mathbf{F_1(a)} := \mathbf{a}^2$ $\quad \mathbf{G_1(b)} := \frac{1}{\mathbf{b}^2}$ — Define two user-defined functions.

$\mathbf{Sample(x,H)} := \frac{3}{\mathbf{x}}\cdot \mathbf{H(x)}$ — This user-defined function uses two arguments. They are "x" - a variable and "H" a function name to be defined later. The function actually used will be given as an argument of the user-defined function "Sample".

$\mathbf{Example_1} := \mathbf{Sample(2, F_1)}$ $\quad \mathbf{Example_1} = 6.00$ $\quad \frac{3}{2}\cdot 2^2 = 6.00$ — This example uses the user-defined function F_1.

$\mathbf{Example_2} := \mathbf{Sample(2, G_1)}$ $\quad \mathbf{Example_2} = 0.38$ $\quad \frac{3}{2}\cdot \frac{1}{2^2} = 0.38$ — This example uses the user-defined function G_1.

$\mathbf{Example_3} := \mathbf{Sample(2, sin)}$ $\quad \mathbf{Example_3} = 1.36$ $\quad \frac{3}{2}\cdot \mathbf{sin}(2) = 1.36$ — This example uses the Mathcad function sin().

$\mathbf{Example_4} := \mathbf{Sample(2, ln)}$ $\quad \mathbf{Example_4} = 1.04$ $\quad \frac{3}{2}\cdot \mathbf{ln}(2) = 1.04$ — This example uses the Mathcad function ln().

Figure 8.9 Function in a function

The second example in Figure 8.9 is a bit more complicated, but much more powerful. In this example, the function H(x) used in the user-defined function "Sample" is also a variable. The function used will not be defined until the user-defined function "Sample" is executed. This means that when you execute the user-defined function "Sample," you need to include a variable argument for "x," and also include a function argument for "H." The function can be a user-defined function or a Mathcad function that uses a single argument.

Custom operator notation

The custom operators allow you to display the names of functions to appear as operators in different forms. There are four different custom operators: prefix, postfix, infix, and treefix. The typical function is displayed as f(x) or f(x,y). The custom operators allow a function to be displayed as fx, xf, xfy, and $x^f y$. The custom operators are located on the Evaluation toolbar.

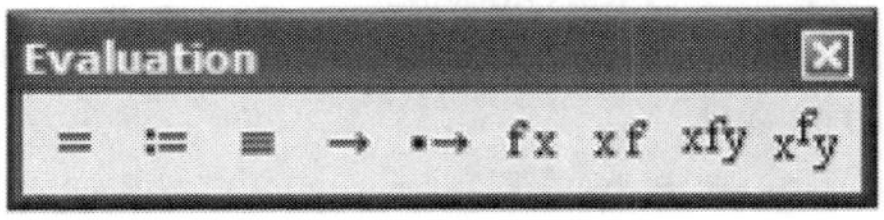

Prefix operator

The prefix operator is similar to the typical function notation except there are no parentheses. When you click the prefix operator you get two placeholders. The first placeholder is for the name of the function. The second placeholder is for the name of the functions argument.

Postfix operator

The postfix operator is very similar to the prefix operator, except it places the function last and the argument first.

Infix operator

The infix operator needs two arguments. It places the function between the x and the y arguments. This allows you to define custom functions that can behave like operators. A simple example would be to define a function "divided by." (Remember that you can include a space in a variable name by using the special text mode, **CTRL+SHIFT+K**.) You can then have displayed "6 divided by 2=3."

Treefix operator

The treefix operator needs two arguments as well. This operator places the function name on top with lines extending down to the arguments.

Figure 8.10 gives simple examples of these four different operators. These operators can be much more complex. An excellent discussion of this topic is found in the December 2001 Mathcad Advisor Newsletter. This can be found on the Mathcad website www.mathcad.com. There is also additional information in Mathcad Help.

Postfix operator (fx)

sin(45**deg**) = 0.71 Normal function notation

sin 45**deg** = 0.71 Prefix notation

Postfix operator (xf)

This example uses the FIF (Feet, Inch, Fraction) function/unit in Mathcad 13

FIF("5' 4 1/2"") = 64.5000 **in** Normal function notation

FIF("5' 4 1/2"") = "5' 4-1/2""**FIF**

"5' 4 1/2""**FIF** = 64.5000 **in** Postfix notation

Infix operator (xfy)

Create a new function called, "to the power of".
Remember, to include spaces in your variable name,
use the special text mode by typing
SHIFT+CTRL+K.

to the power of (**a**, **b**) := $\mathbf{a}^{\mathbf{b}}$

to the power of (2, 5) = 32.00 Normal function notation

2 **to the power of** 5 = 32.00 Infix notation

Treefix operator ($x^f y$)

to the power of = 32.00 Treefix notation

2 5

Figure 8.10 Custom operators

Tip of the day

Mathcad can show a tip each time it is started. This was discussed in Chapter 5—Mathcad settings. You can view and print the entire list of tips if you want. The file is called mtips_EN.txt, and is located in the main Mathcad program directory. This file can be opened in a word processor or in a text editor, and then printed.

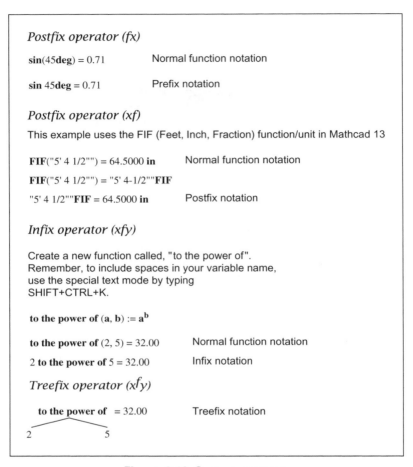

Part II

Hand Tools for your Mathcad Toolbox

You have now built your Mathcad toolbox, and it is time to start filling it with tools. Part II will introduce some simple Mathcad features. The more complex topics will be discussed in Part III. The goal of Part II is to get you comfortable with Mathcad. You will soon see that Mathcad is an easy-to-learn, yet powerful resource.

The chapters in this part will focus on features essential to understand before the more powerful tools are introduced in Part III. The topics covered in this part include: vectors, matrices, simple Mathcad functions, plotting, and simple logic programming.

9 Arrays, vectors, and matrices

An understanding of vectors and matrices can make engineering calculations much more effective. An array is simply either a vector or a matrix. A vector is a matrix with only a single column. Mathcad will perform many complex vector and matrix operations. This chapter will not delve into all of these functions. The purpose of this chapter is to illustrate the benefits that can be had by using simple vectors and matrices in engineering calculations.

Chapter 9 will:

- Tell how to define arrays.
- Introduce the concept of array subscripts.
- Refer to the ORIGIN variable and recommend changing its default value.
- Discuss the concept of range variables and tell how to define them and use them.
- Show how to format vector and matrix output in either matrix or table format.
- Illustrate how to attach units to vectors and matrices.
- Describe how to use simple math operators with vectors and matrices.
- Provide several examples of how vectors and matrices can be used in engineering calculations.
- Demonstrate how to use vectors to evaluate the same equation or function for various input values.

Creating vectors and matrices

It is very easy to create a vector or matrix by using the Insert Matrix dialog box. See Figure 9.1. This dialog box can be accessed in three ways. The first way is by selecting **Matrix** from the **Insert** menu. The second way is by typing the shortcut **CTRL+m**. The third way is by opening the Matrix toolbar and selecting the matrix icon showing a three by three matrix.

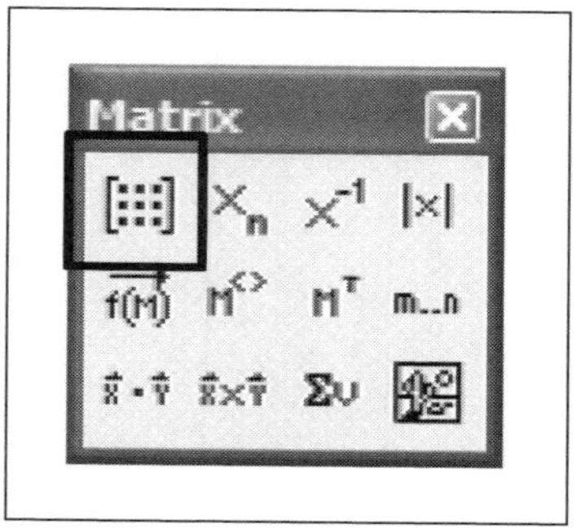

Figure 9.1 Insert matrix icon on Matrix toolbar (*Note*: This worksheet is based on the Mathcad Template "EWM Metric.")

Once the Insert Matrix dialog box is open, change the number of rows and columns to the desired numbers and click **OK**. For example, if you type 4 and 4 in the Rows and Columns boxes you will get a matrix as shown in Figure 9.2.

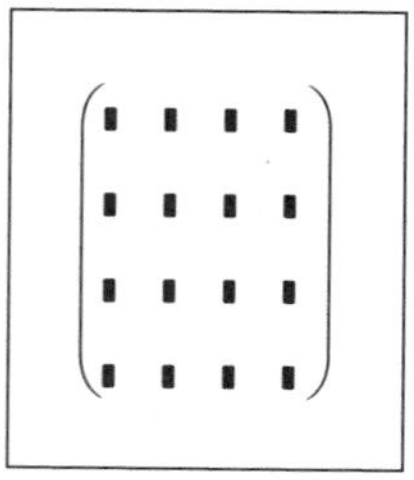

Figure 9.2 Blank 4 × 4 matrix

Now, simply fill in the placeholders with numbers or expressions. Use the tab key or arrow keys to move from placeholder to placeholder. See Figure 9.3 for two sample matrix definitions using numbers and expressions.

$$\text{Matrix\_1} := \begin{pmatrix} 1 & 2 & 3 & 4 \\ 5 & 6 & 7 & 8 \\ 9 & 10 & 11 & 12 \\ 13 & 14 & 15 & 16 \end{pmatrix} \qquad \text{Matrix\_1} = \begin{pmatrix} 1.00 & 2.00 & 3.00 & 4.00 \\ 5.00 & 6.00 & 7.00 & 8.00 \\ 9.00 & 10.00 & 11.00 & 12.00 \\ 13.00 & 14.00 & 15.00 & 16.00 \end{pmatrix}$$

$$\text{Matrix\_2} := \begin{pmatrix} 1+2 & 2+3 & 3+4 & 4+5 \\ 5+6 & 6+7 & 7+8 & 8+9 \\ 9+10 & 10+11 & 11+12 & 12+13 \\ 13+14 & 14+15 & 15+16 & 16+17 \end{pmatrix} \qquad \text{Matrix\_2} = \begin{pmatrix} 3.00 & 5.00 & 7.00 & 9.00 \\ 11.00 & 13.00 & 15.00 & 17.00 \\ 19.00 & 21.00 & 23.00 & 25.00 \\ 27.00 & 29.00 & 31.00 & 33.00 \end{pmatrix}$$

Figure 9.3 Sample matrix definitions

Once you create a vector or matrix, you can add additional rows or columns by using the Insert Matrix dialog box. To do this, select an element in the vector or matrix, and then open the Insert Matrix dialog box. Mathcad will insert additional rows below the selected element, and insert additional columns to the right of the selected element. Tell Mathcad how many additional rows and/or columns you want to add. If you want to add one additional row, but not an additional column, then type 1 for row and 0 for column. If you want to add rows above or columns to the left, then select the entire vector or matrix prior to using the Insert Matrix dialog box. After entering the number of rows and/or columns, click **OK** or **Insert**. If you select **Insert** first, be sure to click **Close** to close the box. If you click **OK** to close the box, additional rows and/or columns will be added.

You can also use the Insert Matrix dialog box to remove rows and columns. To do this, select an element in the row or column that you want to delete. Tell Mathcad how many rows and/or columns you want to delete, and then click on **Delete**. Mathcad will delete the row and/or column of the selected element and additional rows below the element and additional columns to the right of the element. Be sure to click **Close** to close the box. If you click **OK**, additional rows and/or columns will be added.

ORIGIN

We discussed the built-in Mathcad variable ORIGIN in Chapters 5 and 7. The value of ORIGIN tells Mathcad the starting index of your array. The Mathcad default for this variable is 0. This means that a vector or matrix begins indexing with zero. In other words, the first element is the 0th element. Thus, in

Matrix_1 of Figure 9.3, the value of the 0th element of the matrix (Matrix_1(0,0)) would be 1.

For engineering calculations, it is suggested that you change the value of ORIGIN from 0 to 1. This was explained in Chapters 5 and 7. With the value of ORIGIN set at 1, the first element of a matrix is the 1st element. Thus, in Matrix_1 of Figure 9.3, the value of the first element of the matrix (Matrix_1(1,1)) would be 1. For the remainder of this book, the value of ORIGIN will be set at 1. Refer to Chapter 5 to change the value of ORIGIN.

Array subscripts (subscript operator)

We discussed literal subscripts in Chapter 1—Variables. A literal subscript becomes part of the variable name. When using arrays, we use a different subscript called an array subscript. It is also called an array subscript operator. An array subscript allows Mathcad to display the value of a particular element in an array. In this case, the subscript does not become part of the variable name. It is used to refer to a single element in the array. The array subscript is created by using the `[` key. This is referred to as the subscript operator. Thus, if you want Mathcad to display the value of the first element in Matrix_1 Figure 9.3 you would type: `Matrix_1[1,1=` $\mathbf{Matrix\_1}_{1,1} = 1.00$. If you want Mathcad to display the value of the element in the 3rd row, 4th column, you would type `Matrix_1[3,4=` $\mathbf{Matrix\_1}_{3,4} = 12.00$.

In the above example, the name of the variable was Matrix_1. The array subscript is not part of the variable name. It is only used to display an element of the array.

You can also use an array subscript to assign elements of an array. If you type `Matrix_1[1,1:20` then the value of the 1st element in Matrix_1 will be changed from 1 to 20. See Figure 9.4.

$$\mathbf{Matrix\_1}_{1,1} := 20$$

$$\mathbf{Matrix\_1} = \begin{pmatrix} 20.00 & 2.00 & 3.00 & 4.00 \\ 5.00 & 6.00 & 7.00 & 8.00 \\ 9.00 & 10.00 & 11.00 & 12.00 \\ 13.00 & 14.00 & 15.00 & 16.00 \end{pmatrix}$$

Figure 9.4 Changing the value of a single array element

Figure 9.5 shows how to use array subscripts for a vector. Figure 9.6 shows how to use array subscripts to assign new values to vectors and arrays.

$$\textbf{Vector\_1} := \begin{pmatrix} 2 \\ 22 \\ 222 \\ 2222 \end{pmatrix} \qquad \textbf{Vector\_1} = \begin{pmatrix} 2.00 \\ 22.00 \\ 222.00 \\ 2222.00 \end{pmatrix}$$

$\textbf{Vector\_1}_1 = 2.00$

$\textbf{Vector\_1}_2 = 22.00$

$\textbf{Vector\_1}_3 = 222.00$

$\textbf{Vector\_1}_4 = 2222.00$

$\textbf{Vector\_1}_0 = \blacksquare$ With ORIGIN set to one, there is no zero element

There are only 4 elements. The 5th is not recognized.

$\textbf{Vector\_1}_5 = \blacksquare$ $\textbf{Vector\_1}_5 = \blacksquare\blacksquare$

This array index is invalid for this array.

Figure 9.5 Using array subscripts

You can add additional elements to an array, by defining them with array subscripts.

$\textbf{Vector\_1}_6 := 22222$ $\qquad$ $\textbf{Matrix\_1}_{5,5} := 3333$

$$\textbf{Vector\_1} = \begin{pmatrix} 2.00 \\ 22.00 \\ 222.00 \\ 2222.00 \\ 0.00 \\ 22222.00 \end{pmatrix} \qquad \textbf{Matrix\_1} = \begin{pmatrix} 20.00 & 2.00 & 3.00 & 4.00 & 0.00 \\ 5.00 & 6.00 & 7.00 & 8.00 & 0.00 \\ 9.00 & 10.00 & 11.00 & 12.00 & 0.00 \\ 13.00 & 14.00 & 15.00 & 16.00 & 0.00 \\ 0.00 & 0.00 & 0.00 & 0.00 & 3333.00 \end{pmatrix}$$

Figure 9.6 Using array subscripts

Range variables

A range variable is similar to a vector in that it takes on multiple values. It has a range of values. The range of values has a beginning value, an ending value and uniform incremental values between the beginning and ending values. Range variables are used to iterate a calculation over a specific range of values, and to plot a function over a specific range of values. They are often used as subscripts for defining arrays. A range variable looks like this: $\mathrm{RangeVariable}_A := 1, 1.5 .. 5$. This range variable begins with 1.0. The second number in the range variable sets the increment value. Mathcad takes the difference between the first and second numbers and uses this as the incremental value. In the above case, the increment is 0.5. The last number in this range is 5.0. Thus, this range variable has the values: 1.0, 1.5, 2.0, 2.5, 3.0, 3.5, 4.0, 4.5, and 5.0.

To define a range variable, type the variable name followed by the colon [:]. This creates the variable definition. In the placeholder, type the beginning value, and then type a comma. This adds a second placeholder in the expression. Now enter the second value in the placeholder. The second number sets the incremental value. Now type a semicolon [;]. This places two dots in the worksheet, and adds a third placeholder. Enter the ending value in the placeholder. If the second value is less than the beginning value, then the range variable will be decreasing, and the last value sets the lower limit to the range variable. See Figure 9.7 for sample range variables and their displayed results.

The second value sets the increment.

RangeVariable_A := 5, 5.1 .. 6

RangeVariable_A =

5
5.1
5.2
5.3
5.4
5.5
5.6
5.7
5.8
5.9
6

Range variables may also decrease

RangeVariable_B := 10, 8 .. −10

RangeVariable_B =

10.00
8.00
6.00
4.00
2.00
0.00
-2.00
-4.00
-6.00
-8.00
-10.00

If a second value is not given, then an increment of 1 is used.

RangeVariable_C := 1 .. 10

RangeVariable_C =

1.00
2.00
3.00
4.00
5.00
6.00
7.00
8.00
9.00
10.00

Figure 9.7 Sample range variables

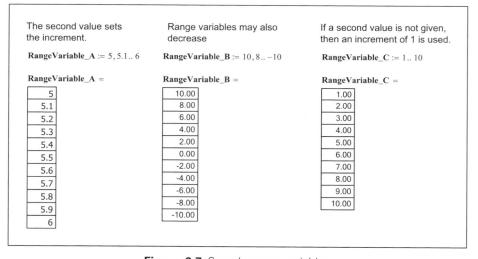

Using range values to create arrays

When you use range variables to iterate a calculation or to create an array, it is important to understand that Mathcad begins at the beginning value and iterates every value in the range. You cannot tell Mathcad to only use part of the range variable.

Range values can be used to define and display arrays. The range variables are used as the array subscripts for defining the array, and they may also be used to define the value of each element. In order to be used in defining an array, the range variables must have positive integer values. The range variable must also begin at a value equal to or greater than ORIGIN. If ORIGIN is zero, then zero may be used as a beginning value. Figures 9.8 and 9.9 give examples of how you can use range variables to create vectors. In both of these figures, the value of each element created is based on the values of the range variable. Figure 9.10 gives an example of using two range variables to create a matrix.

Range variables may be used as arguments for functions. If you create a function with a single argument and then use a range variable for the argument, Mathcad will provide results for every element in the range variable. Figure 9.11 provides an example of this. You may also use multiple range variables in functions with multiple arguments. See Figure 9.12.

In the following example, the range variable "a" is used to create the vector "Sample." The subscript "a" in the definition is an array subscript created using the [key. The definition tells Mathcad to create a vector called "Sample." Each value of the range variable will be used to create the vector.

$\mathbf{a} := 1, 2 .. 9$

$\mathbf{Sample_a} := \mathbf{a}^2 \cdot 2$

Mathcad uses the procedure listed below to calculate the values of the vector "Sample." It uses every value in the range variable from 1 to 10. The element value is the same as the range value.

Element	Range Variable	Value
1	1	1^2*2=2
2	2	2^2*2=8
3	3	3^2*2=18
4	4	4^2*2=32
5	5	5^2*2=50
6	6	6^2*2=72
7	7	7^2*2=98
8	8	8^2*2=128
9	9	9^2*2=162

There are two ways to display the results of the variable "Sample." Typing, "Sample=" will display the entire vector results in a table. Typing, "Sample[a=" will display all the elements of the variable "Sample" associated with the range variable "a." Thus, if you create a new range variable "b" with values 3, 4, 5, & 6 and type, "Sample[b=", then Mathcad will only display elements 3, 4, 5, & 6 of the vector "Sample."

$$\mathbf{Sample} = \begin{pmatrix} 2.00 \\ 8.00 \\ 18.00 \\ 32.00 \\ 50.00 \\ 72.00 \\ 98.00 \\ 128.00 \\ 162.00 \end{pmatrix}$$

$\mathbf{Sample_a} =$

2.00
8.00
18.00
32.00
50.00
72.00
98.00
128.00
162.00

$\mathbf{b} := 3, 4 .. 6$

$\mathbf{Sample_b} =$

18.00
32.00
50.00
72.00

Figure 9.8 Using range variables to create a vector

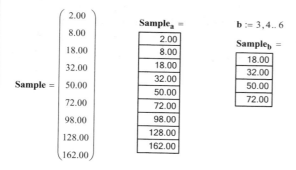

Let's look at the same equation as in Figure 9.8, but using different range variables.

Values less than ORIGIN

$\mathbf{d} := 0, 1 .. 10$

$\mathbf{Sample\_2_d} := \mathbf{d}^2 \cdot 2$

The range variable "d" does not work because ORIGIN is set to 1 and there is not a zero element in the vector.

$\mathbf{Sample\_2_d} := \mathbf{d}^2 \cdot 2$

This array index is invalid for this array.

Non-consecutive numbers

$\mathbf{f} := 2, 4 .. 9$

$\mathbf{Sample\_3_f} := \mathbf{f}^2 \cdot 2$

$$\mathbf{Sample\_3} = \begin{pmatrix} 0.00 \\ 8.00 \\ 0.00 \\ 32.00 \\ 0.00 \\ 72.00 \\ 0.00 \\ 128.00 \end{pmatrix}$$

The range variable "f" has the values 2, 4, 6, 8, and 10. This creates the 2nd, 4th, 6th, and 8th elements of the vector "Sample_3." The elements not specifically defined are assigned the value zero.

Element	Range Variable	Value
1		0
2	2	2^2*2=8
3		0
4	4	4^2*2=32
5		0
6	6	6^2*2=72
7		0
8	8	8^2*2=128
9		0

Non-integers

$\mathbf{h} := 1, 1.5 .. 5$

$\mathbf{Sample\_3_h} := \mathbf{h}^2 \cdot 2$

To create a vector with a range variable, the values in the range variable must be integers to correspond with the element numbers in the vector.

$\mathbf{Sample\_3_h} := \mathbf{h}^2 \cdot 2$

This value must be an integer.

Figure 9.9 Using range variables to create vectors

$\mathbf{k} := 1..4$ $\mathbf{n} := 1..3$

$\mathbf{Sample\_4_{k,n}} := \mathbf{k} + \mathbf{2n}$

$$\mathbf{Sample\_4} = \begin{pmatrix} 3.00 & 5.00 & 7.00 \\ 4.00 & 6.00 & 8.00 \\ 5.00 & 7.00 & 9.00 \\ 6.00 & 8.00 & 10.00 \end{pmatrix}$$

		Column 1 n=1	Column 2 n=2	Column 3 n=3
Row 1	k=1	1+2*1=3	1+2*2=5	1+2*3=7
Row 2	k=2	2+2*1=4	2+2*2=6	2+2*3=8
Row 3	k=3	3+2*1=5	3+2*2=7	3+2*3=9
Row 4	k=4	4+2*1=6	4+2*2=8	4+2*3=10

Figure 9.10 Using range variables to create a matrix

$\mathbf{o} := 0.5, 1..3$ Create range variable

$\mathbf{CircleArea(r)} := \pi \cdot \mathbf{r}^2$ Create user-defined function

The range variable may be used as the argument of the function. Units may also be added to the function argument. The range variables do not need to be integers when using them in a function.

$\mathbf{CircleArea(o \cdot cm)} =$

$\mathbf{cm}^2$
0.79
3.14
7.07
12.57
19.63
28.27

The results are displayed,but this does not allow you to access or reuse individual results.

In order to make each result accessible, you need to assign the results to a vector. But, to do this, the range variables need to be integers.

$\mathbf{p} := 1, 2..6$ New range variable with integer values.

$\mathbf{Area\_Example_p} := \mathbf{CircleArea}\left(\frac{\mathbf{p}}{2} \cdot \mathbf{cm}\right)$

To get the same result as above, the range variable "p" needs to be modified in the argument list.

$$\mathbf{Area\_Example} = \begin{pmatrix} 0.79 \\ 3.14 \\ 7.07 \\ 12.57 \\ 19.63 \\ 28.27 \end{pmatrix} \mathbf{cm}^2$$

$\mathbf{Area\_Example_3} = 7.07\,\mathbf{cm}^2$

$\mathbf{Area\_Example_6} = 28.27\,\mathbf{cm}^2$

Figure 9.11 Using range variables in user-defined functions

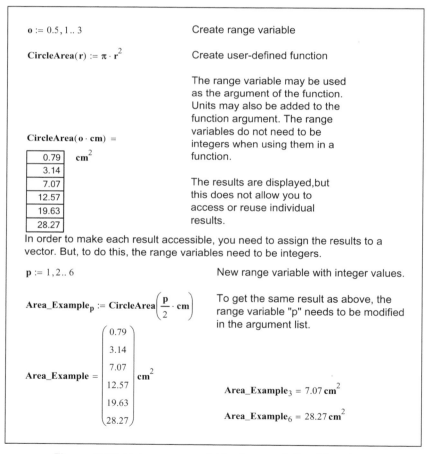

$\mathbf{q} := 1, 2 .. 4$ $\mathbf{r} := 10, 20 .. 40$

$\mathbf{Pressure(P, A)} := \dfrac{\mathbf{P}}{\mathbf{A}}$ $\mathbf{Pressure}\left(1\mathbf{lbf}, 10\mathbf{ft}^2\right) = 0.10\,\mathbf{psf}$

$\mathbf{Pressure}\left(\mathbf{q} \cdot \mathbf{lbf}, \mathbf{r} \cdot \mathbf{ft}^2\right) =$

0.100	psf
0.200	
0.300	
0.400	
0.050	
0.100	
0.150	
0.200	
0.033	
0.067	
0.100	
0.133	
0.025	
0.050	
0.075	
0.100	

In the case of using two range variables, all value of each range variable are used.

Mathcad begins using all values of "q" and the first value of "r." Mathcad then uses all values of "q" again with the second value of "r." The process repeats until all values of "r" have been used.

In this example, 16 values are returned. There is no way to assign a vector variable to this result, because we are using range variables in a way that was not intended by Mathcad.

Chapter 13 provides a better way to organize this calculation using vectors rather than range variables.

Figure 9.12 Using multiple range variables in user-defined functions

Using units in range variables

You can define range variables using units. You simply add units to the beginning value, second value, and ending value. You are required to input a second value when using units. The second and ending values do not need to use the same units, but the units used must be from the same unit dimension. For example, your beginning value can be in feet, and the second value can be in inches and the ending value in feet. This way the increment will be in inches.

Range variables with units attached are primarily used in functions and plotting. See Figures 9.13 and 9.14.

Using consistent units

q := 1ft, 2ft.. 6ft

q =

0.30	m
0.61	
0.91	
1.22	
1.52	
1.83	

q =

1.00	ft
2.00	
3.00	
4.00	
5.00	
6.00	

Using units of feet and inches

r := 1ft, 13in.. 2ft

r =

0.30	m
0.33	
0.36	
0.38	
0.41	
0.43	
0.46	
0.48	
0.51	
0.53	
0.56	
0.58	
0.61	

r =

12.00	in
13.00	
14.00	
15.00	
16.00	
17.00	
18.00	
19.00	
20.00	
21.00	
22.00	
23.00	
24.00	

r =

1.00	ft
1.08	
1.17	
1.25	
1.33	
1.42	
1.50	
1.58	
1.67	
1.75	
1.83	
1.92	
2.00	

Using functions

t := °F(0), °F(10).. °F(100)

t =

255.37	K
260.93	
266.48	
272.04	
277.59	
283.15	
288.71	
294.26	
299.82	
305.37	
310.93	

t =

0.00	°F
10.00	
20.00	
30.00	
40.00	
50.00	
60.00	
70.00	
80.00	
90.00	
100.00	

t =

-17.78	°C
-12.22	
-6.67	
-1.11	
4.44	
10.00	
15.56	
21.11	
26.67	
32.22	
37.78	

Using mixed units of minutes and seconds

u := 1min, 61s.. 2min

u =

60.00	s
61.00	
62.00	
63.00	
64.00	
65.00	
66.00	
67.00	
68.00	
69.00	
70.00	

u =

1.00	min
1.02	
1.03	
1.05	
1.07	
1.08	
1.10	
1.12	
1.13	
1.15	
1.17	

Figure 9.13 Range variables with units

$v := 1cm, 2cm .. 6cm$

The function CircleArea was defined in Figure 9.11 as Area=πr^2.

CircleArea(v) =

3.14	cm^2
12.57	
28.27	
50.27	
78.54	
113.10	

The range variable "v" already has units attached, so the argument does not need to be multiplied by a unit.

The range variable "v" may not be used as an array subscript because it has units attached. In order to use "v" as an array subscript, the units must be divided out.

$Area\_Example_v := CircleArea(v)$

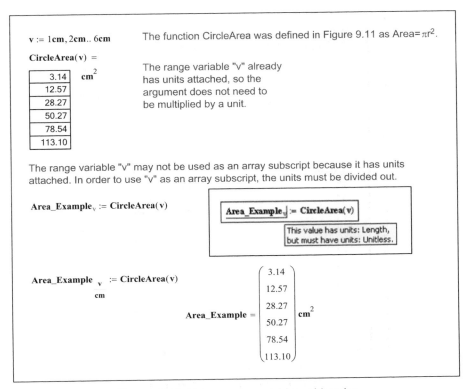

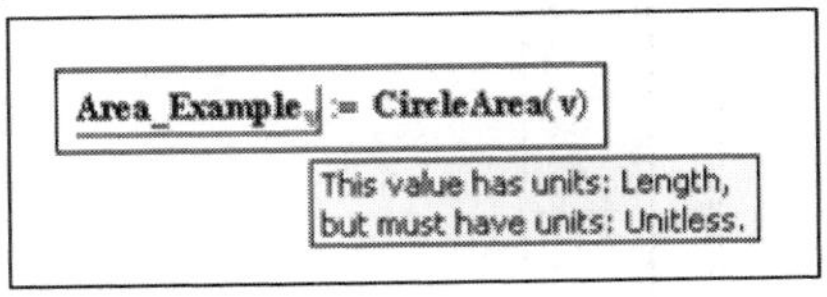

$Area\_Example_{\frac{v}{cm}} := CircleArea(v)$

$$Area\_Example = \begin{pmatrix} 3.14 \\ 12.57 \\ 28.27 \\ 50.27 \\ 78.54 \\ 113.10 \end{pmatrix} cm^2$$

Figure 9.14 Using range variables with units

Calculating increments from the beginning and ending values

If the increments do not allow the range variable to stop exactly on the ending value, Mathcad will stop the range short of the last value. See Figure 9.15. This could cause some unexpected results in your calculations. If it is difficult to calculate an increment that will stop exactly at the last value, you can enter a formula into the second placeholder. The formula is (Last value − First value)/(Number of increments) + First value. See Figure 9.16.

$\mathbf{RangeVariable\_D} := 0, 0.4 .. 3$

$\mathbf{RangeVariable\_D} =$

0
0.4
0.8
1.2
1.6
2
2.4
2.8

The last value of RangeVariable_D does not end at the ending value in the definition. This could cause unexpected results in your calculations.

Figure 9.15 Range variable where increment does not stop at ending value

$\mathbf{RangeVariable\_E} := 1, 1.9 .. 10$

$\mathbf{RangeVariable\_F} := 1, \frac{10-1}{10} + 1 .. 10$

$\mathbf{RangeVariable\_E} =$

1
1.9
2.8
3.7
4.6
5.5
6.4
7.3
8.2
9.1
10

$\mathbf{RangeVariable\_F} =$

1
1.9
2.8
3.7
4.6
5.5
6.4
7.3
8.2
9.1
10

The values for RangeVariable_E and RangeVariable_F are the same. RangeVariable_F used a formula to get 10 increments between the first and last values. The increment for RangeVariable_E needed to be calculated prior to entering the increment.

Figure 9.16 Calculating increments

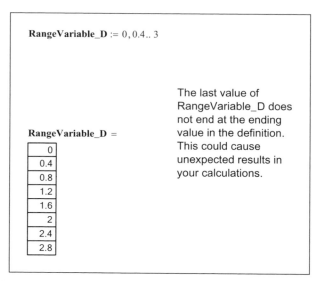

Comparing range variables to vectors

Because range variables and vectors are similar, it is important to understand the difference between them. Here is a comparison between range variables and vectors:

- Range variables must increment (up or down) in uniform steps. Vectors may have numbers in any order.
- Range variables must be real. Vectors may use real or complex numbers.
- You cannot access individual elements of a range variable. Each element of a vector has a unique variable name that can be accessed by using array subscripts.
- Range variables are used to iterate calculations over a range of values. The calculation is performed once for each value in the range. Vectors are often used as arguments for specific Mathcad functions.
- Range variables are often used as subscripts to write or access data in vectors and matrices. Vectors are used to export data.
- Range variables begin at the defined beginning value. Vectors use ORIGIN as the 1st element.

Displaying arrays

There are two ways to display vectors and matrices. In the previous figures, the arrays were displayed in matrix form. The range variables were displayed in table form. Mathcad uses these two methods of displaying vectors and matrices.

The display of arrays is controlled by the "Matrix display style" in the Result Format dialog box. To access this dialog box, select **Result** from the **Format** menu and click on the Display Options tab. See Figure 9.17. The Mathcad default display style is Automatic. When Automatic is selected, matrices smaller than 10 by 10 are displayed in matrix form. When a matrix is 10 by 10 or larger, the default display is in table form. Range variables, by default, are displayed in table form.

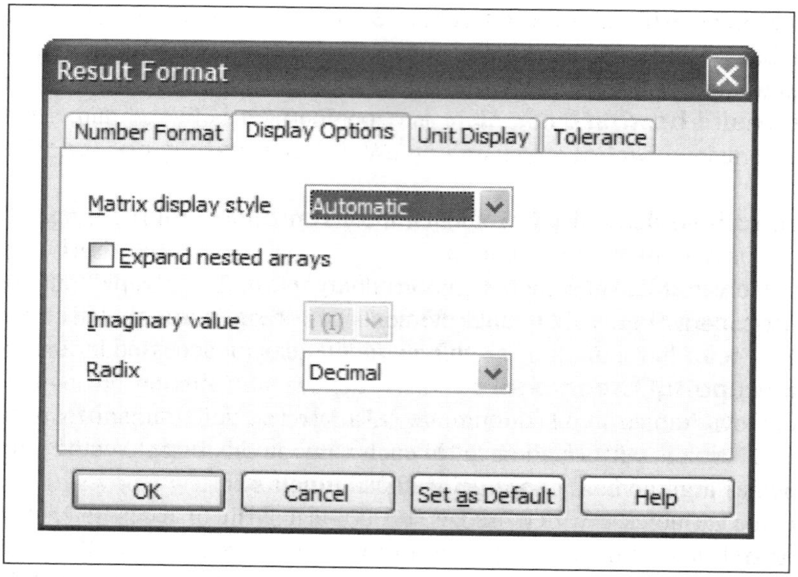

Figure 9.17 Result Format dialog box

If you prefer to have all array output in either matrix form or table form, change the "Matrix display style" from Automatic to either matrix form or table form from the drop-down box adjacent to the "Matrix display style." You may also change the display of individual results by double clicking on the result, clicking on the Display Options tab and selecting the display form from the "Matrix display style" drop-down box. You may also click on the result, and select **Result** from the **Format** menu.

Tip! Remember that when a result is selected, the Result Format dialog box only changes the format of the selected result unless the "Set as Default" button is clicked after setting your result formats.

Table display form

When vectors and matrices are displayed in table form, there are several display options to consider. Some of these options include: columns and row labels, table size, font size, and the display location of the variable name in relationship to the output table.

Column and row labels

Column and row labels appear along the top and left side of an output table. To turn the labels on or off, right-click within the table and select **Properties**. This opens the Component Properties dialog box. To turn the labels on, place a check in the "Show column/row labels" check box. To turn the labels off, clear the check box. The Mathcad default is to have labels turned on for vectors and matrices, and to have labels turned off for range variables. See Figure 9.18 for sample output with and without labels.

$$\text{Matrix\_3} := \begin{pmatrix} 1 & 2 & 3 & 4 & 5 & 6 \\ 2 & 3 & 4 & 5 & 6 & 7 \\ 3 & 4 & 5 & 6 & 7 & 8 \\ 4 & 5 & 6 & 7 & 8 & 9 \\ 5 & 6 & 7 & 8 & 9 & 0 \\ 6 & 7 & 8 & 9 & 0 & 1 \end{pmatrix}$$

Matrix output form.

Matrix_3 =

	1	2	3	4	5	6
1	1.00	2.00	3.00	4.00	5.00	6.00
2	2.00	3.00	4.00	5.00	6.00	7.00
3	3.00	4.00	5.00	6.00	7.00	8.00
4	4.00	5.00	6.00	7.00	8.00	9.00
5	5.00	6.00	7.00	8.00	9.00	0.00
6	6.00	7.00	8.00	9.00	0.00	1.00

Output table with row and column labels turned on.

Matrix_3 =

1.00	2.00	3.00	4.00	5.00	6.00
2.00	3.00	4.00	5.00	6.00	7.00
3.00	4.00	5.00	6.00	7.00	8.00
4.00	5.00	6.00	7.00	8.00	9.00
5.00	6.00	7.00	8.00	9.00	0.00
6.00	7.00	8.00	9.00	0.00	1.00

Output table with row and column labels turned off.

Figure 9.18 Sample output—matrix form and table form

Output table size

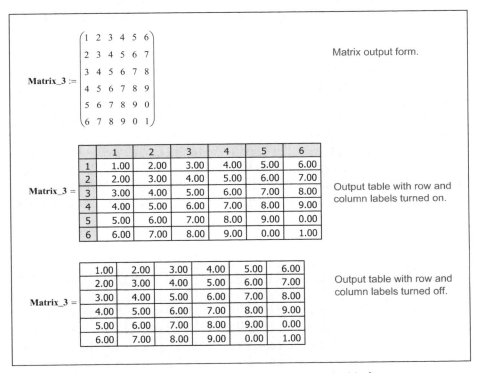

When a vector or matrix becomes too large, Mathcad will only display a portion of the output table. In order to view all the output data, you must click inside the table. This will cause scroll bars to appear at the right side and/or bottom of the table. Use the scroll bars to view the remaining output data. In order to display more of the output table, you can make the table larger by clicking and dragging one of the black anchor boxes located in the corners and at midpoints of the table. See Figure 9.19. If the table crosses the right margin, you will not be able to print the entire table if you have "Print single page width" selected from the Page Setup menu. If this is the case, you can try to make the column widths narrower.

Matrix_3 =

	2	3	4	5	6
2	3.00	4.00	5.00	6.00	7.00
3	4.00	5.00	6.00	7.00	8.00
4	5.00	6.00	7.00	8.00	9.00
5	6.00	7.00	8.00	9.00	0.00

Figure 9.19 Output table with scroll bars

You can only make the columns narrower if the column and row labels are turned on. Move your cursor to the column label and place it on top of a line between columns. The cursor should change to a vertical line with arrows either side. Click and drag to the left to make the columns narrower. You may also do a similar thing to the rows to make them shorter. See Figure 9.20.

Matrix_3 =

	1	2	3	4	5	6
1	1.00	2.00	3.00	4.00	5.00	6.00
2	2.00	3.00	4.00	5.00	6.00	7.00
3	3.00	4.00	5.00	6.00	7.00	8.00
4	4.00	5.00	6.00	7.00	8.00	9.00
5	5.00	6.00	7.00	8.00	9.00	0.00
6	6.00	7.00	8.00	9.00	0.00	1.00

Figure 9.20 Output table with columns narrowed

Output table font size and font properties

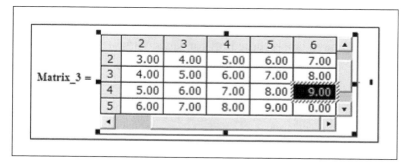

You can also decrease the font size in the output table in order to display more table cells on one page. To change the output table font size, right-click within the table and select **Properties**. This opens the Component Properties dialog box. Click on the **Font** button to open the Font dialog box. From this box, you may change the font style, font properties, and font size. If you want a size different from what is listed, then type the font size in the "Size" box. Remember that the changes you make to this output table will not affect other output tables. See Figure 9.21.

Matrix_3 =

	1	2	3	4	5	6
1	1.00	2.00	3.00	4.00	5.00	6.00
2	2.00	3.00	4.00	5.00	6.00	7.00
3	3.00	4.00	5.00	6.00	7.00	8.00
4	4.00	5.00	6.00	7.00	8.00	9.00
5	5.00	6.00	7.00	8.00	9.00	0.00
6	6.00	7.00	8.00	9.00	0.00	1.00

Figure 9.21 Output table with a smaller font

Variable name location

You can adjust where the variable name is located in relationship to the output table. To change the location of the variable name, right-click in the table, pass your cursor over **Alignment** and select from the menu of options: Top, Center, Bottom, Above, and Below. Remember that you are selecting the location of the variable name in relationship to the table. The Top, Center, and Bottom alignments have the variable name on the left side of the table. See Figure 9.22.

Matrix_3 =

	1	2	3	4	5	6
1	1.00	2.00	3.00	4.00	5.00	6.00
2	2.00	3.00	4.00	5.00	6.00	7.00
3	3.00	4.00	5.00	6.00	7.00	8.00
4	4.00	5.00	6.00	7.00	8.00	9.00
5	5.00	6.00	7.00	8.00	9.00	0.00
6	6.00	7.00	8.00	9.00	0.00	1.00

	1	2	3	4	5	6
1	1.00	2.00	3.00	4.00	5.00	6.00
2	2.00	3.00	4.00	5.00	6.00	7.00
3	3.00	4.00	5.00	6.00	7.00	8.00
4	4.00	5.00	6.00	7.00	8.00	9.00
5	5.00	6.00	7.00	8.00	9.00	0.00
6	6.00	7.00	8.00	9.00	0.00	1.00

Matrix_3 =

	1	2	3	4	5	6
1	1.00	2.00	3.00	4.00	5.00	6.00
2	2.00	3.00	4.00	5.00	6.00	7.00
3	3.00	4.00	5.00	6.00	7.00	8.00
4	4.00	5.00	6.00	7.00	8.00	9.00
5	5.00	6.00	7.00	8.00	9.00	0.00
6	6.00	7.00	8.00	9.00	0.00	1.00

Matrix_3 =

Matrix_3 =

	1	2	3	4	5	6
1	1.00	2.00	3.00	4.00	5.00	6.00
2	2.00	3.00	4.00	5.00	6.00	7.00
3	3.00	4.00	5.00	6.00	7.00	8.00
4	4.00	5.00	6.00	7.00	8.00	9.00
5	5.00	6.00	7.00	8.00	9.00	0.00
6	6.00	7.00	8.00	9.00	0.00	1.00

Figure 9.22 Output table with different variable name locations

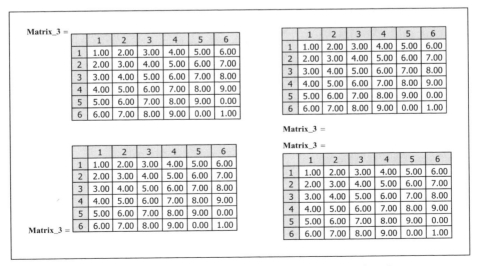

Using units with arrays

Units are just as important with vectors and matrices as they are with other Mathcad variables. Units are essential in engineering calculations. Unfortunately, all data within a vector or matrix must be of the same unit dimension (length, mass, pressure, time, etc.). You cannot have mixed unit dimensions in a vector or matrix.

If all the input values have the same units, then you can multiply the entire matrix by the unit. If the input values have different units (of the same unit dimension), then multiply each individual value by the unit. See Figure 9.23.

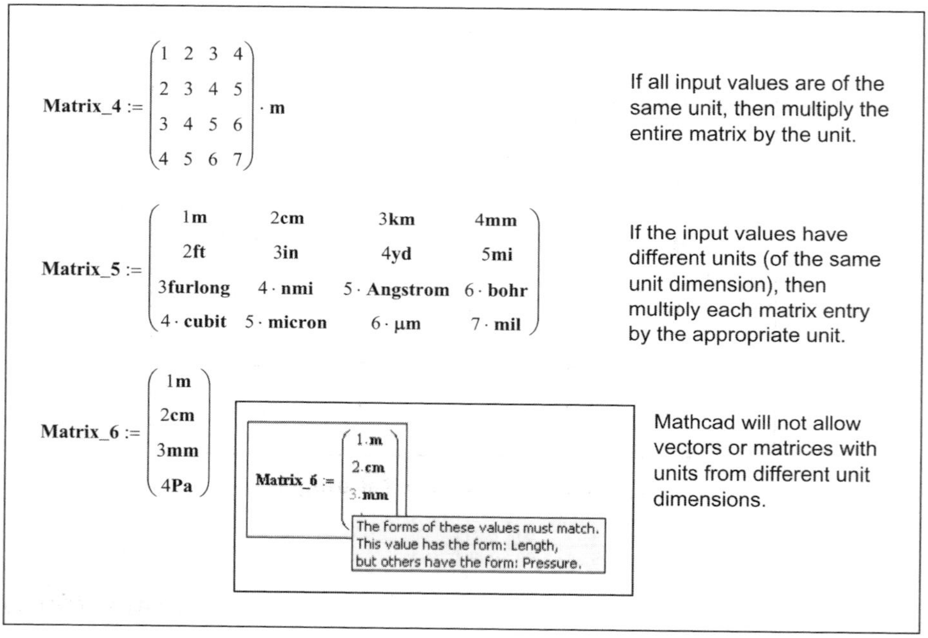

Figure 9.23 Using units with arrays

The input may have different units, but the output (in both matrix form and table form) will be in a single unit system. See Figure 9.24.

$$\mathbf{Matrix\_4} = \begin{pmatrix} 1.00 & 2.00 & 3.00 & 4.00 \\ 2.00 & 3.00 & 4.00 & 5.00 \\ 3.00 & 4.00 & 5.00 & 6.00 \\ 4.00 & 5.00 & 6.00 & 7.00 \end{pmatrix} \mathbf{m}$$

All elements in the output will have the same units.

Matrix_4 = (ft)

	1	2	3	4
1	3.28	6.56	9.84	13.12
2	6.56	9.84	13.12	16.40
3	9.84	13.12	16.40	19.69
4	13.12	16.40	19.69	22.97

Matrix_5 = (m)

	1	2	3	4
1	1.00	0.02	3000.00	0.00
2	0.61	0.08	3.66	8046.72
3	603.50	7408.00	5.00·10^-10	3.18·10^-10
4	1.83	5.00·10^-6	6.00·10^-6	0.00

$$\mathbf{Matrix\_5} = \begin{pmatrix} 3.28 & 0.07 & 9842.52 & 0.01 \\ 2.00 & 0.25 & 12.00 & 26400.00 \\ 1980.00 & 24304.46 & 1.64\times10^{-9} & 1.04\times10^{-9} \\ 6.00 & 0.00 & 0.00 & 0.00 \end{pmatrix} \mathbf{ft}$$

Figure 9.24 All output will have the same units

Calculating with arrays

Mathcad allows you to use many math operators with arrays just as you do with scalar variables, but you need to make sure that you follow the basic rules for matrix math. For example, you can add and subtract arrays just as you would other variables as long as the arrays are the same size. If they are different size arrays, Mathcad will give you an error warning. You can multiply a scalar and an array. The result will be each element of the array multiplied by the scalar. If you multiply two array variables, Mathcad assumes that you want a matrix dot product and gives a result (assuming that the two arrays are compatible with the dot product rules). There is a way to tell Mathcad that you want the matrix cross product. There is also a way to tell Mathcad that you want to multiply two arrays on an element-by-element basis and return a similar size array. These will be discussed shortly.

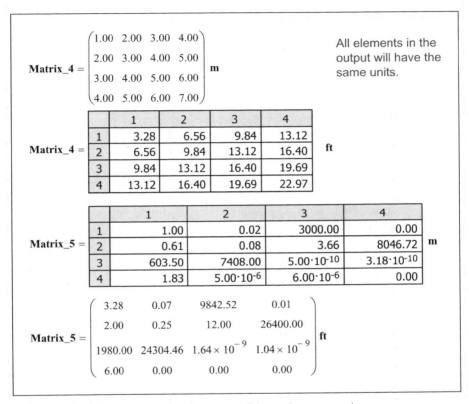

Addition and subtraction

If vectors and matrices are exactly the same size, you may add or subtract them as you would any Mathcad variable. The addition and subtraction is on an element-by-element basis. See Figure 9.25.

Addition and Subtraction: Arrays must be the same size

$$\mathbf{Matrix\_6} := \begin{pmatrix} 11 & 22 \\ 33 & 44 \end{pmatrix} \qquad \mathbf{Matrix\_7} := \begin{pmatrix} 2 & 3 \\ 4 & 5 \end{pmatrix} \qquad \mathbf{Matrix\_8} := \begin{pmatrix} 1 & 2 & 3 \\ 2 & 3 & 4 \end{pmatrix}$$

$$\mathbf{AA} := (1 \quad 2)$$

$$\mathbf{Matrix\_9} := \mathbf{Matrix\_6} + \mathbf{Matrix\_7}$$

$$\mathbf{Matrix\_9} = \begin{pmatrix} 13.00 & 25.00 \\ 37.00 & 49.00 \end{pmatrix}$$

The addition operator adds element by element.

$$\mathbf{Matrix\_10} := \mathbf{Matrix\_6} - \mathbf{Matrix\_7}$$

$$\mathbf{Matrix\_10} = \begin{pmatrix} 9.00 & 19.00 \\ 29.00 & 39.00 \end{pmatrix}$$

The subtraction operator subtracts element by element.

The following does not work because the size of the arrays are different.

$$\mathbf{Matrix\_11} := \mathbf{Matrix\_6} + \mathbf{Matrix\_8}$$

Matrix_11 := Matrix_6 + Matrix_8

These array dimensions do not match.

Figure 9.25 Addition and subtraction

Multiplication

There are several different ways to multiply arrays. These are discussed below.

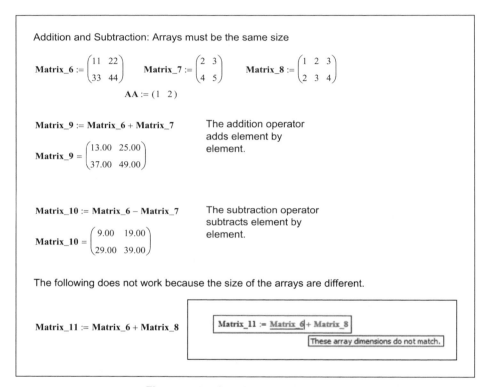

Scalar multiplication

You can multiply any vector or matrix by a scalar number. Mathcad multiplies each element in the array by the scalar and returns the results. The scalar can be before or after the array. See Figure 9.26.

You can multiply any vector or matrix by a scalar.

Refer to Figure 9.25 for the definition of $Matrix_6$, $Matrix_7$, and $Matrix_8$.

$$\mathbf{Matrix\_12} := \mathbf{2} \cdot \mathbf{Matrix\_6} \qquad \mathbf{Matrix\_12} = \begin{pmatrix} 22.00 & 44.00 \\ 66.00 & 88.00 \end{pmatrix}$$

$$\mathbf{Matrix\_13} := \mathbf{3} \cdot \mathbf{Matrix\_7} \qquad \mathbf{Matrix\_13} = \begin{pmatrix} 6.00 & 9.00 \\ 12.00 & 15.00 \end{pmatrix}$$

$$\mathbf{Matrix\_14} := \mathbf{Matrix\_8} \cdot 4 \qquad \mathbf{Matrix\_14} = \begin{pmatrix} 4.00 & 8.00 & 12.00 \\ 8.00 & 12.00 & 16.00 \end{pmatrix}$$

Figure 9.26 Scalar multiplication

Dot product multiplication

When you multiply two arrays, Mathcad assumes that you want the matrix dot product of the two arrays. (The dot product is calculated by multiplying each element of the first vector by the corresponding element of the complex conjugate of the second vector, and then summing the result. Refer to a text on matrix math for a discussion of the matrix dot product.) In order for the dot product to work, the number of columns in the first matrix must match the number of rows in the second matrix. In other words, the matrices must be of the size $m^x n$ and $n^x p$. See Figure 9.27.

The multiplication operator returns the matrix dot product of the arrays. The arrays must be of the size m×n and n×p. The number or rows of the second matrix must match the number of columns in the first matrix.

$$\mathbf{Matrix\_6} = \begin{pmatrix} 11.00 & 22.00 \\ 33.00 & 44.00 \end{pmatrix} \qquad \mathbf{Matrix\_7} = \begin{pmatrix} 2.00 & 3.00 \\ 4.00 & 5.00 \end{pmatrix} \qquad \mathbf{Matrix\_8} = \begin{pmatrix} 1.00 & 2.00 & 3.00 \\ 2.00 & 3.00 & 4.00 \end{pmatrix}$$

$$\mathbf{Matrix\_15} := \mathbf{Matrix\_6} \cdot \mathbf{Matrix\_7} \qquad \mathbf{Matrix\_15} = \begin{pmatrix} 110.00 & 143.00 \\ 242.00 & 319.00 \end{pmatrix}$$

$$\mathbf{Matrix\_16} := \mathbf{Matrix\_7} \cdot \mathbf{Matrix\_8} \qquad \mathbf{Matrix\_16} = \begin{pmatrix} 8.00 & 13.00 & 18.00 \\ 14.00 & 23.00 & 32.00 \end{pmatrix}$$

The first matrix must have two columns.

$$\mathbf{Matrix\_17} := \begin{pmatrix} 1 \\ 2 \end{pmatrix} \cdot \mathbf{Matrix\_7}$$

$$\mathbf{Matrix\_17} := \begin{pmatrix} 1 \\ 2 \end{pmatrix} \cdot \mathbf{Matrix\_7}$$

These array dimensions do not match.

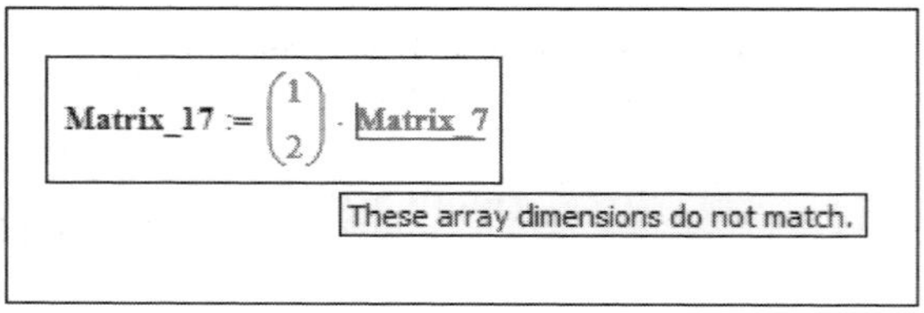

Figure 9.27 Array dot product multiplication

Vector cross product multiplication

Mathcad can perform a vector cross product on two column vectors. Each vector must have three elements. The result is a vector perpendicular to the plane for the first two vectors. The direction is according to the right-hand rule. (Refer to a text on matrix math for a discussion of the vector cross product.) Use the matrix toolbar to insert the vector cross product operator, or type **CTRL+8**. See Figure 9.28.

$$\textbf{Vector\_2} := \begin{pmatrix} 1 \\ 1 \\ 50 \end{pmatrix} \qquad \textbf{Vector\_3} := \begin{pmatrix} 2 \\ 3 \\ 60 \end{pmatrix}$$

$$\textbf{Vector\_4} := \textbf{Vector\_2} \times \textbf{Vector\_3} \qquad \textbf{Vector\_4} = \begin{pmatrix} -90.00 \\ 40.00 \\ 1.00 \end{pmatrix}$$

Figure 9.28 Vector cross product

Element-by-element multiplication (vectorize)

In order to do an element-by-element multiplication, you need to use the vectorize operator. This will tell Mathcad to ignore the normal matrix rules and perform the operation on each element. To vectorize the multiplication operation, select the entire expression, and then type **CTRL+MINUS SIGN**.

This places an arrow above the expression and tells Mathcad to perform the operation on an element-by-element basis. When using a dot product multiplication, the number of columns in the first array needs to match the number of rows in the second array. When using the vectorize operation, the arrays must be exactly the same size because the multiplication is being done on an element-by-element basis (similar to addition and subtraction). See Figure 9.29 for examples of using the vectorize operator.

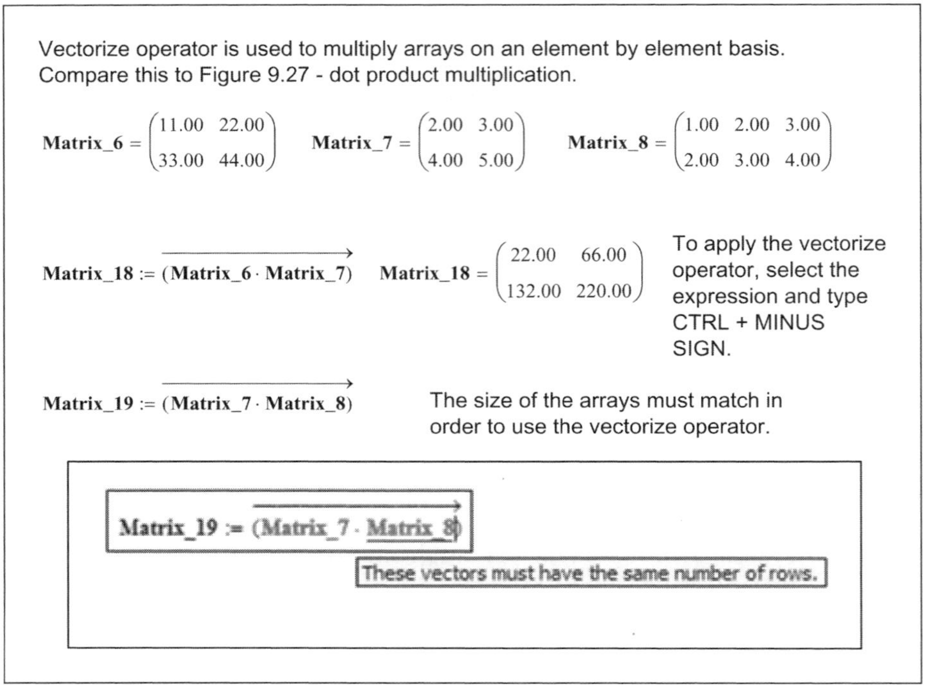

Vectorize operator is used to multiply arrays on an element by element basis. Compare this to Figure 9.27 - dot product multiplication.

$$\mathbf{Matrix\_6} = \begin{pmatrix} 11.00 & 22.00 \\ 33.00 & 44.00 \end{pmatrix} \quad \mathbf{Matrix\_7} = \begin{pmatrix} 2.00 & 3.00 \\ 4.00 & 5.00 \end{pmatrix} \quad \mathbf{Matrix\_8} = \begin{pmatrix} 1.00 & 2.00 & 3.00 \\ 2.00 & 3.00 & 4.00 \end{pmatrix}$$

$$\mathbf{Matrix\_18} := \overrightarrow{(\mathbf{Matrix\_6} \cdot \mathbf{Matrix\_7})} \quad \mathbf{Matrix\_18} = \begin{pmatrix} 22.00 & 66.00 \\ 132.00 & 220.00 \end{pmatrix}$$

To apply the vectorize operator, select the expression and type CTRL + MINUS SIGN.

$$\mathbf{Matrix\_19} := \overrightarrow{(\mathbf{Matrix\_7} \cdot \mathbf{Matrix\_8})}$$

The size of the arrays must match in order to use the vectorize operator.

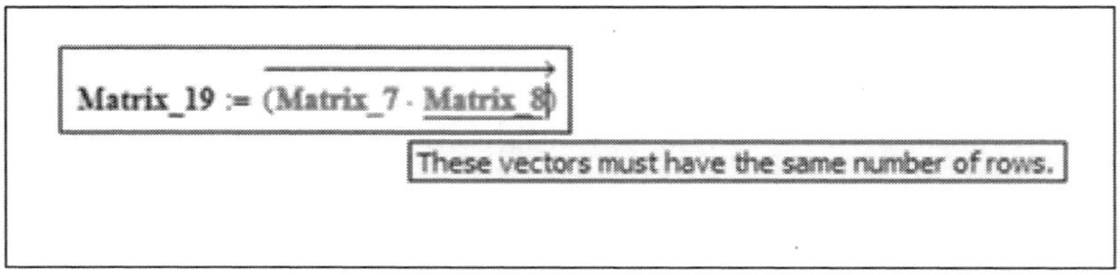

Figure 9.29 Array element-by-element multiplication

Division

For the case of X/Y, the result is dependant on whether X or Y are scalars or arrays. The result is also dependant on whether Y is a square matrix. Here are the rules:

- If Y is a square matrix, the result is the dot product of $X*Y^{-1}$, where Y^{-1} is the inverse of the square matrix Y.
- If either X or Y is a scalar, then the division is done element-by-element.
- If both X and Y are arrays, then both arrays must be of the same size. If they are not square matrices, then Mathcad does an element-by-element division. If they are square matrices, then the result is the dot product $X*Y^{-1}$.
- If both X and Y are square matrices, then in order to do an element-by-element division, you must use the vectorize operator. To do this, select the expression and type **CTRL+MINUS SIGN**.

See Figures 9.30–9.33 for examples of array division.

$$\mathbf{Matrix\_6} = \begin{pmatrix} 11.00 & 22.00 \\ 33.00 & 44.00 \end{pmatrix} \quad \mathbf{Matrix\_7} = \begin{pmatrix} 2.00 & 3.00 \\ 4.00 & 5.00 \end{pmatrix} \quad \mathbf{Matrix\_8} = \begin{pmatrix} 1.00 & 2.00 & 3.00 \\ 2.00 & 3.00 & 4.00 \end{pmatrix}$$

$$\mathbf{Matrix\_20} := \frac{\mathbf{Matrix\_6}}{2} \qquad \mathbf{Matrix\_20} = \begin{pmatrix} 5.50 & 11.00 \\ 16.50 & 22.00 \end{pmatrix}$$

Each element of Matrix_6 is divided by the scalar.

$$\mathbf{Matrix\_21a} := \frac{2}{\mathbf{Matrix\_7}} \qquad \mathbf{Matrix\_21a} = \begin{pmatrix} -5.00 & 3.00 \\ 4.00 & -2.00 \end{pmatrix}$$

Because Matrix_7 is a square matrix, the result is equal to 2*Matrix_7^{-1}.

$$\mathbf{Matrix\_7}^{-1} = \begin{pmatrix} -2.50 & 1.50 \\ 2.00 & -1.00 \end{pmatrix} \qquad 2 \cdot \mathbf{Matrix\_7}^{-1} = \begin{pmatrix} -5.00 & 3.00 \\ 4.00 & -2.00 \end{pmatrix}$$

$$\mathbf{Matrix\_21b} := \frac{2}{\mathbf{Matrix\_8}} \qquad \mathbf{Matrix\_21b} = \begin{pmatrix} 2.00 & 1.00 & 0.67 \\ 1.00 & 0.67 & 0.50 \end{pmatrix}$$

Because Matrix_8 is not a square matrix, the result is element-by-element division.

Figure 9.30 Scalar division

$$\mathbf{Matrix\_6} = \begin{pmatrix} 11.00 & 22.00 \\ 33.00 & 44.00 \end{pmatrix} \quad \mathbf{Matrix\_7} = \begin{pmatrix} 2.00 & 3.00 \\ 4.00 & 5.00 \end{pmatrix} \quad \mathbf{Matrix\_8} = \begin{pmatrix} 1.00 & 2.00 & 3.00 \\ 2.00 & 3.00 & 4.00 \end{pmatrix}$$

$$\mathbf{Matrix\_22} := \frac{\mathbf{Matrix\_6}}{\mathbf{Matrix\_7}} \qquad \mathbf{Matrix\_22} = \begin{pmatrix} 16.50 & -5.50 \\ 5.50 & 5.50 \end{pmatrix}$$

Because Matrix_7 is a square matrix, the result is equal to Matrix_6 * Matrix_7^{-1}.

$$\mathbf{Matrix\_7}^{-1} = \begin{pmatrix} -2.50 & 1.50 \\ 2.00 & -1.00 \end{pmatrix} \qquad \mathbf{Matrix\_6} \cdot \mathbf{Matrix\_7}^{-1} = \begin{pmatrix} 16.50 & -5.50 \\ 5.50 & 5.50 \end{pmatrix}$$

$$\mathbf{Matrix\_23} := \frac{\mathbf{Matrix\_8}}{\begin{pmatrix} 2 & 4 & 6 \\ 4 & 6 & 8 \end{pmatrix}} \qquad \mathbf{Matrix\_23} = \begin{pmatrix} 0.50 & 0.50 & 0.50 \\ 0.50 & 0.50 & 0.50 \end{pmatrix}$$

Both matrices must be of the same size. Because the bottom matrix is not a square matrix, the result is element-by-element division.

Figure 9.31 Array division

$$\textbf{Matrix\_7} = \begin{pmatrix} 2.00 & 3.00 \\ 4.00 & 5.00 \end{pmatrix} \qquad \textbf{Matrix\_8} = \begin{pmatrix} 1.00 & 2.00 & 3.00 \\ 2.00 & 3.00 & 4.00 \end{pmatrix}$$

$$\frac{\textbf{Matrix\_7}}{\textbf{Matrix\_8}} = \blacksquare$$

This division will not work because the matrices are different sizes.

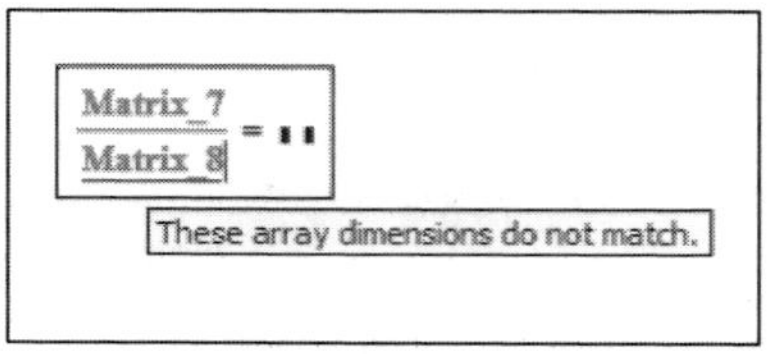

Figure 9.32 Array division

Array element-by-element division using the vectorize operator when both matrices are square.

$$\textbf{Matrix\_6} = \begin{pmatrix} 11.00 & 22.00 \\ 33.00 & 44.00 \end{pmatrix} \qquad \textbf{Matrix\_7} = \begin{pmatrix} 2.00 & 3.00 \\ 4.00 & 5.00 \end{pmatrix}$$

To apply the vectorize operator, select the expression and type CTRL+MINUS SIGN. This allows an element-by-element division to occur.

$$\textbf{Matrix\_24a} := \overrightarrow{\frac{\textbf{Matrix\_6}}{\textbf{Matrix\_7}}} \qquad \textbf{Matrix\_24a} = \begin{pmatrix} 5.50 & 7.33 \\ 8.25 & 8.80 \end{pmatrix}$$

Figure 9.33 Array division

Calculation summary

It is not the intent of this chapter to provide a complete discussion of all the many different ways that Mathcad can be used to process and manipulate vectors and matrices. Mathcad has some very useful and powerful matrix features such as transpose, inverse, determinant, and statistical functions. An excellent discussion of vectors and matrices can be found in Mathcad Help. Two sections can be referenced. The first section is found as a topic in the Contents. The second section is contained under Operators. This section discusses the many different operators that can be used with vectors and matrices. The Mathcad Tutorial also has an excellent discussion of arrays.

Engineering Examples

We have spent several pages introducing vectors and matrices. Let's now give some examples of how to use arrays in your engineering equations.

Engineering Example 9.1 shows how vectors can be used in user-defined function.

Engineering Example 9.2 shows how vectors can be use in expressions for many different input cases.

Engineering Example 9.3 shows how to use arrays as input values.

Engineering Example 9.1 Using vectors in a user-defined function

$\mathbf{Density} := 997.1 \frac{\mathbf{kg}}{\mathbf{m}^3}$ — Input variable for the user-defined function

$\mathbf{NetPressure(d)} := \mathbf{Density} \cdot \mathbf{g} \cdot \mathbf{d}$ — Define user-defined function.

$\mathbf{g} = 9.81 \frac{\mathbf{m}}{\mathbf{s}^2}$

$\mathbf{Depths} := \begin{pmatrix} 1 \\ 2 \\ 4 \\ 5 \\ 7 \\ 17 \\ 33 \end{pmatrix} \cdot \mathbf{m}$ — Create a vector of various input depth values

$\mathbf{kPa} := 1000\mathbf{Pa}$ — Define custom unit for kPa

$\mathbf{Pressures} := \mathbf{NetPressure(Depths)}$

The variable "Pressures" is defined using the user defined function "NetPressure". The argument for the user-defined function is the vector variable "Depths". This causes the variable "Pressures" to become a vector variable.

Pressures = (kPa)

	1
1	9.778
2	19.556
3	39.113
4	48.891
5	68.447
6	166.23
7	322.681

Note: The array output was changed from matrix form to an output table.

$\mathbf{Pressures}_2 = 19.56\,\mathbf{kPa}$

$\mathbf{Pressures}_6 = 166.23\,\mathbf{kPa}$

Depths = (m)

1.00
2.00
4.00
5.00
7.00
17.00
33.00

Pressures = (kPa)

9.78
19.56
39.11
48.89
68.45
166.23
322.68

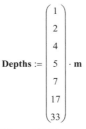

Engineering Example 9.2 Using vectors in expressions

Suppose you want to find the force caused by 4 different mass elements and 4 different accelerations.

You could assign the first value of mass and acceleration and calculate the result, and then change the input variable to the second values, or add new definitions below.

$\mathbf{Mass} := 5\mathbf{kg}$ $\quad \mathbf{Acceleration} := 3\dfrac{\mathbf{m}}{\mathbf{sec}^2}$

$\mathbf{Force} := \mathbf{Mass} \cdot \mathbf{Acceleration}$ $\quad \mathbf{Force} = 15.00\ \mathbf{N}$

$\mathbf{Mass} := 6\mathbf{lb}$ $\quad \mathbf{Acceleration} := 5\dfrac{\mathbf{ft}}{\mathbf{sec}^2}$ $\quad \mathbf{Force} := \mathbf{Mass} \cdot \mathbf{Acceleration}$ $\quad \mathbf{Force} = 4.15\ \mathbf{N}$

An easier way is to create vectors for the mass values and for the acceleration values. The same equation, now gives output for the four input conditions.

$$\mathbf{Mass} := \begin{pmatrix} 5\mathbf{kg} \\ 6\mathbf{lb} \\ 3\mathbf{gm} \\ 2 \cdot \mathbf{oz} \end{pmatrix} \qquad \mathbf{Acceleration} := \begin{pmatrix} 3\dfrac{\mathbf{m}}{\mathbf{sec}^2} \\ 5\dfrac{\mathbf{ft}}{\mathbf{sec}^2} \\ 2\dfrac{\mathbf{m}}{\mathbf{sec}^2} \\ 4\dfrac{\mathbf{ft}}{\mathbf{sec}^2} \end{pmatrix}$$

Note the use of mixed input units, but same derived unit dimensions.

$\mathbf{Force_1} := \mathbf{Mass} \cdot \mathbf{Acceleration}$ $\quad \mathbf{Force_2} := \overrightarrow{(\mathbf{Mass} \cdot \mathbf{Acceleration})}$

Use the vectorize operator to cause an element-by-element multiplication.

$\mathbf{Force_1} = 19.22\ \mathbf{N}$

The above did not perform an element by element multiplication.

$$\mathbf{Force_2} = \begin{pmatrix} 15.00 \\ 4.15 \\ 0.01 \\ 0.07 \end{pmatrix} \mathbf{N}$$

$\mathbf{Force_2}_{2} = 4.15\ \mathbf{N}$

Note: In this example,"Force_2" is the name of the variable using a literal subscript. The additional subscript is an array subscript.

Engineering Example 9.3 Using matrices in functions and expressions

The input values do not need to be only arrays. They can also be contained in a matrix.

$$\mathbf{MassInput} := \begin{pmatrix} 5 & 7 \\ 6 & 8 \end{pmatrix} \cdot \mathbf{kg} \qquad \mathbf{AccelerationInput} := \begin{pmatrix} 3 & 7 \\ 5 & 9 \end{pmatrix} \frac{\mathbf{m}}{\mathbf{sec}^2}$$

Example using a function:

$$\mathbf{ForceFunction(m, a)} := \mathbf{m} \cdot \mathbf{a}$$

Define a function

$$\mathbf{ForceOutput_1} := \mathbf{ForceFunction(MassInput, AccelerationInput)}$$

Not accurate.

$$\mathbf{ForceOutput_1} = \begin{pmatrix} 50.00 & 98.00 \\ 58.00 & 114.00 \end{pmatrix} \mathbf{N}$$

$$\mathbf{ForceOutput_2} := \overrightarrow{\mathbf{ForceFunction(MassInput, AccelerationInput)}}$$

You must use the vectorize operator to get the correct result.

$$\mathbf{ForceOutput_2} = \begin{pmatrix} 15.00 & 49.00 \\ 30.00 & 72.00 \end{pmatrix} \mathbf{N}$$

Example using an expression:

$$\mathbf{ForceOutput_3} := \mathbf{MassInput} \cdot \mathbf{AccelerationInput}$$

Not accurate.

$$\mathbf{ForceOutput_3} = \begin{pmatrix} 50.00 & 98.00 \\ 58.00 & 114.00 \end{pmatrix} \mathbf{N}$$

$$\mathbf{ForceOutput_4} := \overrightarrow{(\mathbf{MassInput} \cdot \mathbf{AccelerationInput})}$$

You must use the vectorize operator to get the correct result.

$$\mathbf{ForceOutput_4} = \begin{pmatrix} 15.00 & 49.00 \\ 30.00 & 72.00 \end{pmatrix} \mathbf{N}$$

Summary

Arrays are a very useful tool in engineering calculations. Mathcad can do very advanced matrix computations. This chapter focused mostly on using arrays to perform multiple iterations of engineering expressions.

In Chapter 9 we:

- Learned how to create, and modify the size of vectors and matrices.
- Set the ORIGIN built-in variable to 1.
- Learned how to attach units to variables.
- Demonstrated different ways to have array output data displayed.
- Learned how to make output tables smaller so that more information can be displayed on a single page.
- Learned how to add, subtract, multiply, and divide arrays.
- Illustrated how arrays can be used in engineering calculations.

Practice

1. Open the Insert Matrix dialog box. Create a 4×5 matrix. Fill-in the matrix with some numbers.
2. In the matrix created above, insert a new column between the 3rd and 4th columns. Insert a new row between the 2nd and 3rd rows. Fill-in new data.
3. In the above matrix, delete the 3rd row. Delete the 5th column.
4. Create 10 different arrays. Practice inserting and deleting rows and columns in the different arrays. Have at least two arrays larger than 20×20.
5. Assign variable names to the above arrays. Display each array in matrix form, in an output table with row and column labels, and in an output table without row and column labels. If the entire matrix does not display in table form, then enlarge the output table. Practice changing the column widths, row heights, and table font. In the output tables, use all five different alignment options for locating the variable name.
6. Select 20 elements from the above variables, and use the subscript operator to display the elements.
7. Create 10 range variables. Use various increments, some using fractions, decimals, and negative numbers. Use some formulas in the range variables.
8. Create three arrays of the following sizes: 1×3, 2×2, 2×3, and 3×3. Attach units of length to each matrix. Assign each array a variable name.
9. Using the arrays created above; create expressions to perform the following calculations. Practice using different forms for displaying results.

   [a] Six addition expressions.
   [b] Six subtraction expressions.
   [c] Six dot product expressions.

[d] Six element-by-element multiplication expressions.
[e] Twelve division expressions. For square matrices, create a dot product solution and an element-by-element solution.

10. Define a simple function with only one argument (see Figure 9.11). Create a range variable that has at least six elements. Create an expression that uses the function with the range variable as the argument. Display the result in matrix form and in table form. Use the subscript operator to display at least three individual elements of the output vector.
11. Define another simple function with two arguments. Create two input vectors. Create an expression that uses the function and the input vectors (see Engineering Examples 9.1 and 9.2). Display the result in matrix form and in table form. Use the subscript operator to display at least three individual elements of the output vector.
12. Use the function and expression created above, but instead of using input vectors, use input matrices.
13. Verify that the results in the above three practice exercises are correct. Do you need to use the vectorize operator to get the correct results?

10 Selected Mathcad functions

By now, you should be familiar with many Mathcad functions. There have been many functions discussed in this book. Chapter 3 discussed how to use several different functions.

There are hundreds of Mathcad functions. The following is a partial list of the different categories that have functions: Bessel Functions, Complex Numbers, Curve Fitting, Data Analysis, Differential Equation Solving, Finance, Fourier Transforms, Graphing, Hyperbolic Functions, Image Processing, Interpolation, Logs, Number Theory, Probability, Solving, Sorting, Statistics, Trigonometry, Vectors, and Waves.

There are many functions that many of us have never heard of, and will never use. Several books have been written discussing the many Mathcad functions. As explained earlier, the purpose of this book is to teach the application of Mathcad to engineering calculations; therefore, a great deal of time will not be spent discussing specific Mathcad functions. The Mathcad Users Guide is an excellent resource to learn the details of specific functions. The Mathcad Help is also useful to learn about the many Mathcad functions.

Of all the many Mathcad functions, which ones are most important for engineering calculations? The answer to this question depends upon your own perspective. A function that is important for one person may not be important for another.

With that said, this chapter will introduce and discuss a few functions that (in the author's opinion) will be beneficial to most engineers doing engineering calculations. These functions were chosen because they are easy to learn, and add power to engineering calculations.

Chapter 10 will:

- Review the basic concept of built-in functions.
- Discuss the calculation toolbar.
- Introduce the following functions:
    - ***max***
    - ***min***
    - ***mean***
    - ***median***
    - ***floor & Floor***
    - ***ceil and Ceil***
    - ***trunc and Trunc***
    - ***round and Round***
    - ***Vector Sum***
    - ***Summation***
    - ***Range Sum***
    - ***if***
    - ***linterp***

Review of built-in functions

In Chapter 3, we learned that every built-in Mathcad function is set-up in a similar way. The name of the function is given, followed by a pair of parenthesis. The information required within the parenthesis is called the argument. Mathcad processes the argument(s) based on rules that are defined for the specific function.

To insert a function into a worksheet, use the Insert Function dialog box. To open this dialog box, select **Function** from the **Insert** menu. You may also type **CTRL+E**, or you may also click the measuring cup icon on the Standard toolbar.

Refer to Chapter 3 for a more detailed description of functions.

Toolbars

There are many useful functions and operators located as icons on many of the Mathcad toolbars. You can access many of these toolbars from the

Math toolbar. To open the Math toolbar, hover your mouse over **Toolbars** on the **View** menu and click **Math**.

In addition to math operators, the Calculator toolbar is a good place to look for icons, which insert simple math operators and functions. This toolbar contains simple trigonometry functions, log functions, factorial, and absolute value.

Selected functions

max and *min* functions

The ***max*** and ***min*** function are useful to select the maximum or minimum values from a list of values. The ***max*** function takes the form ***max***(A,B,C, . . .). Mathcad returns the largest value from A, B, C, etc. The ***min*** function takes the same form, ***min***(A,B,C, . . .). In its simplest form, you can just type the list of values in the function. See Figure 10.1.

The *max* function selects the maximum value from a list of arguments. The *min* function selects the minimum value.

$\mathbf{Test\_max_1} := \mathbf{max}(1,3,5,7,9,8,6,4,2,-4)$ $\quad\quad \mathbf{Test\_max_1} = 9.00$

$\mathbf{Test\_min_1} := \mathbf{min}(1,3,5,7,9,8,6,4,2,-4)$ $\quad\quad \mathbf{Test\_min_1} = -4.00$

Figure 10.1 ***max*** and ***min*** functions with a list

A, B, C, etc. can also be variable names. See Figure 10.2.

The *max* and *min* functions can include a list of variable names.

$\mathbf{Var_1} := 3 \quad\quad \mathbf{Var_2} := 5 \quad\quad \mathbf{Var_3} := 7$

$\mathbf{max}(\mathbf{Var_1}, \mathbf{Var_2}, \mathbf{Var_3}) = 7.00$

$\mathbf{min}(\mathbf{Var_1}, \mathbf{Var_2}, \mathbf{Var_3}) = 3.00$

$\mathbf{Text\_max_2} := \mathbf{max}(\mathbf{Var_1}, \mathbf{Var_2}, \mathbf{Var_3})$ $\quad\quad \mathbf{Text\_max_2} = 7.00$

$\mathbf{Test\_min_2} := \mathbf{min}(\mathbf{Var_1}, \mathbf{Var_2}, \mathbf{Var_3})$ $\quad\quad \mathbf{Test\_min_2} = 3.00$

Figure 10.2 ***max*** and ***min*** functions with variables

A, B, C, etc. can also be a list of arrays. See Figure 10.3.

The *max* and *min* function can also be used to select the maximum and minimum values from vectors and matrices.

$$\mathbf{Matrix_{10.2}} := \begin{pmatrix} 1 & 2 & 3 \\ 4 & 5 & 6 \\ 7 & 8 & 9 \end{pmatrix} \qquad \mathbf{Vector_{10.2}} := \begin{pmatrix} 13 \\ 16 \\ 19 \end{pmatrix}$$

$\mathbf{Test\_max_3} := \mathbf{max}(\mathbf{Matrix_{10.2}})$ $\quad$ $\mathbf{Test\_max_4} := \mathbf{max}(\mathbf{Vector_{10.2}})$

$\mathbf{Test\_max_3} = 9.00$ $\quad$ $\mathbf{Test\_max_4} = 19.00$

$\mathbf{Test\_min_3} := \mathbf{min}(\mathbf{Matrix_{10.2}})$ $\quad$ $\mathbf{Test\_min_4} := \mathbf{min}(\mathbf{Vector_{10.2}})$

$\mathbf{Test\_min_3} = 1.00$ $\quad$ $\mathbf{Test\_min_4} = 13.00$

max and *min* can also select from multiple arrays.

$\mathbf{Test\_max_5} := \mathbf{max}(\mathbf{Matrix_{10.2}}, \mathbf{Vector_{10.2}})$ $\quad$ $\mathbf{Test\_min_5} := \mathbf{min}(\mathbf{Matrix_{10.2}}, \mathbf{Vector_{10.2}})$

$\mathbf{Test\_max_5} = 19.00$ $\quad$ $\mathbf{Test\_min_5} = 1.00$

Figure 10.3 ***max*** and ***min*** functions with arrays

Units can also be attached to A, B, C, etc. as long as they are of the same unit dimension.

See Figure 10.4.

The *max* and *min* functions may also include units from the same unit dimension. Mathcad converts all units to consistent SI units, and then selects the maximum or minimum values. The result can be displayed in any unit.

$\mathbf{Test\_max_6 := max(1m, 3m, 5m, 7m, 9cm, 8m, 6m, 4m, 25ft)}$ $\quad \mathbf{Test\_max_6} = 8.00\ \mathbf{m}$

$\mathbf{Test\_min_6 := min(1m, 3m, 5m, 7m, 9cm, 8m, 6m, 4m, 2m)}$ $\quad \mathbf{Test\_min_6} = 0.09\ \mathbf{m}$

$\mathbf{Test\_max_7 := max(1min, 30sec, 0.5day, 5hr)}$ $\quad \mathbf{Test\_max_7} = 12.00\ \mathbf{hr}$

$\mathbf{Test\_min_7 := min(1min, 30sec, 0.5day, 5hr)}$ $\quad \mathbf{Test\_min_7} = 30.00\ \mathbf{s}$

$$\mathbf{Matrix_{10.4}} := \begin{pmatrix} 3\mathbf{ft} & 5\mathbf{in} & 10\mathbf{mm} \\ 1\mathbf{m} & 56\mathbf{cm} & 1\mathbf{yd} \\ 0.0005\mathbf{mile} & 5000\mathbf{mil} & 0.0005\mathbf{furlong} \end{pmatrix}$$

$$\mathbf{Matrix_{10.4}} = \begin{pmatrix} 0.91 & 0.13 & 0.01 \\ 1.00 & 0.56 & 0.91 \\ 0.80 & 0.13 & 0.10 \end{pmatrix} \mathbf{m}$$

$\mathbf{Test\_max_8 := max(Matrix_{10.4})}$ $\quad \mathbf{Test\_max_8} = 1.00\ \mathbf{m}$

$\mathbf{Test\_min_8 := min(Matrix_{10.4})}$ $\quad \mathbf{Test\_min_8} = 10.00\ \mathbf{mm}$

Figure 10.4 ***max*** and ***min*** functions with units

If A, B, C, etc. include a complex number, then the ***max*** function returns the largest real part of any value, and i times the largest imaginary part of any value. For the ***min*** function, Mathcad returns the smallest real part of any value, and i times the smallest imaginary part of any value. See Figure 10.5.

If the argument list for the *max* and *min* functions include complex numbers, then Mathcad selects the largest or smallest real values and the largest or smallest imaginary part, even though they may be from different elements.

$$\mathbf{Vector_{10.5a}} := \begin{pmatrix} 1+9i \\ 8 \\ 3+3i \\ 4+4i \end{pmatrix} \qquad \mathbf{Vector_{10.5b}} := \begin{pmatrix} 1+9i \\ 8 \\ 3-3i \\ 4+4i \end{pmatrix}$$

$\mathbf{Test\_max_9} := \mathbf{max}(\mathbf{Vector_{10.5a}})$ $\mathbf{Test\_max_9} = 8.00 + 9.00i$

$\mathbf{Test\_min_9} := \mathbf{min}(\mathbf{Vector_{10.5a}})$ $\mathbf{Test\_min_9} = 1.00$ In this case, 0*i is the smallest imaginary part.

$\mathbf{Test\_max_{10}} := \mathbf{max}(\mathbf{Vector_{10.5b}})$ $\mathbf{Test\_max_{10}} = 8.00 + 9.00i$

$\mathbf{Test\_min_{10}} := \mathbf{min}(\mathbf{Vector_{10.5b}})$ $\mathbf{Test\_min_{10}} = 1.00 - 3.00i$

Figure 10.5 ***max*** and ***min*** functions with complex numbers

A, B, C, etc. can even be strings. For strings, "z" is larger than "a," thus the string "cat" is larger than the string "alligator." You cannot mix strings and numbers. See Figure 10.6.

The *max* and *min* functions can also select from a list of string variables.

Alpha_max := **max**("cat" , "alligator") **Alpha_max** = "cat" For string variables, letter "b" > letter "a".

Alpha_min := **min**("cat" , "alligator") **Alpha_min** = "alligator"

Figure 10.6 ***max*** and ***min*** functions with string variables

mean *and* ***median*** *functions*

Mathcad has many statistical and data analysis functions. We will discuss only two of these functions. The ***mean*** function is useful for calculating averages of a list of values. The ***median*** function returns the value above and below which there are an equal number of values.

The ***mean*** function takes the form ***mean***(A,B,C,. . .). The arguments A, B, C, etc. can be scalars, arrays or complex numbers. The arguments can also have units attached. The ***mean*** function sums all the elements in the argument list and divides by the number of elements. If one or all of the arguments are arrays, Mathcad sums all the elements in the arrays and counts all the elements in the arrays. If any of the arguments are complex numbers, Mathcad returns i times the sum of the imaginary parts divided by the total number of all elements (not just the complex numbers). When using units, Mathcad converts all units to SI units before taking the average. It then displays the average in the desired unit system. See Figure 10.7.

The *mean* function calculates the average of a list of variables.

$$\mathbf{Matrix_{10.7a}} := \begin{pmatrix} 3 & 5 & 7 \\ 9 & 11 & 13 \\ 15 & 17 & 19 \end{pmatrix} \qquad \mathbf{Vector_{10.7a}} := \begin{pmatrix} 5 \\ 15 \\ 30 \end{pmatrix}$$

$$\mathbf{Test\_mean_1} := \mathbf{mean}(\mathbf{Matrix_{10.7a}}) \qquad \mathbf{Test\_mean_1} = 11.00$$

$$\mathbf{Test\_mean_2} := \mathbf{mean}(\mathbf{Vector_{10.7a}}) \qquad \mathbf{Test\_mean_2} = 16.67$$

$$\mathbf{Test\_mean_3} := \mathbf{mean}(\mathbf{Matrix_{10.7a}}, \mathbf{Vector_{10.7a}}) \qquad \mathbf{Test\_mean_3} = 12.42$$

The arguments may include units. Mathcad converts all units to consistent SI units, and then takes the mean.

$$\mathbf{Matrix_{10.7b}} := \begin{pmatrix} 3\mathbf{ft} & 5\mathbf{in} & 10\mathbf{mm} \\ 1\mathbf{m} & 56\mathbf{cm} & 1\mathbf{yd} \\ 0.0005\mathbf{mile} & 5000\mathbf{mil} & 0.0005\mathbf{furlong} \end{pmatrix} \qquad \mathbf{Matrix_{10.7b}} = \begin{pmatrix} 0.91 & 0.13 & 0.01 \\ 1.00 & 0.56 & 0.91 \\ 0.80 & 0.13 & 0.10 \end{pmatrix} \mathbf{m}$$

$$\mathbf{Test\_mean_4} := \mathbf{mean}(\mathbf{Matrix_{10.7b}}) \qquad \mathbf{Test\_mean_4} = 0.51\ \mathbf{m}$$

If any of the arguments are complex numbers, Mathcad takes the average of the real parts and the average of the imaginary parts (including the numbers with 0i).

$$\mathbf{Vector_{10.7c}} := \begin{pmatrix} 1 + 9i \\ 8 \\ 3 + 3i \\ 4 + 4i \end{pmatrix} \qquad \mathbf{Vector_{10.7d}} := \begin{pmatrix} 5 + 9i \\ 8 \\ 3 - 3i \\ 4 + 4i \end{pmatrix}$$

$$\mathbf{Test\_mean_5} := \mathbf{mean}(\mathbf{Vector_{10.7c}}) \qquad \mathbf{Test\_mean_5} = 4.00 + 4.00i$$

$$\mathbf{Test\_mean_6} := \mathbf{mean}(\mathbf{Vector_{10.7d}}) \qquad \mathbf{Test\_mean_6} = 5.00 + 2.50i$$

$$\mathbf{Test\_mean_7} := \mathbf{mean}(\mathbf{Vector_{10.7c}}, \mathbf{Vector_{10.7d}}) \qquad \mathbf{Test\_mean_7} = 4.50 + 3.25i$$

Figure 10.7 ***mean*** function

The ***median*** function returns the median of all the elements in the argument list. The median is the value above and below which there are an equal number of values. If there are an even number of values, the median is the arithmetic mean of the two central values. The arguments A, B, C, etc. can be scalars or arrays, but not complex numbers. The arguments can also have units attached. Mathcad first sorts the arguments from lowest to highest prior to taking the median. See Figure 10.8.

The *median* function returns the value above and below which there are an equal number of values.

If the lists of values used to the left are rearranged, it is easier to see how Mathcad selects the median.

$\mathbf{Test\_median_1} := \mathbf{median}(3,5,7,9,8,6,4)$ $\quad \mathbf{Test\_median_1} = 6.00$ $\quad \mathbf{median}(3,4,5,6,7,8,9) = 6.00$

$\mathbf{Test\_median_2} := \mathbf{median}(5\mathbf{m},8\mathbf{m},6\mathbf{m},9\mathbf{m})$ $\quad \mathbf{Test\_median_2} = 7.00\ \mathbf{m}$ $\quad \mathbf{median}(5,6,8,9) = 7.00$

$\mathbf{Test\_median_3} := \mathbf{median}(5\mathbf{m},8\mathbf{m},6\mathbf{cm},9\mathbf{mm})$ $\quad \mathbf{Test\_median_3} = 2.53\ \mathbf{m}$ $\quad \mathbf{median}(0.009,0.06,5,8) = 2.53$

$$\frac{0.06+5}{2} = 2.53$$

If there are an even number of values, then Mathcad takes the arithmetic mean of the two central values.

$$\mathbf{Matrix_{10.7a}} = \begin{pmatrix} 3.00 & 5.00 & 7.00 \\ 9.00 & 11.00 & 13.00 \\ 15.00 & 17.00 & 19.00 \end{pmatrix} \qquad \mathbf{Vector_{10.7a}} = \begin{pmatrix} 5.00 \\ 15.00 \\ 30.00 \end{pmatrix}$$

$\mathbf{Test\_median_4} := \mathbf{median}(\mathbf{Matrix_{10.7a}})$ $\quad \mathbf{Test\_median_4} = 11.00$

$\mathbf{Test\_median_5} := \mathbf{median}(\mathbf{Vector_{10.7a}})$ $\quad \mathbf{Test\_median_5} = 15.00$

$\mathbf{Test\_median_6} := \mathbf{median}(\mathbf{Matrix_{10.7a}},\mathbf{Vector_{10.7a}})$ $\quad \mathbf{Test\_median_6} = 12.00$

Check

$\mathbf{median}(3,5,5,7,9,11,13,15,15,17,19,30) = 12.00$

$$\mathbf{Matrix_{10.7b}} = \begin{pmatrix} 0.91 & 0.13 & 0.01 \\ 1.00 & 0.56 & 0.91 \\ 0.80 & 0.13 & 0.10 \end{pmatrix} \mathbf{m}$$

$\mathbf{Test\_median_7} := \mathbf{median}(\mathbf{Matrix_{10.7b}})$ $\quad \mathbf{Test\_median_7} = 0.56\ \mathbf{m}$

Check

$\mathbf{median}(0.01,0.10,0.13,0.13,0.56,0.80,0.91,0.91,1.00) = 0.56$

Figure 10.8 ***median*** function

Truncation and rounding functions

We will discuss four truncation and rounding functions (***floor***, ***ceil***, ***trunc***, and ***round***). Each of these functions has two forms. One form is the lower case form, and the other form is the upper case form. We will discuss the lower case forms first.

The function ***floor***(z) returns the greatest integer less than z.

The function ***ceil***(z) returns the smallest integer greater than z.

The function ***trunc***(z) returns the integer part of z by removing the fractional part. If z is greater than zero, then this function is identical to ***floor***(z). If z is less than zero, then this function is identical to ***ceil***(z).

The function ***round***(z,[n]) returns z rounded to n decimal places. The argument n must be an integer. If n is omitted (or equal to zero), it returns z rounded to the nearest integer. If n is less than zero, it returns z rounded to n places to the left of the decimal point.

The argument z can be a real or complex scalar or vector. It cannot be an array. See Figures 10.9 and 10.10 for examples of the lower case truncate and round functions. The argument z in these lower case functions may not have units attached, but there is still a way to use units with these functions. If your list of values has units attached, divide the list of values by the unit you want to use, then multiply the function by the same unit. See Figure 10.11 for an example of using units with the lower case truncate and round functions.

Use the following four values to examine the results of the following four functions: *floor, ceil, trunc*, and *round.*

$Var_{10.9a} := 3.49$	$Var_{10.9b} := 3.51$	$Var_{10.9c} := -3.49$	$Var_{10.9d} := -3.51$
$Test\_floor_1 := floor(Var_{10.9a})$	$Test\_floor_2 := floor(Var_{10.9b})$	$Test\_floor_3 := floor(Var_{10.9c})$	$Test\_floor_4 := floor(Var_{10.9d})$
$Test\_floor_1 = 3.00$	$Test\_floor_2 = 3.00$	$Test\_floor_3 = -4.00$	$Test\_floor_4 = -4.00$
$Test\_ceil_1 := ceil(Var_{10.9a})$	$Test\_ceil_2 := ceil(Var_{10.9b})$	$Test\_ceil_3 := ceil(Var_{10.9c})$	$Test\_ceil_4 := ceil(Var_{10.9d})$
$Test\_ceil_1 = 4.00$	$Test\_ceil_2 = 4.00$	$Test\_ceil_3 = -3.00$	$Test\_ceil_4 = -3.00$
$Test\_trunc_1 := trunc(Var_{10.9a})$	$Test\_trunc_2 := trunc(Var_{10.9b})$	$Test\_trunc_3 := trunc(Var_{10.9c})$	$Test\_trunc_4 := trunc(Var_{10.9d})$
$Test\_trunc_1 = 3.00$	$Test\_trunc_2 = 3.00$	$Test\_trunc_3 = -3.00$	$Test\_trunc_4 = -3.00$
$Test\_round_1 := round(Var_{10.9a})$	$Test\_round_2 := round(Var_{10.9b})$	$Test\_round_3 := round(Var_{10.9c})$	$Test\_round_4 := round(Var_{10.9d})$
$Test\_round_1 = 3.00$	$Test\_round_2 = 4.00$	$Test\_round_3 = -3.00$	$Test\_round_4 = -4.00$

Figure 10.9 Lower case truncate and round functions

The lower case truncate and round functions may have vectors as arguments, but not matrices.

$$\mathbf{Var_{10.10a}} := \begin{pmatrix} 4.49 \\ 49.50 \\ -150.01 \\ -1499.99 \\ 5000.01 \end{pmatrix}$$

$\mathbf{Test\_floor_5} := \mathbf{floor}(\mathbf{Var_{10.10a}})$ $\mathbf{Test\_ceil_5} := \mathbf{ceil}(\mathbf{Var_{10.10a}})$ $\mathbf{Test\_trunc_5} := \mathbf{trunc}(\mathbf{Var_{10.10a}})$ $\mathbf{Test\_round_5} := \mathbf{round}(\mathbf{Var_{10.10a}})$

$$\mathbf{Test\_floor_5} = \begin{pmatrix} 4.00 \\ 49.00 \\ -151.00 \\ -1500.00 \\ 5000.00 \end{pmatrix} \quad \mathbf{Test\_ceil_5} = \begin{pmatrix} 5.00 \\ 50.00 \\ -150.00 \\ -1499.00 \\ 5001.00 \end{pmatrix} \quad \mathbf{Test\_trunc_5} = \begin{pmatrix} 4.00 \\ 49.00 \\ -150.00 \\ -1499.00 \\ 5000.00 \end{pmatrix} \quad \mathbf{Test\_round_5} = \begin{pmatrix} 4.00 \\ 50.00 \\ -150.00 \\ -1500.00 \\ 5000.00 \end{pmatrix}$$

The round function has the form round(z,[n]), where n is an integer telling Mathcad at which decimal place to round. If n is less than zero, Mathcad rounds to n places to the left of the decimal point. If n is left blank (zero), Mathcad rounds to an integer.

n=1 rounds to 1 decimal place n=-1 rounds to the 10's n=-2 rounds to the 100's

$\mathbf{Test\_round_6} := \mathbf{round}(\mathbf{Var_{10.10a}}, 1)$ $\mathbf{Test\_round_7} := \mathbf{round}(\mathbf{Var_{10.10a}}, -1)$ $\mathbf{Test\_round_8} := \mathbf{round}(\mathbf{Var_{10.10a}}, -2)$

$$\mathbf{Test\_round_6} = \begin{pmatrix} 4.50 \\ 49.50 \\ -150.00 \\ -1500.00 \\ 5000.00 \end{pmatrix} \quad \mathbf{Test\_round_7} = \begin{pmatrix} 0.00 \\ 50.00 \\ -150.00 \\ -1500.00 \\ 5000.00 \end{pmatrix} \quad \mathbf{Test\_round_8} = \begin{pmatrix} 0.00 \\ 0.00 \\ -200.00 \\ -1500.00 \\ 5000.00 \end{pmatrix}$$

Figure 10.10 Lower case truncate and round functions with vectors

Figure 10.11 is the same as figure 10.10 except that units are used in 10.11.

Even though the lower case truncate and round functions do not allow the use of units, you can still use units with these functions. To do this, divide the list of units by the unit you want to truncate or round to, then multiply the function by the same unit.

$$\mathbf{Var_{10.11}} := \begin{pmatrix} 4.49 \\ 49.50 \\ -150.01 \\ -1499.99 \\ 5000.01 \end{pmatrix} \cdot \mathbf{m} \qquad \mathbf{Var_{10.11}} = \begin{pmatrix} 4.49 \\ 49.50 \\ -150.01 \\ -1499.99 \\ 5000.01 \end{pmatrix} \mathbf{m} \qquad \mathbf{Var_{10.11}} = \begin{pmatrix} 14.73 \\ 162.40 \\ -492.16 \\ -4921.23 \\ 1.64 \times 10^{4} \end{pmatrix} \mathbf{ft}$$

In order to fit all examples in this figure, the results are not attached to a variable. This is not good engineering practice, because the results will not be available for later use.

$$\mathbf{floor}\left(\frac{\mathbf{Var_{10.11}}}{\mathbf{m}}\right) \cdot \mathbf{m} = \begin{pmatrix} 4.00 \\ 49.00 \\ -151.00 \\ -1500.00 \\ 5000.00 \end{pmatrix} \mathbf{m} \qquad \mathbf{trunc}\left(\frac{\mathbf{Var_{10.11}}}{\mathbf{m}}\right) \cdot \mathbf{m} = \begin{pmatrix} 4.00 \\ 49.00 \\ -150.00 \\ -1499.00 \\ 5000.00 \end{pmatrix} \mathbf{m}$$

$$\mathbf{ceil}\left(\frac{\mathbf{Var_{10.11}}}{\mathbf{m}}\right) \cdot \mathbf{m} = \begin{pmatrix} 5.00 \\ 50.00 \\ -150.00 \\ -1499.00 \\ 5001.00 \end{pmatrix} \mathbf{m} \qquad \mathbf{round}\left(\frac{\mathbf{Var_{10.11}}}{\mathbf{m}}\right) \cdot \mathbf{m} = \begin{pmatrix} 4.00 \\ 50.00 \\ -150.00 \\ -1500.00 \\ 5000.00 \end{pmatrix} \mathbf{m}$$

$$\mathbf{floor}\left(\frac{\mathbf{Var_{10.11}}}{\mathbf{ft}}\right) \cdot \mathbf{ft} = \begin{pmatrix} 14.000 \\ 162.000 \\ -493.000 \\ -4922.000 \\ 1.640 \times 10^{4} \end{pmatrix} \mathbf{ft} \qquad \mathbf{round}\left(\frac{\mathbf{Var_{10.11}}}{\mathbf{ft}}\right) \cdot \mathbf{ft} = \begin{pmatrix} 15.00 \\ 162.00 \\ -492.00 \\ -4921.00 \\ 1.64 \times 10^{4} \end{pmatrix} \mathbf{ft}$$

The rounding unit may be different than the input units. You can use feet, even though the input was in meters.

The round function has the form round(z,[n]), where n is an integer telling Mathcad at which decimal place to round. If n is less than zero, Mathcad rounds to n places to the left of the decimal point. If n is left blank (zero), Mathcad rounds to an integer.

n=1 rounds to 1 decimal place

$$\mathbf{round}\left(\frac{\mathbf{Var_{10.11}}}{\mathbf{m}}, 1\right) \cdot \mathbf{m} = \begin{pmatrix} 4.50 \\ 49.50 \\ -150.00 \\ -1500.00 \\ 5000.00 \end{pmatrix} \mathbf{m}$$

n=-1 rounds to the 10's

$$\mathbf{round}\left(\frac{\mathbf{Var_{10.11}}}{\mathbf{m}}, -1\right) \cdot \mathbf{m} = \begin{pmatrix} 0.00 \\ 50.00 \\ -150.00 \\ -1500.00 \\ 5000.00 \end{pmatrix} \mathbf{m}$$

n=-2 rounds to the 100's

$$\mathbf{round}\left(\frac{\mathbf{Var_{10.11}}}{\mathbf{ft}}, -2\right) \cdot \mathbf{ft} = \begin{pmatrix} 0.00 \\ 200.00 \\ -500.00 \\ -4900.00 \\ 1.64 \times 10^{4} \end{pmatrix} \mathbf{ft}$$

Figure 10.11 Lower case truncate and round functions with units

The upper case forms of these equations are a bit more complicated. The upper case functions introduce an additional argument "y." This argument must be a real, nonzero scalar or vector. The argument y tells Mathcad to truncate or round to a multiple of y. The lower case functions are equivalent to the upper case functions with y equal to one. (The n in round(z,[n]) being zero.) If y is equal to three, then Mathcad truncates or rounds the argument z to a multiple of three. The upper case functions also allow the use of units. Let's look at a few examples. See Figures 10.12, 10.13 (without units) and Figure 10.14 (with units).

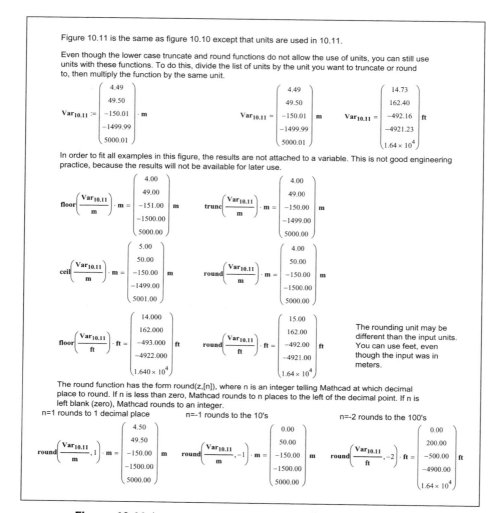

The upper case truncate and round functions introduce a second argument y. The functions have the form Floor(z,y). The argument y tells Mathcad to truncate or round to a multiple of y. Let's look at the same variables as in Figure 10.9 and look at two values of y.

$\text{Var}_{10.9a} = 3.49$	$\text{Var}_{10.9b} = 3.51$	$\text{Var}_{10.9c} = -3.49$

Using y=2 means that all results will be a multiple of 2 (or an even number.)

$\text{Test\_Floor}_9 := \text{Floor}(\text{Var}_{10.9a}, 2)$	$\text{Test\_Floor}_{10} := \text{Floor}(\text{Var}_{10.9b}, 2)$	$\text{Test\_Floor}_{11} := \text{Floor}(\text{Var}_{10.9c}, 2)$
$\text{Test\_Floor}_9 = 2.00$	$\text{Test\_Floor}_{10} = 2.00$	$\text{Test\_Floor}_{11} = -4.00$
$\text{Test\_Ceil}_9 := \text{Ceil}(\text{Var}_{10.9a}, 2)$	$\text{Test\_Ceil}_{10} := \text{Ceil}(\text{Var}_{10.9b}, 2)$	$\text{Test\_Ceil}_{11} := \text{Ceil}(\text{Var}_{10.9c}, 2)$
$\text{Test\_Ceil}_9 = 4.00$	$\text{Test\_Ceil}_{10} = 4.00$	$\text{Test\_Ceil}_{11} = -2.00$
$\text{Test\_Trunc}_9 := \text{Trunc}(\text{Var}_{10.9a}, 2)$	$\text{Test\_Trunc}_{10} := \text{Trunc}(\text{Var}_{10.9b}, 2)$	$\text{Test\_Trunc}_{11} := \text{Trunc}(\text{Var}_{10.9c}, 2)$
$\text{Test\_Trunc}_9 = 2.00$	$\text{Test\_Trunc}_{10} = 2.00$	$\text{Test\_Trunc}_{11} = -2.00$
$\text{Test\_Round}_9 := \text{Round}(\text{Var}_{10.9a}, 2)$	$\text{Test\_Round}_{10} := \text{Round}(\text{Var}_{10.9b}, 2)$	$\text{Test\_Round}_{11} := \text{Round}(\text{Var}_{10.9c}, 2)$
$\text{Test\_Round}_9 = 4.00$	$\text{Test\_Round}_{10} = 4.00$	$\text{Test\_Round}_{11} = -4.00$

Using y=0.1 means that all results will be a multiple of 0.1, or to the first decimal place.

$\text{Test\_Floor}_{12} := \text{Floor}(\text{Var}_{10.9a}, 0.1)$	$\text{Test\_Floor}_{13} := \text{Floor}(\text{Var}_{10.9b}, 0.1)$	$\text{Test\_Floor}_{14} := \text{Floor}(\text{Var}_{10.9c}, 0.2)$
$\text{Test\_Floor}_{12} = 3.40$	$\text{Test\_Floor}_{13} = 3.50$	$\text{Test\_Floor}_{14} = -3.60$
$\text{Test\_Ceil}_{12} := \text{Ceil}(\text{Var}_{10.9a}, 0.1)$	$\text{Test\_Ceil}_{13} := \text{Ceil}(\text{Var}_{10.9b}, 0.1)$	$\text{Test\_Ceil}_{14} := \text{Ceil}(\text{Var}_{10.9c}, 0.1)$
$\text{Test\_Ceil}_{12} = 3.50$	$\text{Test\_Ceil}_{13} = 3.60$	$\text{Test\_Ceil}_{14} = -3.40$
$\text{Test\_Trunc}_{12} := \text{Trunc}(\text{Var}_{10.9a}, 0.1)$	$\text{Test\_Trunc}_{13} := \text{Trunc}(\text{Var}_{10.9b}, 0.1)$	$\text{Test\_Trunc}_{14} := \text{Trunc}(\text{Var}_{10.9c}, 0.1)$
$\text{Test\_Trunc}_{12} = 3.40$	$\text{Test\_Trunc}_{13} = 3.50$	$\text{Test\_Trunc}_{14} = -3.40$
$\text{Test\_Round}_{12} := \text{Round}(\text{Var}_{10.9a}, 0.1)$	$\text{Test\_Round}_{13} := \text{Round}(\text{Var}_{10.9b}, 0.1)$	$\text{Test\_Round}_{14} := \text{Round}(\text{Var}_{10.9c}, 0.1)$
$\text{Test\_Round}_{12} = 3.50$	$\text{Test\_Round}_{13} = 3.50$	$\text{Test\_Round}_{14} = -3.50$

Figure 10.12 Upper case truncate and round functions

The upper case truncate and round functions may have vectors as arguments, but not matrices. Let's look at the same variables from Figure 10.10 and use 3 values of y.

$$\mathbf{Var_{10.10a}} = \begin{pmatrix} 4.49 \\ 49.50 \\ -150.01 \\ -1499.99 \\ 5000.01 \end{pmatrix}$$

In order to fit all examples in this figure, the results are not attached to a variable. This is not good engineering practice, because the results will not be available for later use.

y=2

$$\mathbf{Floor}(\mathbf{Var_{10.10a}}, 2) = \begin{pmatrix} 4.00 \\ 48.00 \\ -152.00 \\ -1500.00 \\ 5000.00 \end{pmatrix}$$

$$\mathbf{Ceil}(\mathbf{Var_{10.10a}}, 2) = \begin{pmatrix} 6.00 \\ 50.00 \\ -150.00 \\ -1498.00 \\ 5002.00 \end{pmatrix}$$

$$\mathbf{Trunc}(\mathbf{Var_{10.10a}}, 2) = \begin{pmatrix} 4.00 \\ 48.00 \\ -150.00 \\ -1498.00 \\ 5000.00 \end{pmatrix}$$

$$\mathbf{Round}(\mathbf{Var_{10.10a}}, 2) = \begin{pmatrix} 4.00 \\ 50.00 \\ -150.00 \\ -1500.00 \\ 5000.00 \end{pmatrix}$$

y=0.1

$$\mathbf{Floor}(\mathbf{Var_{10.10a}}, 0.1) = \begin{pmatrix} 4.40 \\ 49.50 \\ -150.10 \\ -1500.00 \\ 5000.00 \end{pmatrix}$$

$$\mathbf{Ceil}(\mathbf{Var_{10.10a}}, 0.1) = \begin{pmatrix} 4.50 \\ 49.50 \\ -150.00 \\ -1499.90 \\ 5000.10 \end{pmatrix}$$

$$\mathbf{Trunc}(\mathbf{Var_{10.10a}}, 0.1) = \begin{pmatrix} 4.40 \\ 49.50 \\ -150.00 \\ -1499.90 \\ 5000.00 \end{pmatrix}$$

$$\mathbf{Round}(\mathbf{Var_{10.10a}}, 0.1) = \begin{pmatrix} 4.50 \\ 49.50 \\ -150.00 \\ -1500.00 \\ 5000.00 \end{pmatrix}$$

y=0.01

$$\mathbf{Floor}(\mathbf{Var_{10.10a}}, 0.01) = \begin{pmatrix} 4.49 \\ 49.50 \\ -150.01 \\ -1499.99 \\ 5000.01 \end{pmatrix}$$

$$\mathbf{Ceil}(\mathbf{Var_{10.10a}}, 0.01) = \begin{pmatrix} 4.49 \\ 49.50 \\ -150.01 \\ -1499.99 \\ 5000.01 \end{pmatrix}$$

$$\mathbf{Trunc}(\mathbf{Var_{10.10a}}, 0.01) = \begin{pmatrix} 4.49 \\ 49.50 \\ -150.01 \\ -1499.99 \\ 5000.01 \end{pmatrix}$$

$$\mathbf{Round}(\mathbf{Var_{10.10a}}, 0.01) = \begin{pmatrix} 4.49 \\ 49.50 \\ -150.01 \\ -1499.99 \\ 5000.01 \end{pmatrix}$$

Figure 10.13 Upper case truncate and round functions with vectors

The upper case truncate and round functions allow the use of units. You do not need to divide out the units as you do with the lower case functions.

$$\mathbf{Var_{10.14}} := \begin{pmatrix} 4.49 \\ 49.50 \\ 150.01 \\ 1499.99 \\ 5000.01 \end{pmatrix} \mathbf{m} \qquad \mathbf{Var_{10.14}} = \begin{pmatrix} 14.73 \\ 162.40 \\ 492.16 \\ 4921.23 \\ 1.64 \times 10^4 \end{pmatrix} \mathbf{ft}$$

$$\mathbf{Test\_Floor_{15}} := \mathbf{Floor}(\mathbf{Var_{10.14}}, 1\mathbf{m}) \qquad \mathbf{Test\_Round_{15}} := \mathbf{Round}(\mathbf{Var_{10.14}}, 1\mathbf{m})$$

$$\mathbf{Test\_Floor_{15}} = \begin{pmatrix} 4.00 \\ 49.00 \\ 150.00 \\ 1499.00 \\ 5000.00 \end{pmatrix} \mathbf{m} \qquad \mathbf{Test\_Round_{15}} = \begin{pmatrix} 4.00 \\ 50.00 \\ 150.00 \\ 1500.00 \\ 5000.00 \end{pmatrix} \mathbf{m}$$

Floor and Round to the nearest meter

$$\mathbf{Test\_Floor_{16}} := \mathbf{Floor}(\mathbf{Var_{10.14}}, 1\mathbf{ft}) \qquad \mathbf{Test\_Round_{16}} := \mathbf{Round}(\mathbf{Var_{10.14}}, 1\mathbf{ft})$$

$$\mathbf{Test\_Floor_{16}} = \begin{pmatrix} 4.27 \\ 49.38 \\ 149.96 \\ 1499.92 \\ 4999.94 \end{pmatrix} \mathbf{m} \qquad \mathbf{Test\_Round_{16}} = \begin{pmatrix} 4.57 \\ 49.38 \\ 149.96 \\ 1499.92 \\ 4999.94 \end{pmatrix} \mathbf{m}$$

Floor and Round to the nearest foot.
Displayed in meters.

$$\mathbf{Test\_Floor_{16}} = \begin{pmatrix} 14.00 \\ 162.00 \\ 492.00 \\ 4921.00 \\ 1.64 \times 10^4 \end{pmatrix} \mathbf{ft} \qquad \mathbf{Test\_Round_{16}} = \begin{pmatrix} 15.00 \\ 162.00 \\ 492.00 \\ 4921.00 \\ 1.64 \times 10^4 \end{pmatrix} \mathbf{ft}$$

Floor and Round to the nearest foot.
Displayed in feet.

Figure 10.14 Upper case truncate and round functions with units

Summation operators

Mathcad has three ways of summing data. These are the Vector Sum operator, the Summation operator, and the Range Sum operator. These are technically not functions, but they are useful in engineering calculations, so they are included in this chapter.

The simplest summation operator to use is the Vector Sum. This operator adds all the elements in a vector. This operator is useful when you want to add a variable series of numbers. If you include all the numbers you want to add in a vector, this function will give the sum of all the elements. To insert the Vector

Sum operator, type **CTRL+4** or use the summation icon on the Matrix toolbar. See Figure 10.15 for some examples.

$\sum$ Type CTRL+4 or use the Matrix toolbar to get the Vector Sum operator. Type the name of the vector in the placeholder.

$$\mathbf{Vector_{10.15}} := \begin{pmatrix} 5 \\ 4 \\ 3 \\ 2 \\ 1 \end{pmatrix} \qquad \sum \mathbf{Vector_{10.15}} = 15.00$$

Suppose you want to input various data, and then sum them.

$\mathbf{Input}_1 := 50\mathbf{Pa}$

$\mathbf{Input}_2 := 100\mathbf{Pa}$

$\mathbf{Input}_3 := 40\mathbf{Pa}$

$\mathbf{Input}_4 := 120\mathbf{Pa}$

Note that these variables are not literal subscripts (typed with the period key). They are array subscripts typed with the [key.

$$\mathbf{Input} = \begin{pmatrix} 50.00 \\ 100.00 \\ 40.00 \\ 120.00 \end{pmatrix} \mathbf{Pa}$$

Give the sum of the input variables.

$$\sum \mathbf{Input} = 310.00\ \mathbf{Pa}$$

Figure 10.15 Vector Sum operator

The Summation operator allows you to sum an expression or a function over a range of values. The operator has four placeholders. $\sum$ The placeholder below the sigma and to the left of the equal sign contains the index of the summation. This index is independent of any variable name outside of the operator. It can be any variable name, but since it is independent to the operator, it is best to keep it a single letter. The placeholder to the right of the sigma is the expression that is to be summed. The expression usually contains the index, but it is not necessary. The remaining two placeholders give the beginning and ending limits of the index. Let's look at some examples. See Figures 10.16 and 10.17.

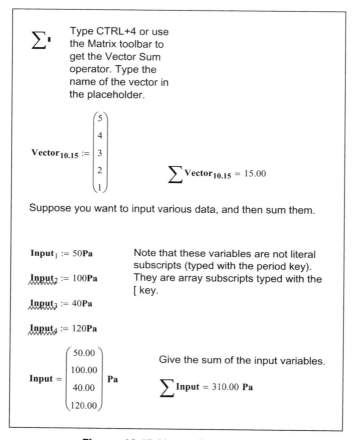

$\sum_{\blacksquare=\blacksquare}^{\blacksquare} \blacksquare$ Type CTRL+SHIFT+4 or use the Calculus toolbar.

Define the index, the beginning limit, the ending limit, and the expression.

$\sum_{i=1}^{4} 2^i = 30.00$ This is equivalent to: $2^1 + 2^2 + 2^3 + 2^4 = 30.00$

$f(x) := 2^x$ Define a function to use in the summation.

$\mathbf{BeginningLimit} := 1$ The beginning limit and ending limit may be variables.

$\mathbf{EndingLimit} := 4$

$\sum_{j=\mathbf{BeginningLimit}}^{\mathbf{EndingLimit}} f(j) = 30.00$ This is the same as above, except the function and limits were defined previously.

The summation expression does not need to contain the index, as shown below.

$\sum_{k=1}^{5} 2 = 10.00$ This is equivalent to: $2 + 2 + 2 + 2 + 2 = 10.00$

Figure 10.16 Summation operator

You can also use vectors and arrays with the Summation operator.

$\mathbf{Vector}_{10.17} := \begin{pmatrix} 5 \\ 4 \\ 3 \\ 2 \\ 1 \end{pmatrix}$ $\mathbf{Matrix}_{10.17} := \begin{pmatrix} 1 & 2 \\ 3 & 4 \end{pmatrix}$

$\mathbf{Summation\_1} := \sum_{l=1}^{3} \left(\mathbf{Vector}_{10.17} \cdot l\right)$

This is equivalent to:

$\mathbf{Summation\_1} = \begin{pmatrix} 30.00 \\ 24.00 \\ 18.00 \\ 12.00 \\ 6.00 \end{pmatrix}$ $\begin{pmatrix} 5 \cdot 1 + 5 \cdot 2 + 5 \cdot 3 \\ 4 \cdot 1 + 4 \cdot 2 + 4 \cdot 3 \\ 3 \cdot 1 + 3 \cdot 2 + 3 \cdot 3 \\ 2 \cdot 1 + 2 \cdot 2 + 2 \cdot 3 \\ 1 \cdot 1 + 1 \cdot 2 + 1 \cdot 3 \end{pmatrix} = \begin{pmatrix} 30.00 \\ 24.00 \\ 18.00 \\ 12.00 \\ 6.00 \end{pmatrix}$

$\mathbf{Summation\_2} := \sum_{m=1}^{3} \left(\mathbf{Matrix}_{10.17} \cdot m\right)$

This is equivalent to:

$\mathbf{Summation\_2} = \begin{pmatrix} 6.00 & 12.00 \\ 18.00 & 24.00 \end{pmatrix}$ $\begin{pmatrix} 1 \cdot 1 + 1 \cdot 2 + 1 \cdot 3 & 2 \cdot 1 + 2 \cdot 2 + 2 \cdot 3 \\ 3 \cdot 1 + 3 \cdot 2 + 3 \cdot 3 & 4 \cdot 1 + 4 \cdot 2 + 4 \cdot 3 \end{pmatrix} = \begin{pmatrix} 6.00 & 12.00 \\ 18.00 & 24.00 \end{pmatrix}$

Figure 10.17 Summation operator with vectors

The Range Sum operator is similar to the Summation operator, except you need to have a range variable defined before using the Range Sum operator. This operator has only two placeholders. $\sum_{\blacksquare} \blacksquare$ The placeholder below the sigma is for the name of a previously defined range variable. The placeholder to the right of the sigma is the expression that is to be summed. The expression usually contains the range variable, but it is not necessary. Let's look at some examples. See Figures 10.18 and 10.19.

$\sum_{\blacksquare} \blacksquare$ Type $ or use the Calculus toolbar

These examples are exactly the same as in Figure 10.16, except the Range Sum operator is used.

In order to use the Range Sum operator, the range must be defined previous to using the operator.

$i := 1, 2 .. 4$ Define a range variable

$\sum_{i} 2^{i} = 30.00$ This is the same as in Figure 10.16, except the limits are defined by the range variable i.

$gg(x) := 2^{x}$ Define a function to use in the summation.

$\mathbf{BeginningCounter} := 1$ Define limits for the range variable

$\mathbf{EndingCounter} := 4$

$j := \mathbf{BeginningCounter} .. \mathbf{EndingCounter}$ Define a range variable using previously defined variables.

$\sum_{j} gg(j) = 30.00$ This is the same as above, except the function was defined previously.

The summation expression does not need to contain the range variable.

$k := 1 .. 5$

$\sum_{k} 2 = 10.00$ This is equivalent to: $2 + 2 + 2 + 2 + 2 = 10.00$

Figure 10.18 Range Sum operator

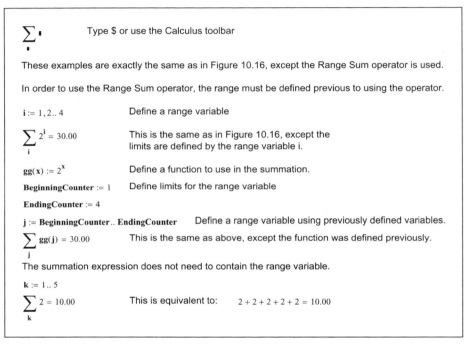

These examples are exactly the same as in Figure 10.17, except the Range Sum operator is used

You can also use vectors and arrays with the Range Sum operator.

$$\mathbf{Vector_{10.17}} = \begin{pmatrix} 5.00 \\ 4.00 \\ 3.00 \\ 2.00 \\ 1.00 \end{pmatrix} \qquad \mathbf{Matrix_{10.17}} = \begin{pmatrix} 1.00 & 2.00 \\ 3.00 & 4.00 \end{pmatrix}$$

$\mathbf{l} := 1..3$

$$\mathbf{Summation\_3} := \sum_{\mathbf{l}} \left(\mathbf{Vector_{10.17}} \cdot \mathbf{l}\right)$$

This is equivalent to:

$$\mathbf{Summation\_3} = \begin{pmatrix} 30.00 \\ 24.00 \\ 18.00 \\ 12.00 \\ 6.00 \end{pmatrix} \qquad \begin{pmatrix} 5\cdot 1 + 5\cdot 2 + 5\cdot 3 \\ 4\cdot 1 + 4\cdot 2 + 4\cdot 3 \\ 3\cdot 1 + 3\cdot 2 + 3\cdot 3 \\ 2\cdot 1 + 2\cdot 2 + 2\cdot 3 \\ 1\cdot 1 + 1\cdot 2 + 1\cdot 3 \end{pmatrix} = \begin{pmatrix} 30.00 \\ 24.00 \\ 18.00 \\ 12.00 \\ 6.00 \end{pmatrix}$$

$\mathbf{M} := 1..3$

$$\mathbf{Summation\_4} := \sum_{\mathbf{M}} \left(\mathbf{Matrix_{10.17}} \cdot \mathbf{M}\right)$$

This is equivalent to:

$$\mathbf{Summation\_4} = \begin{pmatrix} 6.00 & 12.00 \\ 18.00 & 24.00 \end{pmatrix} \qquad \begin{pmatrix} 1\cdot 1 + 1\cdot 2 + 1\cdot 3 & 2\cdot 1 + 2\cdot 2 + 2\cdot 3 \\ 3\cdot 1 + 3\cdot 2 + 3\cdot 3 & 4\cdot 1 + 4\cdot 2 + 4\cdot 3 \end{pmatrix} = \begin{pmatrix} 6.00 & 12.00 \\ 18.00 & 24.00 \end{pmatrix}$$

Figure 10.19 Range Sum operator with vectors

Figure 10.20 shows the use of the Range Sum and Summation operator in an engineering example.

Use the summation operator to calculate the total floor mass of a structure.

$\mathbf{TopFloor} := 4$ — Define the top floor

$\mathbf{Counter} := 1..\,\mathbf{TopFloor}$ — Create a range variable for each floor

$$\mathbf{Area} := \begin{pmatrix} 1400 \\ 1200 \\ 1200 \\ 1000 \end{pmatrix} \cdot \mathbf{m}^2 \qquad \mathbf{Mass} := \begin{pmatrix} 700 \\ 700 \\ 700 \\ 500 \end{pmatrix} \cdot \frac{\mathbf{kg}}{\mathbf{m}^2}$$

Create two vectors with the area and mass/m^2 of each floor.

$$\mathbf{TotalMass} := \sum_{\mathbf{Counter}} \left(\mathbf{Area}_{\mathbf{Counter}} \cdot \mathbf{Mass}_{\mathbf{Counter}}\right)$$

This equation takes the first element of Area and the first element of Mass and multiplies them. The counter then increments and the 2nd elements are multiplied. The process continues for the 3rd and 4th elements. It then takes the sum of the results.

$$\mathbf{TotalMass} = 3.16 \times 10^6\ \mathbf{kg}$$

Check

$$\mathbf{test} := \overrightarrow{\left[\begin{pmatrix} 1400 \\ 1200 \\ 1200 \\ 1000 \end{pmatrix} \mathbf{m}^2 \cdot \begin{pmatrix} 700 \\ 700 \\ 700 \\ 500 \end{pmatrix} \cdot \frac{\mathbf{kg}}{\mathbf{m}^2}\right]} \qquad \mathbf{test} = \begin{pmatrix} 9.80 \times 10^5 \\ 8.40 \times 10^5 \\ 8.40 \times 10^5 \\ 5.00 \times 10^5 \end{pmatrix} \mathbf{kg}$$

Note: Use CTRL+MINUS SIGN to get the vectorize operator.

$$\sum \mathbf{test} = 3.16 \times 10^6\ \mathbf{kg}$$

OK. Same result

Figure 10.20 Range Sum operator—engineering example

if function

The ***if*** function allows Mathcad to make a determination between two or more choices.

The ***if*** function is very similar to the ***if*** function in Microsoft Excel. It takes the form ***if***(Cond,x,y). Cond is an expression, typically involving a logical or Boolean operator. The function returns x if Cond is true, and y otherwise.

Let's look at some engineering examples. See Figures 10.21 and 10.22.

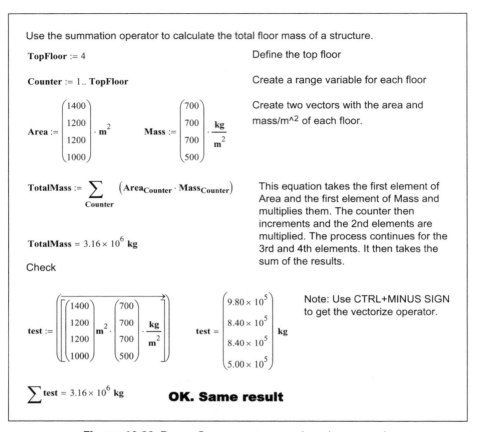

The *if* function takes the form *if*(cond,x,y). If cond is true then x. If cond is false then y.

Assume that some Mathcad expressions returned two results.

$\mathbf{Result_1} := 10 \cdot \mathbf{Hz}$

$\mathbf{Result_2} := 0 \cdot \mathbf{Hz}$

IfTrue := "Result1 is greater than Result2" — Create two strings to use in the if function.

IfFalse := "Result1 is less than Result 2"

Is Result_1 greater than Result_2?

$\mathbf{Result_3} := \mathbf{if}(\mathbf{Result_1} > \mathbf{Result_2}, \mathbf{IfTrue}, \mathbf{IfFalse})$

$\mathbf{Result_3}$ = "Result1 is greater than Result2"

$$\mathbf{Result_4} := \mathbf{if}\left(\mathbf{Result_2} = 0, \text{"Division by Zero"}, \frac{\mathbf{Result_1}}{\mathbf{Result_2}}\right)$$

$\mathbf{Result_4}$ = "Division by Zero"

Figure 10.21 *if* function

In this example, there is an input length. A specific formula needs to use the input length, but the length cannot be less than 25 ft.

InputLength := 20**ft**

Length := **if**(**InputLength** < 25**ft**, 25**ft**, **InputLength**)

Length = 25.00 **ft**

Figure 10.22 *if* function

linterp *function*

Mathcad has several interpolation and regression functions. The ***linterp*** function allows straight-line interpolation between points. It is a straight-line interpolation, and is the easiest to use. You may have a specific need to use some of the more advanced functions, but for our discussion, we will use the liner interpolation function.

The ***linterp*** function has the form ***linterp***(vx,vy,x). The value vx is a vector of real data values in ascending order. The value vy is a vector of real data values having the same number of elements as vector vx. The value x is the value of the independent variable at which to interpolate the value. It is best if the value x is contained within the data range of vx. If x is below the first value of vx, the Mathcad extrapolates a straight line between the first two data points. If x is above the last value of vx, Mathcad extrapolates a straight line between the last two data points.

The ***linterp*** function draws a straight line between each data point and uses straight-line interpolation between the pairs of points. The ***linterp*** function is very useful if you have a table or graph of data and need to interpolate between data points. If your data is scattered, you may want to consider using a regression function instead.

Let's first look at a simple example. See Figure 10.23.

The *linterp* function takes the form linterp(vx,vy,x), where vx and vy are vectors in ascending order, and x is the independent variable.

$$\mathbf{vx} := \begin{pmatrix} 1 \\ 2 \end{pmatrix} \qquad \mathbf{vy} := \begin{pmatrix} 4 \\ 8 \end{pmatrix}$$

The vectors vx and vy are the same length and have corresponding data.

$$\mathbf{linterp}(\mathbf{vx}, \mathbf{vy}, 1.5) = 6.00$$

Mathcad calculates the value of y for x=1.5.

Figure 10.23 ***linterp*** function

In this next example, there is a longer list of data values. Mathcad uses linear interpolation between each pair of data points. See Figure 10.24.

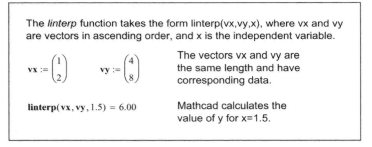

Time Velocity

$$\mathbf{Time} := \begin{pmatrix} 1 \\ 2 \\ 3 \\ 4 \\ 5 \end{pmatrix} \cdot \mathbf{sec} \qquad \mathbf{Velocity} := \begin{pmatrix} 2.1 \\ 3.9 \\ 5.9 \\ 7.8 \\ 10.1 \end{pmatrix} \cdot \frac{\mathbf{m}}{\mathbf{s}}$$

The vectors Time and Velocity must be the same length and have corresponding data. Mathcad uses linear interpolation between each data point.

$$\mathbf{linterp(Time, Velocity, 1.5s)} = 3.00\,\frac{\mathbf{m}}{\mathbf{s}}$$

$$\mathbf{linterp(Time, Velocity, 2.2s)} = 4.30\,\frac{\mathbf{m}}{\mathbf{s}}$$

$$\mathbf{linterp(Time, Velocity, 0.0s)} = 0.30\,\frac{\mathbf{m}}{\mathbf{s}}$$

$$\mathbf{linterp(Time, Velocity, 6s)} = 12.40\,\frac{\mathbf{m}}{\mathbf{s}}$$

Use the *linterp* function to interpolate the velocity at different moments in time.

You may also assign a vector of values to the x argument.

$$\mathbf{xValues} := \begin{pmatrix} 1.5 \\ 2.2 \\ 0 \\ 6 \end{pmatrix} \cdot \mathbf{s} \qquad \mathbf{linterp(Time, Velocity, xValues)} = \begin{pmatrix} 3.00 \\ 4.30 \\ 0.30 \\ 12.40 \end{pmatrix} \frac{\mathbf{m}}{\mathbf{s}}$$

In order to reuse the interpolated values, assign the output to a variable.

$$\mathbf{InterpolatedResults} := \mathbf{linterp(Time, Velocity, xValues)}$$

$$\mathbf{InterpolatedResults} = \begin{pmatrix} 3.00 \\ 4.30 \\ 0.30 \\ 12.40 \end{pmatrix} \frac{\mathbf{m}}{\mathbf{s}} \qquad \mathbf{xValues} = \begin{pmatrix} 1.50 \\ 2.20 \\ 0.00 \\ 6.00 \end{pmatrix} \mathbf{s}$$

You can now access individual results from the vector "Interpolated Results."

$$\mathbf{InterpolatedResults}_1 = 3.00\,\frac{\mathbf{m}}{\mathbf{s}}$$

$$\mathbf{InterpolatedResults}_2 = 4.30\,\frac{\mathbf{m}}{\mathbf{s}}$$

Figure 10.24 ***linterp*** function

Figure 10.25 is a bit more complicated engineering example. It uses interpolation to calculate pressures at different heights above the ground. It then uses some of the array information discussed in Chapter 9 to calculate force and overturning moments.

This example uses linear interpolation to calculate the pressure on areas located at various heights above the ground. It uses pressure data taken from a table that lists pressures at various heights. The vectors Ht_{Data} and $Pressure_{Data}$ are the same length and have corresponding data.

Ht_{Data} and $Pressure_{Data}$ are taken from tables.

Area is the surface area of an object. Ht is the height above the ground to the centroid of the corresponding area.

$$\mathbf{Ht_{Data}} := \begin{pmatrix} 0 \\ 10 \\ 20 \\ 30 \\ 40 \\ 50 \\ 60 \\ 70 \end{pmatrix} \cdot \mathbf{ft} \qquad \mathbf{Pressure_{Data}} := \begin{pmatrix} 24 \\ 24 \\ 25 \\ 27 \\ 30 \\ 34 \\ 39 \\ 45 \end{pmatrix} \cdot \mathbf{psf} \qquad \mathbf{Area} := \begin{pmatrix} 10 \\ 12 \\ 9 \\ 11 \\ 20 \end{pmatrix} \cdot \mathbf{ft}^2 \qquad \mathbf{Ht} := \begin{pmatrix} 12 \\ 22 \\ 27 \\ 33 \\ 45 \end{pmatrix} \cdot \mathbf{ft}$$

$$\mathbf{Pressure} := \mathbf{linterp}(\mathbf{Ht_{Data}}, \mathbf{Pressure_{Data}}, \mathbf{Ht})$$

Each value of Ht is interpolated between values of Ht_{Data} and $Pressure_{Data}$.

$$\mathbf{Pressure} = \begin{pmatrix} 24.20 \\ 25.40 \\ 26.40 \\ 27.90 \\ 32.00 \end{pmatrix} \mathbf{psf}$$

Pressure is a vector of the interpolated pressures at the heights in the vector Ht.

The following part of this example uses vector information discussed in Chapter 9.

Calculate the force on each area, which is the area times the pressure

$$\mathbf{Force} := \overrightarrow{(\mathbf{Area} \cdot \mathbf{Pressure})}$$

$$\mathbf{Force} = \begin{pmatrix} 242.00 \\ 304.80 \\ 237.60 \\ 306.90 \\ 640.00 \end{pmatrix} \mathbf{lbf}$$

Use the vectorize operator to ensure that there is an element-by element multiplication.

Calculate the overturning moment, which is the force times the height.

$$\mathbf{OTM} := \overrightarrow{(\mathbf{Force} \cdot \mathbf{Ht})} \qquad \mathbf{OTM} = \begin{pmatrix} 2904.00 \\ 6705.60 \\ 6415.20 \\ 10127.70 \\ 28800.00 \end{pmatrix} \mathbf{ft} \cdot \mathbf{lbf}$$

Calculate the sum of the overturning moments. Use the Vector Sum operator contained on the matrix toolbar.

$$\mathbf{SumOTM} := \sum \mathbf{OTM} \qquad \mathbf{SumOTM} = 54952.50\,\mathbf{ft} \cdot \mathbf{lbf}$$

Figure 10.25 ***linterp*** function—engineering example

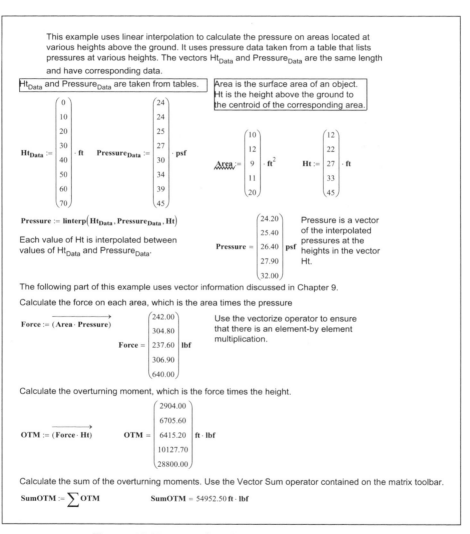

Summary

We have reviewed only a handful of the hundreds of Mathcad functions. Hopefully the functions discussed in this chapter will be helpful to your engineering calculations. As you become more familiar with Mathcad, you will add other functions to your Mathcad toolbox. We will discuss additional functions in later chapters. In the meantime, open the Insert Function dialog box and look for functions that appear interesting to you. Use the Mathcad Help to learn about these functions.

In Chapter 10 we:

- Reviewed the basics of Mathcad functions.
- Learned about the following functions:
  - max and min functions
  - mean and median functions
  - truncation and rounding functions
  - summation operators
  - if function
  - interpolation functions
- Encouraged you to search for and learn about additional functions that you can add to your Mathcad toolbox.

Practice

1. Use the ***max*** and ***min*** functions to find the maximum and minimum values of the following:

   [a] 10 m, 1 km, 3000 mm, 1 mi, 2000 yd, 10 furlong, 1 nmi.
   [b] 100 s, 1.25 min, 0.15 day, 0.05 week, 0.005 year.
   [c] 1000 Pa, 1 psi, 1 atm, 10 torr, 0.002 ksi, 0.002 MPa, 1in_Hg.
   [d] 1 bhp, 10 ehp, 10.1 kW, 19 mhp, 10 hpUK, 10 hhp, 10000 W.

2. Use the ***mean*** and ***median*** functions to determine the mean and median values from the list of values used in practice exercise 1.
3. Place the list of values from practice exercise 1 into four different vectors.
4. Use the ***max*** and ***min*** functions and the ***mean*** and ***median*** functions with the vectors created in practice exercise 3.

5. Use the ***floor***, ***ceil***, and ***trunc*** functions with the vectors created in practice exercise 3. Select two different units to use for each vector. For example, truncate to units of meters and feet for the first vector.
6. Use the ***round*** function with the vectors created in practice exercise 3. Round to the following: 2 decimal places, 1 decimal place, 0 decimal places, 10's, and 100's. Select two units for each case. For example, round to units of meters and feet for the first vector.
7. Use the Vector Sum operator to calculate the sum of the vectors created in practice exercise 3. Display the result in three different units.
8. Use the Summation operator and the Range Sum operator to calculate the total area of ten squares with sides incrementing from 1 m to 10 m.
9. Use the following variables for this practice exercise: R1 = 3Amp, R2 = 4Amp, R3 = 10Volt, R4 = 20Volt. Create four variables using the ***if*** function to meet the following conditions:

Condition	First	Operator	Second	If true	If false
1	R1	<	R2	R3	R4
2	R1	=	R2	R3	R4
3	R2	=	0	"Division by zero"	R1/R2
4	R3	>=	R4	R1	R2

[a] Vary the values of the variables, to see how the results are affected.

10. Use the ***linterp*** function to interpolate the following values:

Time	Distance	Interpolate the distance for this time:
0 s	0.0 m	0.5 s
1 s	2.5 m	1.2 s
2 s	10.0 m	2.4 s
3 s	22.5 m	3.3 s
4 s	40.0 m	4.6 s
5 s	62.5 m	5.1 s

11
Plotting

Plots are an important part of engineering calculations because they allow visualization of data and equations. They are an important part of equation solving because they can help you select an initial guess for a solution. Plots also allow you to visualize trends in engineering data. Chapter 11 will deal with 2D plots. Reference will be made to 3D plots, but they will not be discussed in much detail. This chapter will focus on using plots as a tool for visualization and solving of equations.

Chapter 11 will:

- Show how to create simple 2D X-Y plots and Polar plots.
- Show how to set plot ranges.
- Instruct how to graph multiple functions in the same plot.
- Discuss the use of range variables to control plots.
- Tell how to plot data points.
- Describe the steps necessary to format plots, including the use of log scale, grid lines, scaling, numbering, and setting defaults.
- Discuss the use of titles and labels.
- Show how to get numeric readout of plotted coordinates.
- Show how the use of plots can help find the solutions to various engineering problems.
- Use engineering examples to illustrate the concepts.

Creating a simple X-Y QuickPlot

To create a simple X-Y QuickPlot, type @, or hover the mouse over **Graph** on the **Insert** menu and click **X-Y Plot**. This places a blank X-Y plot operator on the worksheet.

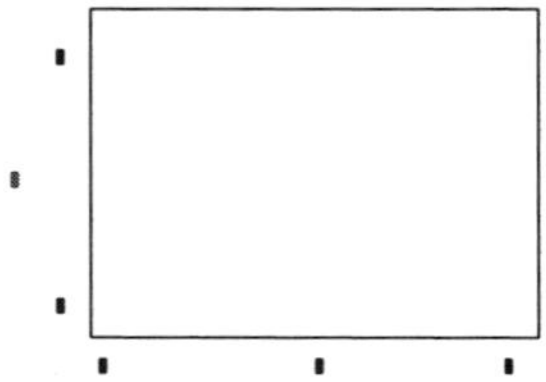

Click on the bottom middle placeholder. This is where you type the x-axis variable. Type the name of a previously undefined variable. The variable is allowed to be "x" but can be any Mathcad variable name. Next, click on the middle left placeholder, and type an expression using the variable named on the x-axis. Click outside the operator to view the X-Y plot. Mathcad automatically selects the range for both the x-axis and the y-axis. There is a way to change the plot range, which will be discussed shortly. Another shortcut is to only type the expression in the left placeholder. Mathcad automatically adds the independent variable in the bottom placeholder. See Figure 11.1.

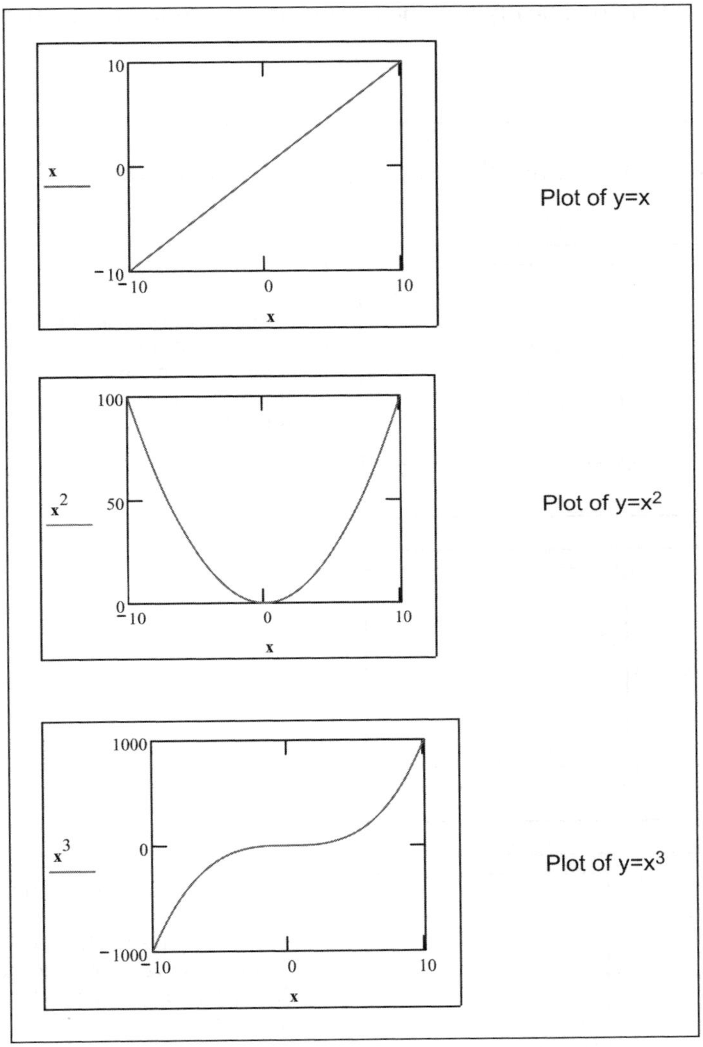

Figure 11.1 X-Y QuickPlot of equations

Another way to create a QuickPlot is to define a function prior to creating the plot. Open the X-Y plot operator by typing [@]. Click the bottom placeholder and type a variable name for the x-axis. This variable name does not need to be the same one used as the argument to define the function. On the left placeholder, type the name of the function. Use the variable name from the x-axis as the argument of the function. Here again, Mathcad selects the range for both the x-axis and the y-axis. You may skip the step of typing a variable for the x-axis.

Mathcad will automatically add the argument used in the y-axis function. See Figure 11.2.

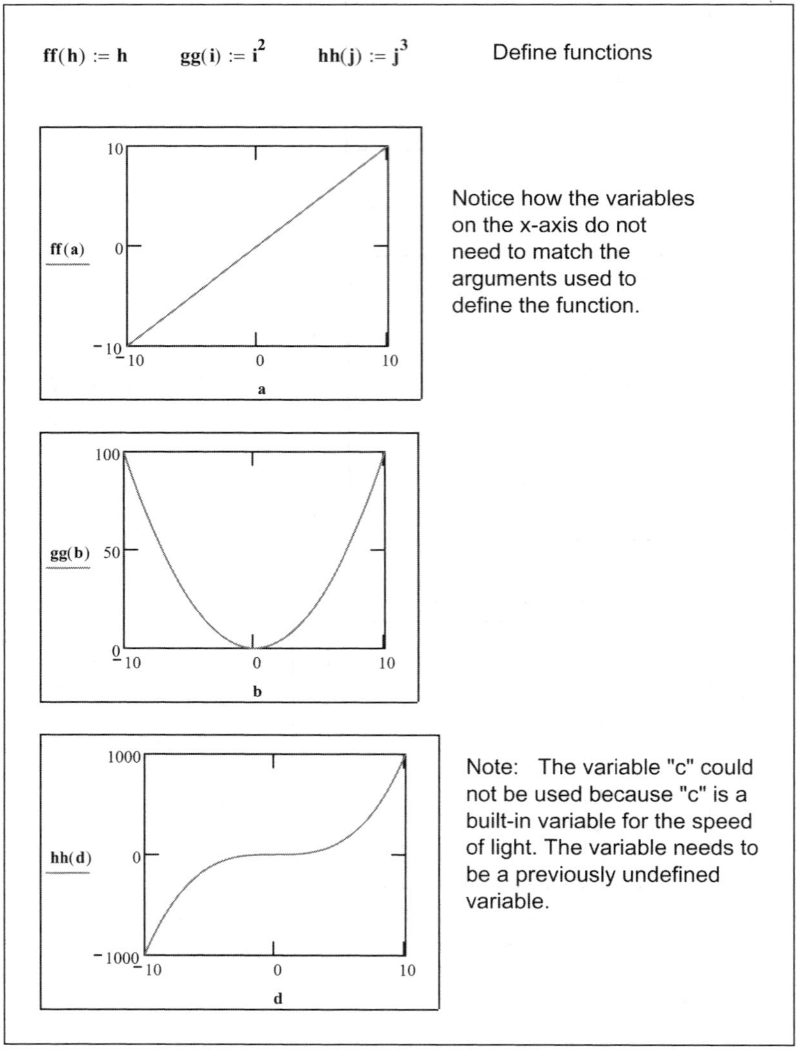

Figure 11.2 X-Y QuickPlot of functions

If you use a previously defined variable, then Mathcad will not plot a graph over a range of values. It will only plot the value of the variable used. In some cases, this may only be a single point. For a QuickPlot, it is important to use only undefined variables. We will shortly discuss the use of range variables in plots. This is a case where previously defined variable can be used.

Creating a simple Polar plot

Creating a simple Polar QuickPlot is very similar to creating a simple X-Y QuickPlot. Open the Polar plot operator by typing CTRL+7, or hover the mouse over **Graph** on the **Insert** menu, and select **Polar Plot**.

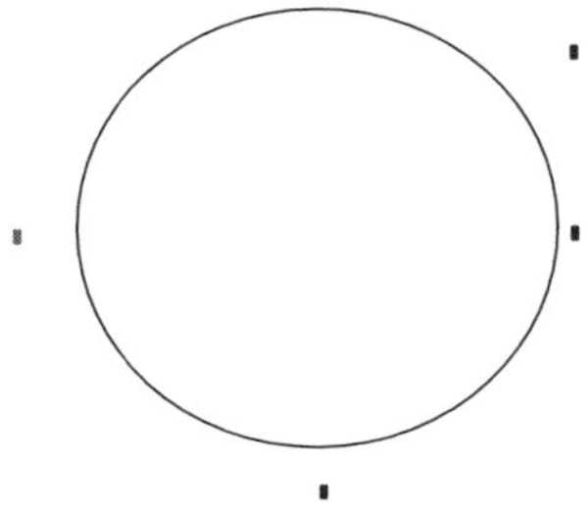

Click on the bottom middle placeholder. This is where you type the angular variable. Unless you specify otherwise, Mathcad assumes the variable to be in radians. Type the name of a previously undefined variable. The variable can be any Mathcad variable name. Next, click on the middle left placeholder and type an expression using the angular variable defined on the x-axis. This sets the properties of the radial axis. Click outside of the operator to view the plot. For every angle from 0 to 2π, Mathcad plots a radial value. Mathcad automatically selects the radial range. See Figure 11.3.

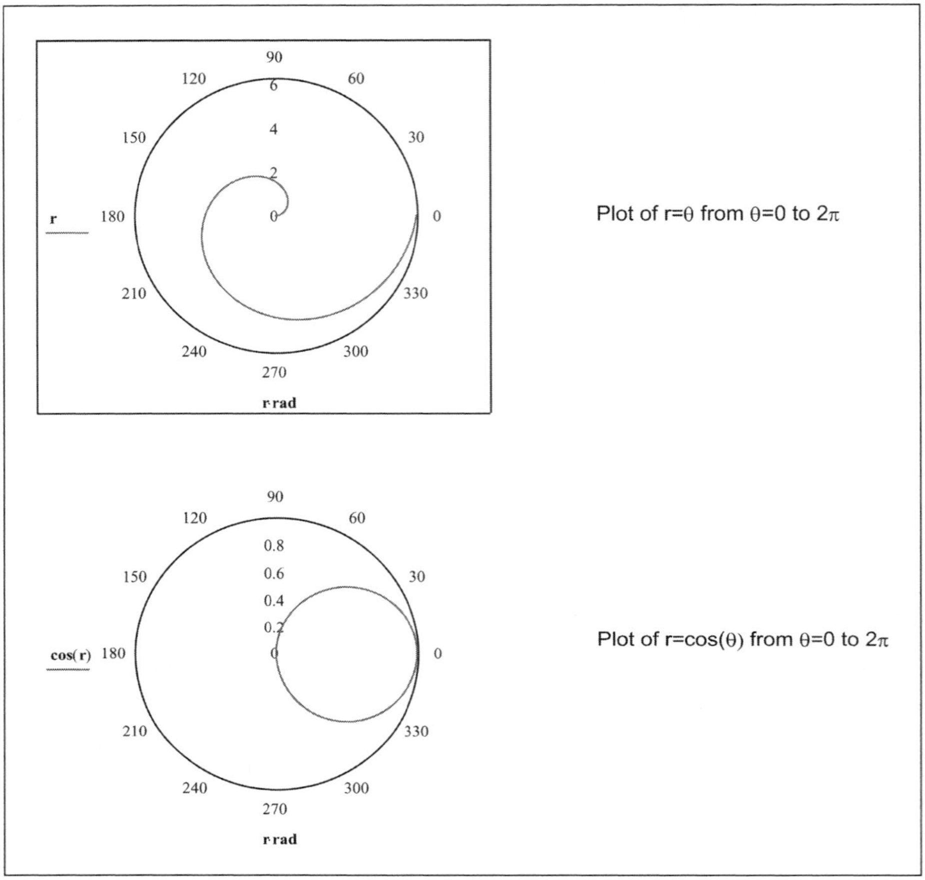

Figure 11.3 Polar QuickPlot of functions

You may also use a previously defined function in a simple Polar QuickPlot. Open the Polar plot operator by typing **CTRL+7**. Type the name of the angular variable in the bottom placeholder. Type the name of the function on the left placeholder using the angular variable name as the argument of the function. See Figure 11.4.

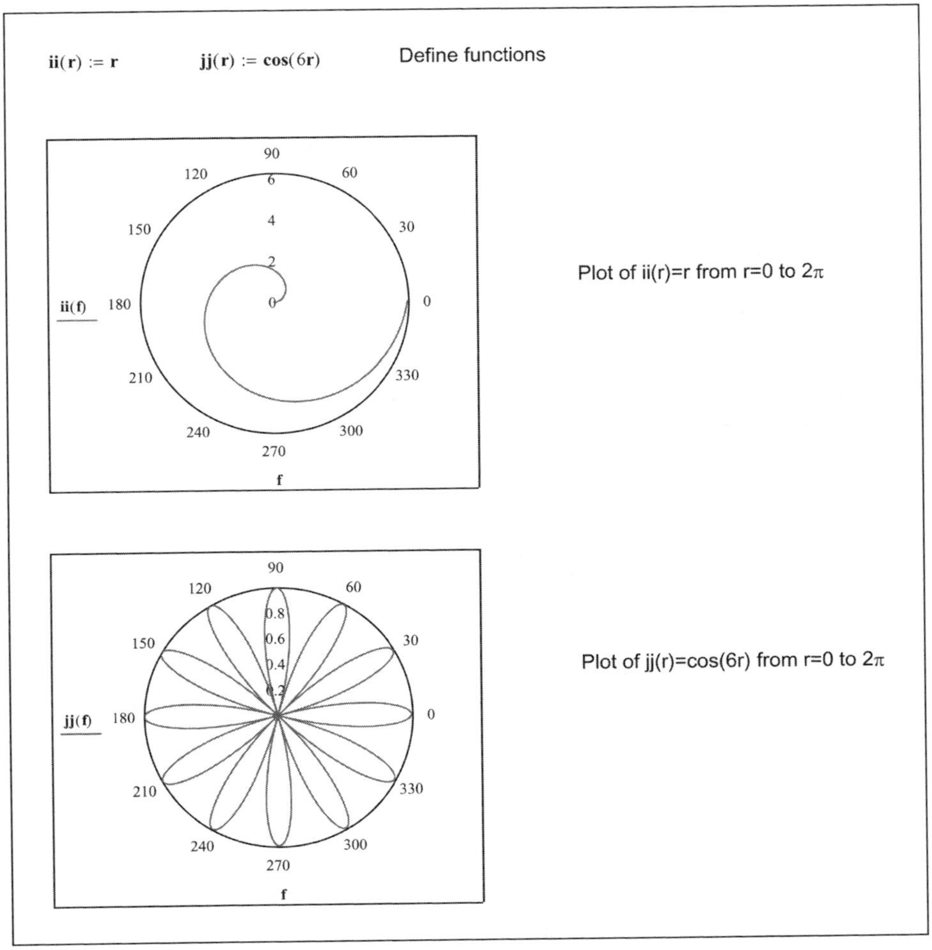

Figure 11.4 Polar QuickPlot of functions

Using range variables

Range variables are used to tell Mathcad what range of values to use when graphing an expression or function. When you create simple X-Y QuickPlots and Polar QuickPlots, Mathcad sets the range of values. By using range variables, you can control the range of values. To graph using a range variable, create the range variable before using the plot operator. Next, open the plot operator, and type the name of the range variable in the placeholder on the x-axis. In the left placeholder, type the name of a function or expression using the range variable. See Figures 11.5 and 11.6. When using range variables, you are actually telling

$\mathbf{D(t)} := \frac{1}{2} \cdot 9.81 \cdot \mathbf{t}^2$ Define function

$\mathbf{t_1} := 0, 1 .. 21$ Define range variables

$\mathbf{t_2} := 0, 1 .. 100$

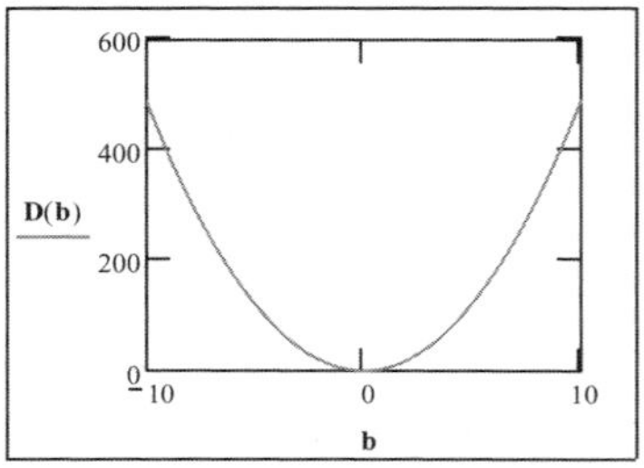

Using a simple X-Y QuickPlot with Mathcad default ranges.

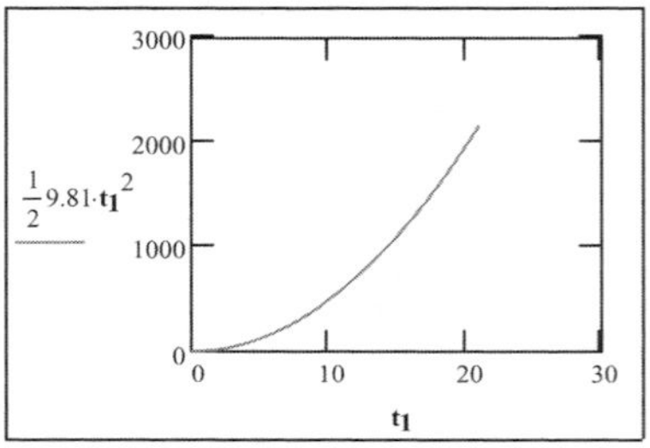

Using range variable t_1 with an expression.

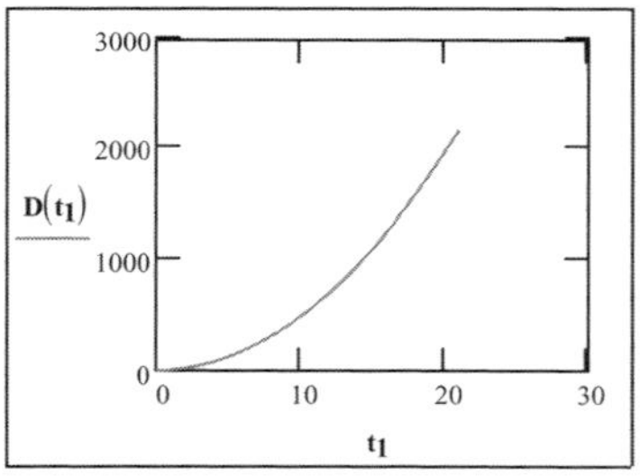

Function "D(t)" using the range variable t_1.

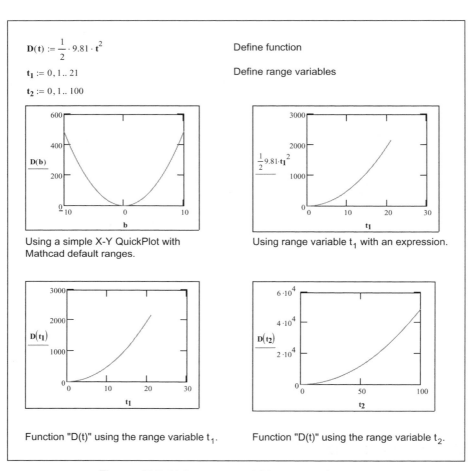

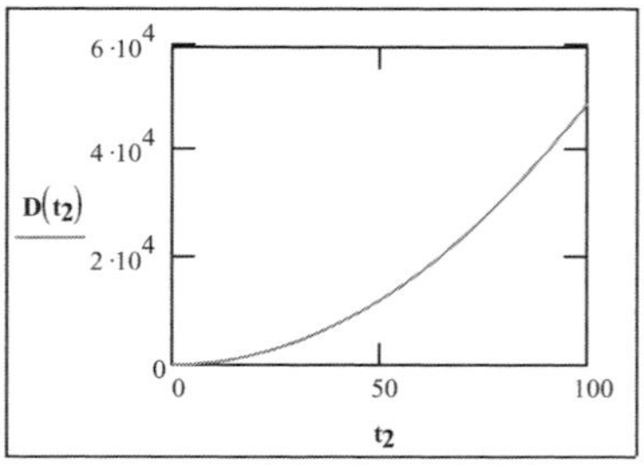

Function "D(t)" using the range variable t_2.

Figure 11.5 Using range variables to set plot range

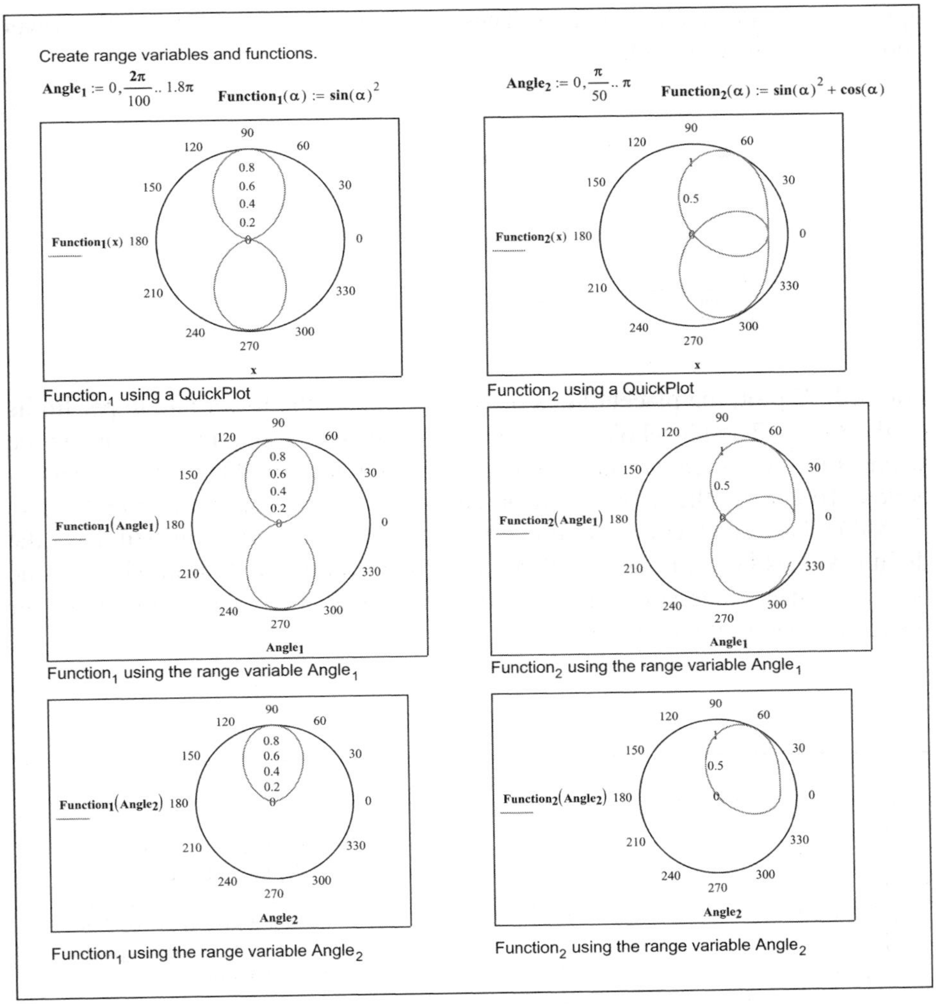

Figure 11.6 Using range variables to set plot range

Mathcad to plot each point in the range variable and to draw a line between the points. This will be illustrated later.

Setting plotting ranges

In addition to using range variables, there is another way to set the plot range. You may have noticed additional placeholders when you opened an X-Y plot operator or a Polar plot operator. These additional placeholders set the lower and upper limits of the plot.

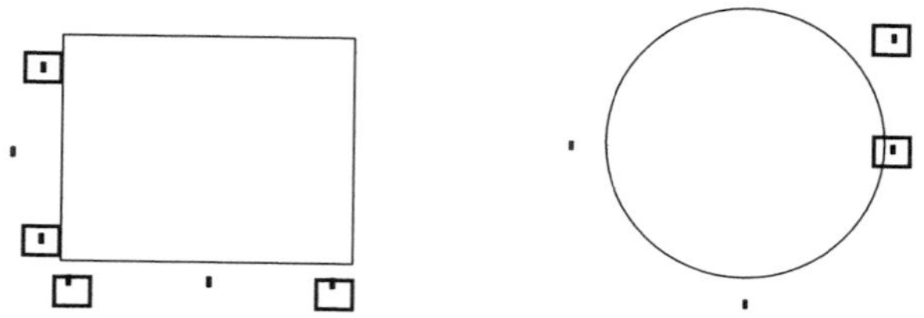

For an X-Y plot, the placeholders on the bottom set the lower and upper limits on the x-axis. The placeholders on the left set the lower and upper limits on the y-axis. Once you create a QuickPlot, these placeholders will have default values added. To change the default values, click on the limit placeholder and delete the value. Next, add a new plot limit. You can tell which plot limits still have the default values because there will be small brackets on the bottom sides of the default values. Once you change the default values, the brackets are no longer displayed. See Figures 11.7 and 11.8.

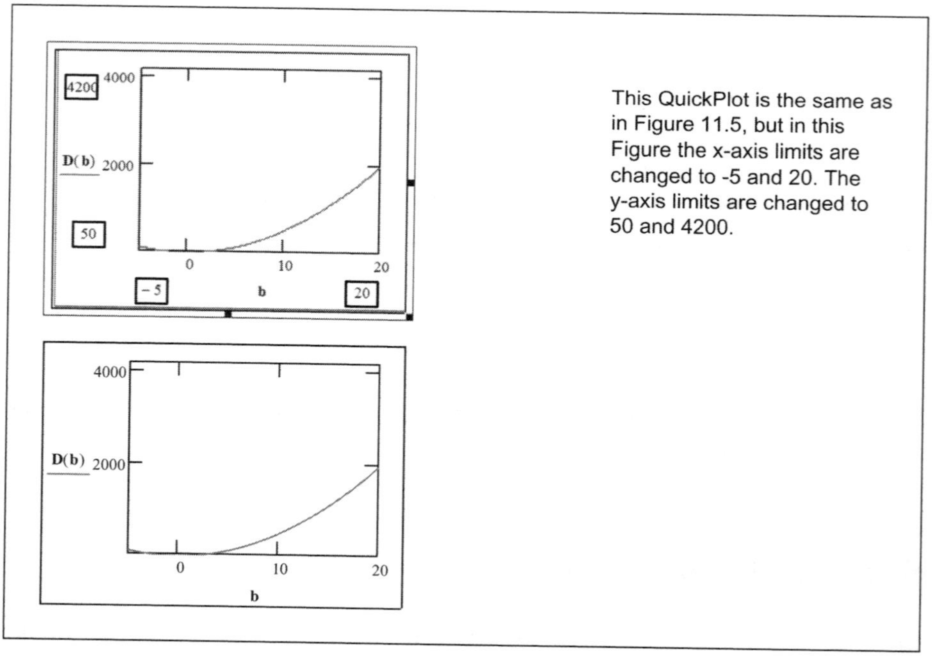

Figure 11.7 Setting plot range

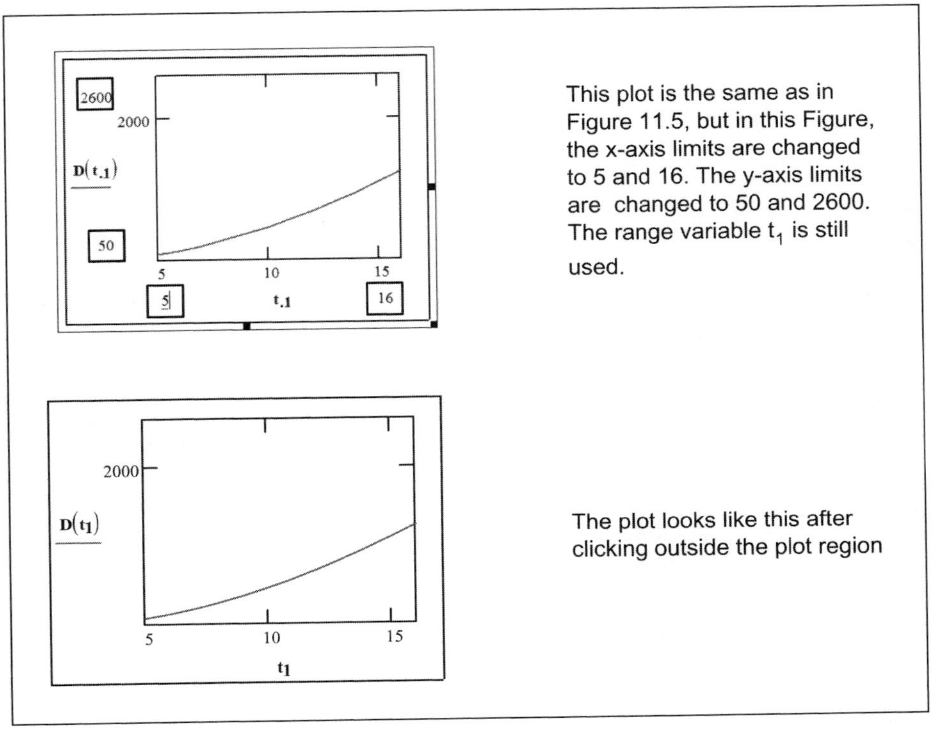

Figure 11.8 Setting plot range

For a Polar plot, the limits of the radial axis are set by the two placeholders on the right. You can experiment with changing the lower placeholder, but it seems to work best when this placeholder is left at zero. See Figures 11.9 and 11.10.

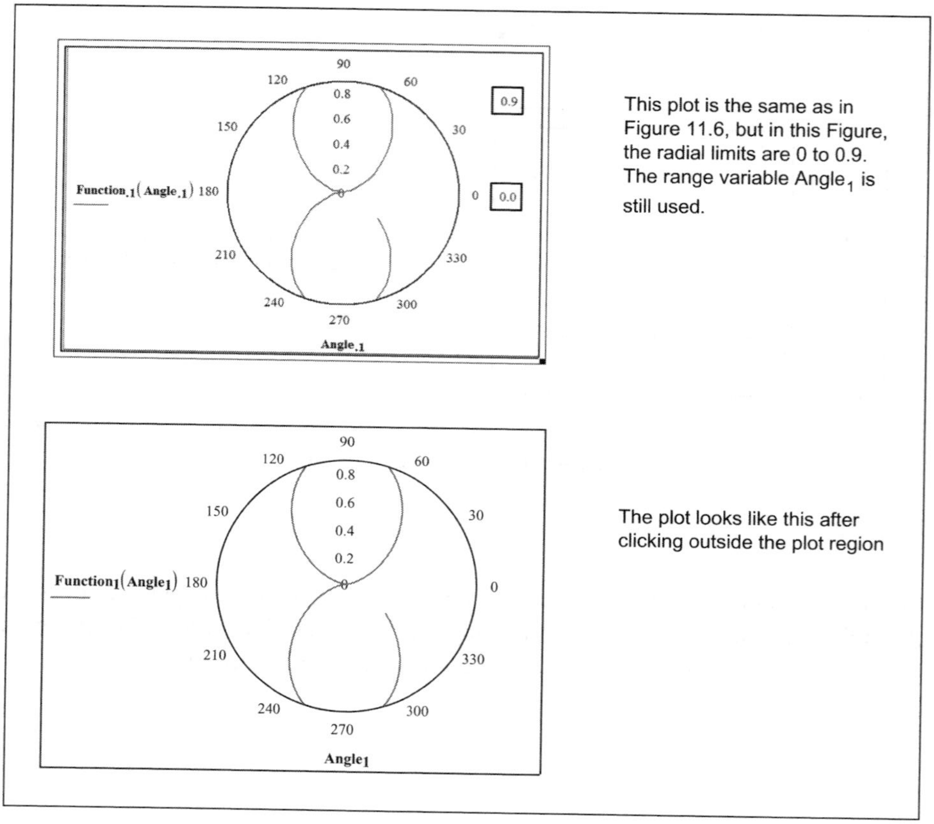

Figure 11.9 Setting plot range

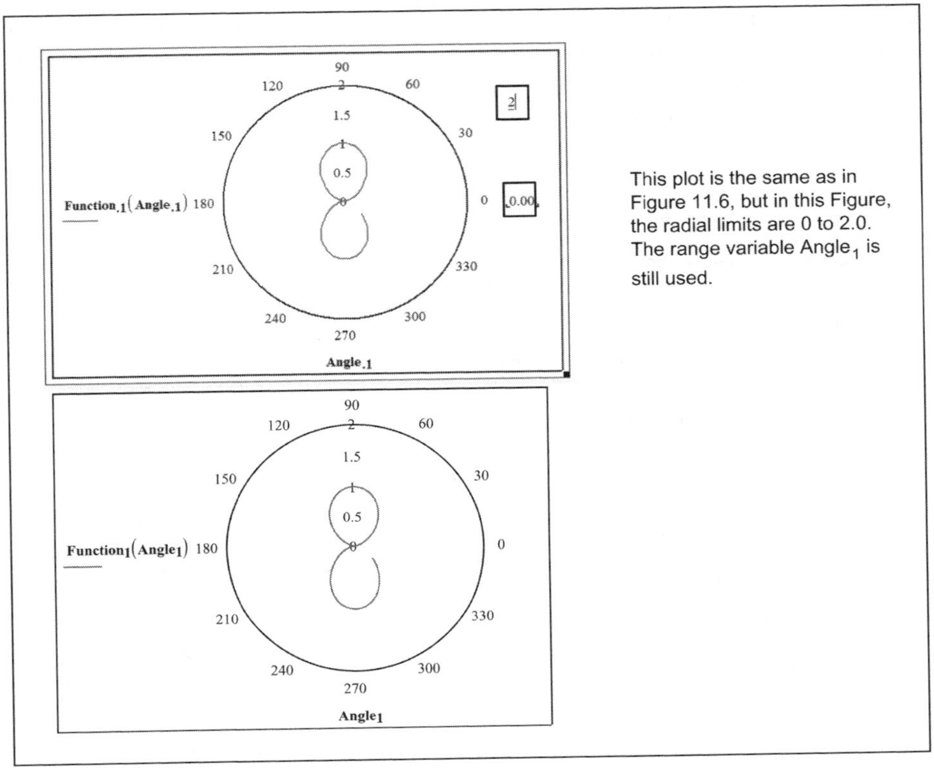

Figure 11.10 Setting plot range

Graphing with units

When you plot functions and data with units attached, the numbers displayed on the axis correspond to the base unit or the derived unit for the unit dimension. If you are plotting distance, the numbers on the axis will correspond to meters (if SI is the default unit system). If you are plotting pressure, the numbers on the axis will correspond to Pa (if SI is the default unit system). In Figure 11.11 we want to plot pressure in psf, but the worksheet default unit system is SI. This causes the depth to be plotted in meters and the pressure to be plotted as Pa. How can the pressure be plotted in psf, psi, or other pressure units? How can the depth be plotted in feet?

In Chapter 4, we discussed using units in empirical equations. Plots are similar. To plot functions and data with units attached, divide the function or data by the units you want plotted. This creates unitless data with the values you want displayed.

In Figures 11.11 and 11.12, the range variable "i" had units attached to it. If the range variable did not have units attached, the function argument would need to have units added. The function would then need to be divided by the pressure units to be plotted. The y-axis function would be Pressure(i*ft)/psf.

The density of water is about 62.24 pound force per cubic foot.
Graph the water pressure (in psf) at various depths (in feet).

Density := 62.24**pcf**

Pressure(Depth) := **Density** · **Depth** **Pressure(**10**ft)** = 622.40 **psf**

i := 0**ft**, 1**ft**.. 10**ft** Set plot limits using a range variable.

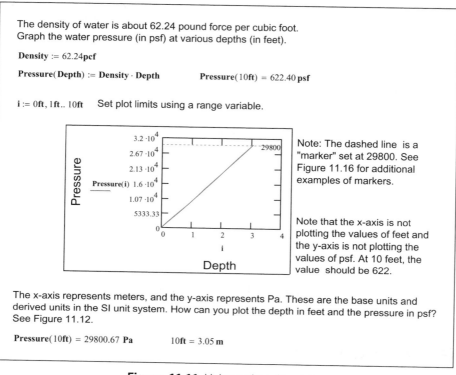

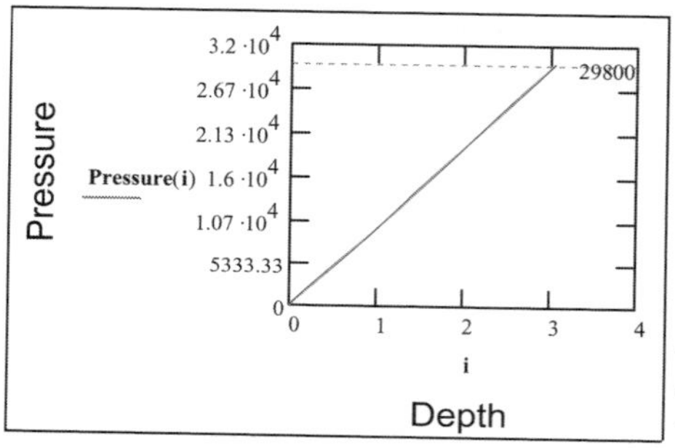

Note: The dashed line is a "marker" set at 29800. See Figure 11.16 for additional examples of markers.

Note that the x-axis is not plotting the values of feet and the y-axis is not plotting the values of psf. At 10 feet, the value should be 622.

The x-axis represents meters, and the y-axis represents Pa. These are the base units and derived units in the SI unit system. How can you plot the depth in feet and the pressure in psf? See Figure 11.12.

Pressure(10**ft)** = 29800.67 **Pa** 10**ft** = 3.05 **m**

Figure 11.11 Using units with plots

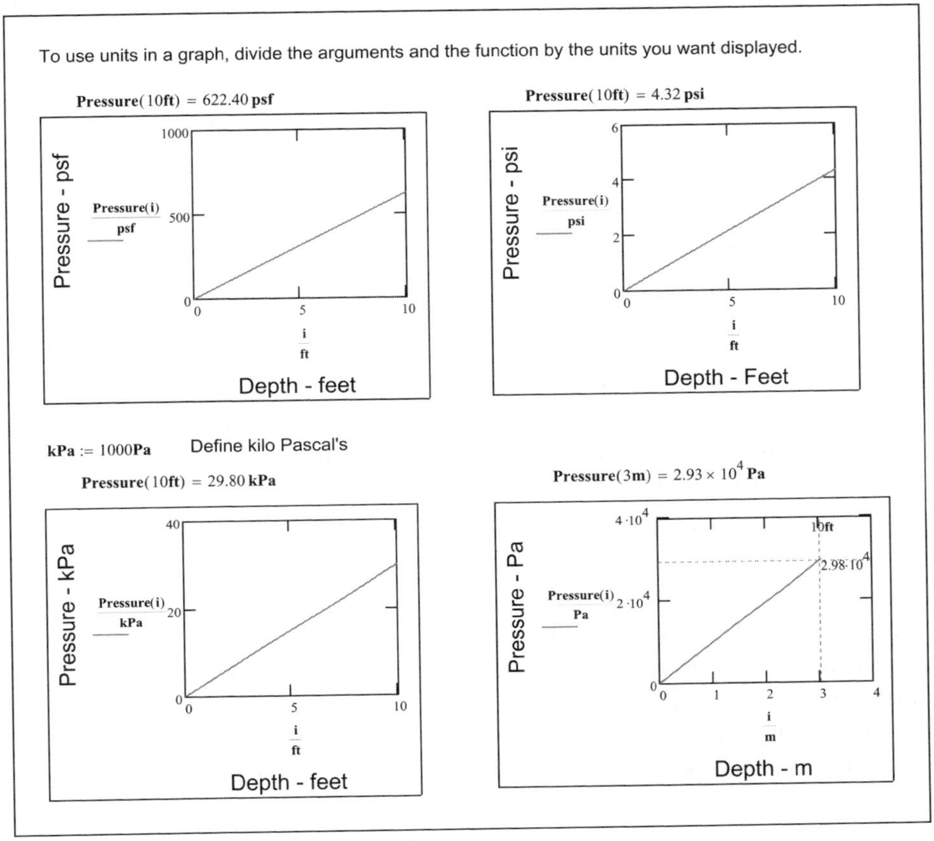

Figure 11.12 Using units with plots

Graphing multiple functions

You can graph up to 16 multiple functions or expressions on the same plot. To graph multiple expressions using the same x-axis variable, type a comma after entering the function or expression in the left middle placeholder. This places a new placeholder below the original placeholder. You can now type a new function or expression. You can repeat the process until you have up to 16 functions or expressions. Each plot is called a trace. See Figure 11.13.

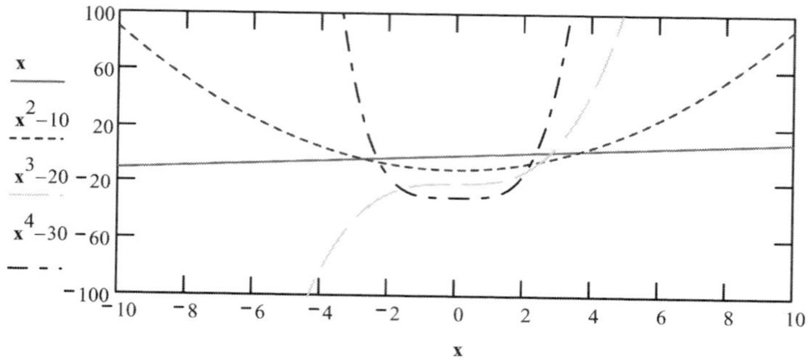

This figure is a plot of 4 expressions, each using the same x-axis variable. The x-axis limits are set at -10 and 10. The y-axis limits are set at -100 and 100.

Figure 11.13 Multiple plots

You can also use multiple variables on the x-axis and then plot corresponding expressions on the y-axis. To do this, type a comma after entering the variable name on the x-axis. This places a new placeholder adjacent to the original placeholder. On the y-axis, create new placeholders as noted above, and use the corresponding x-axis variable name in your expression. See Figure 11.14.

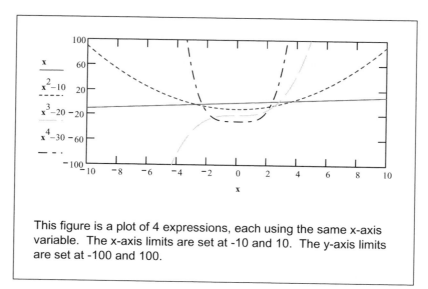

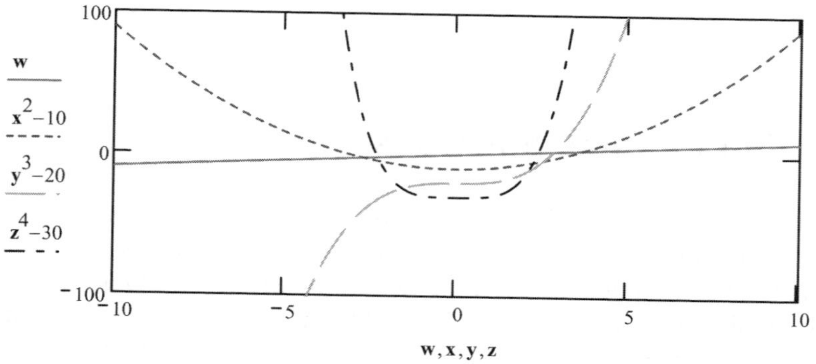

Figure 11.14 Multiple plots with multiple variables

Beginning with Mathcad 12, you can create a secondary y-axis on the right-hand vertical axis of an X-Y plot. This secondary y-axis can be used to graph additional traces at a different scale than the primary y-axis. To create a secondary y-axis, double-click on the plot. This opens a plot formatting dialog box. Place a check in the box "Enable secondary Y axis." This will add new placeholders on the right side of the plot. You can now type expressions, functions, and range limits in the new placeholders. See Figure 11.15.

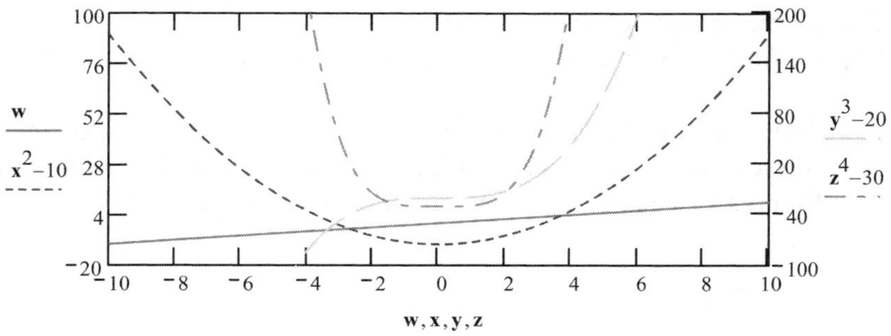

The primary and secondary y-axis plots can have different limits.

Figure 11.15 Mulitple plots using secondary Y-axis

Formatting plots

Mathcad allows you to customize many aspects of your plots. You can add grid lines; change the spacing of the grid lines; plot with equal x- and y-axes; change the color, line weight, or line type of each trace or add symbols to the trace. You can even plot in log scale.

To make customizations, double-click on the plot. This opens a plot formatting dialog box. The features in this box will depend on whether you are working with an X-Y plot or a Polar plot. Each version of Mathcad seems to have a slightly different dialog box, but they all control similar features.

Axes tab

This tab will be either the X-Y Axes tab, or the Polar Axes tab depending on the type of plot you have open. This tab controls the appearance of the axes and grids. This allows you to:

- Change an axis to log scale.
- Add grid lines to an axis.
- Display or not display numbers on the axes.

- Tell Mathcad whether the axis limits are set at the data limits or whether the axis limits are set to the next major tick mark beyond the end of the data (Auto scale).
- Add horizontal, vertical, or radial marker lines at values that you set. The marker lines are dashed horizontal or vertical lines that can be set at specified locations. Each plot axis may have two marker locations.
- Change the spacing of the grid lines.
- Change from axes on the sides of the plot to axes in the center of the plot.

Mathcad Help has an excellent description of each of these various features. To access these descriptions click the **Help** button within the Formatting dialog box. See Figures 11.16 and 11.17 for examples of these different features.

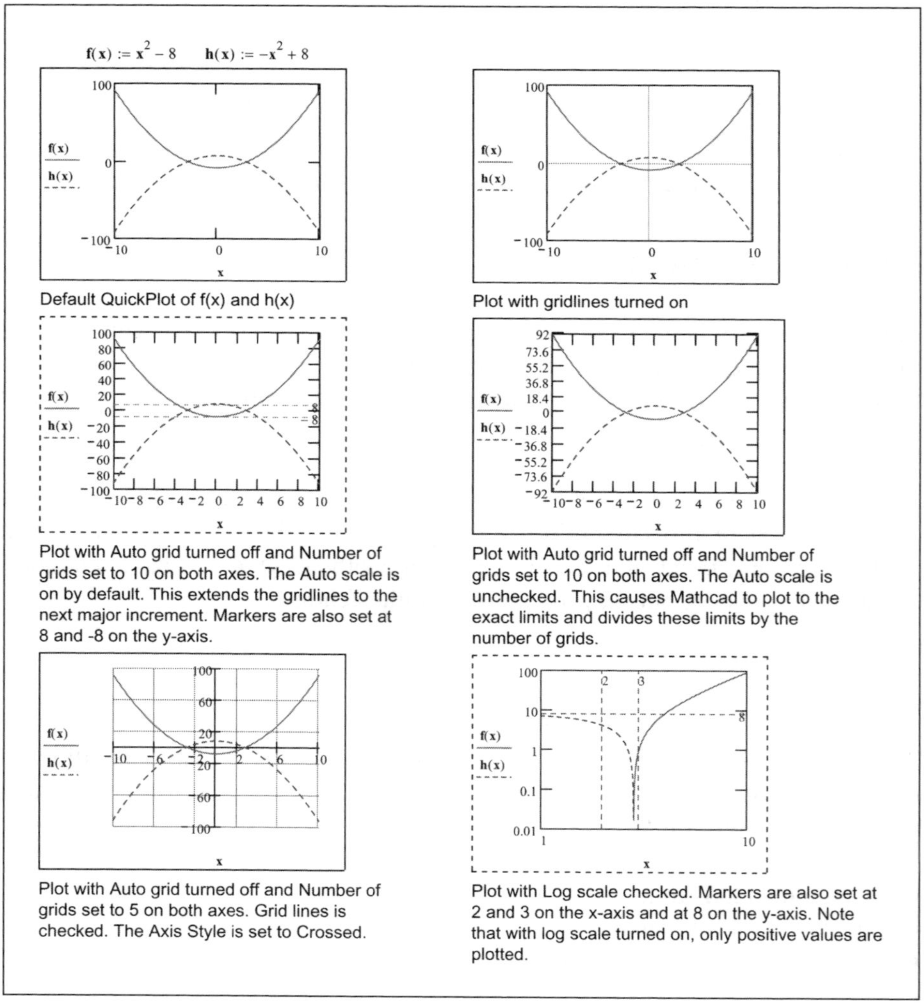

Default QuickPlot of f(x) and h(x)

Plot with gridlines turned on

Plot with Auto grid turned off and Number of grids set to 10 on both axes. The Auto scale is on by default. This extends the gridlines to the next major increment. Markers are also set at 8 and -8 on the y-axis.

Plot with Auto grid turned off and Number of grids set to 10 on both axes. The Auto scale is unchecked. This causes Mathcad to plot to the exact limits and divides these limits by the number of grids.

Plot with Auto grid turned off and Number of grids set to 5 on both axes. Grid lines is checked. The Axis Style is set to Crossed.

Plot with Log scale checked. Markers are also set at 2 and 3 on the x-axis and at 8 on the y-axis. Note that with log scale turned on, only positive values are plotted.

Figure 11.16 Formatting examples

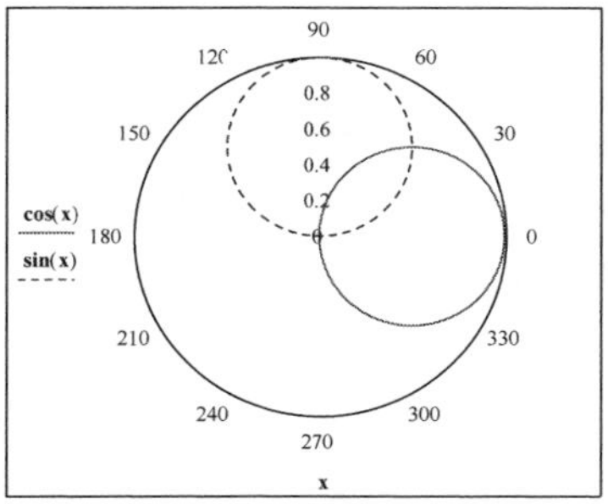

Default QuickPlot of cos(x) and sin(x)

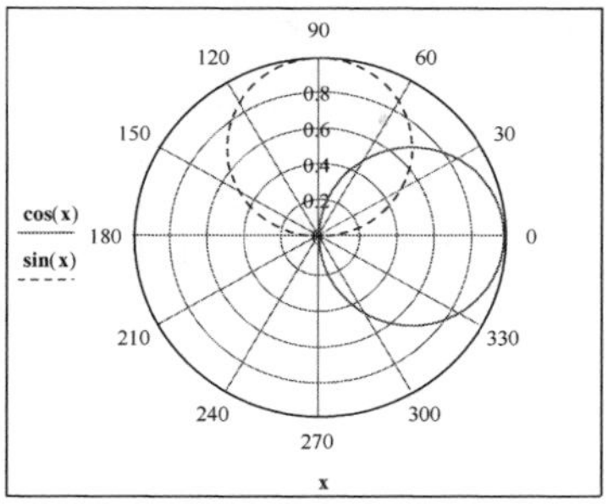

Plot with Angular and Radial grid lines turned on

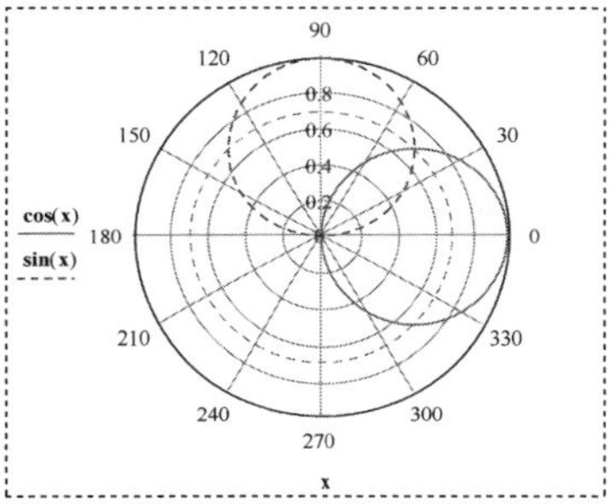

Plot with Angular and Radial grid lines turned on. Radial marker set at 0.7.

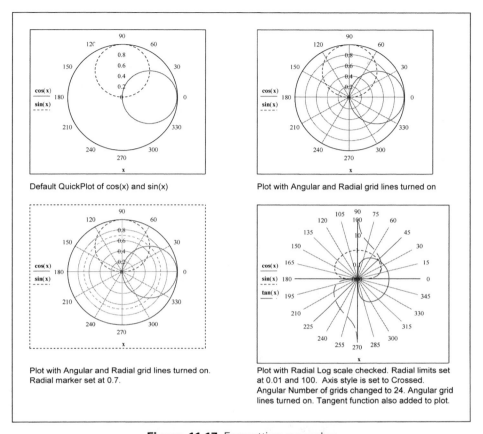

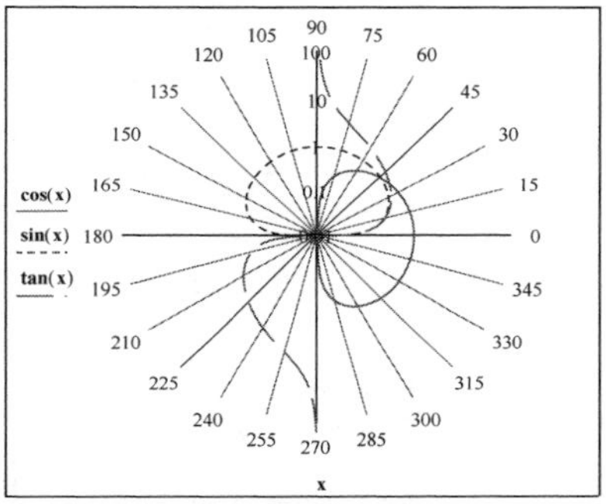

Plot with Radial Log scale checked. Radial limits set at 0.01 and 100. Axis style is set to Crossed. Angular Number of grids changed to 24. Angular grid lines turned on. Tangent function also added to plot.

Figure 11.17 Formatting examples

Traces tab

This tab allows you to change how each trace appears. You can change line type, line weight, and line color. You can also change the type of plot from a line plot to various forms of bar graphs, or change to just plotting points. You can also add a legend giving titles to the different traces. You can also add different symbols for the data points. Data points will be discussed later in the chapter.

See Figures 11.18–11.20 for a display of these various features.

The best way to learn these features is to try them. See the practice exercises at the end of the chapter.

Use the same functions as in Figure 11.16.

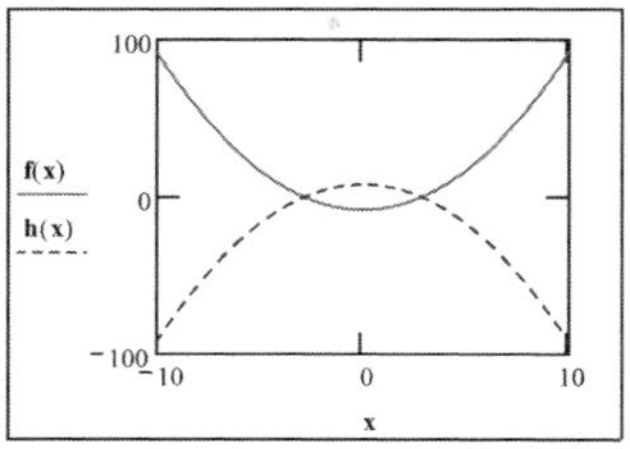

QuickPlot with defaults

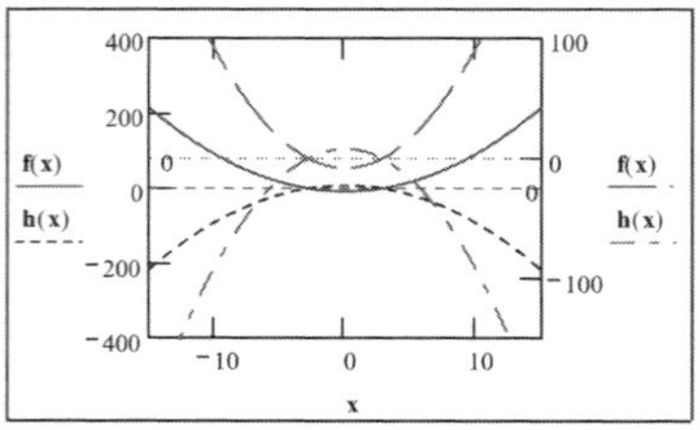

Plot with secondary y axis using same functions, but with different y-axis limits.

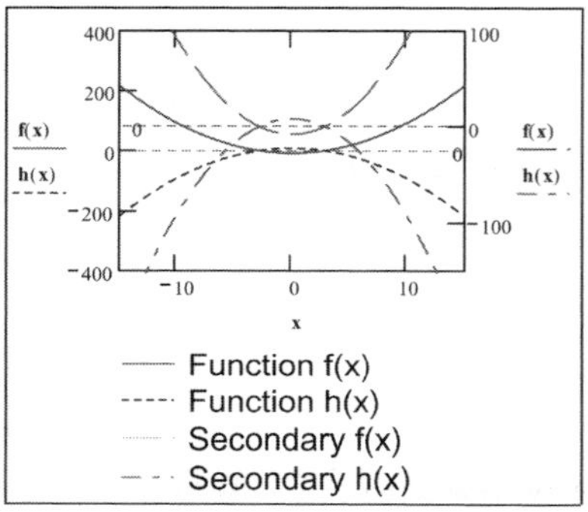

Legend turned on. Positioned below plot.

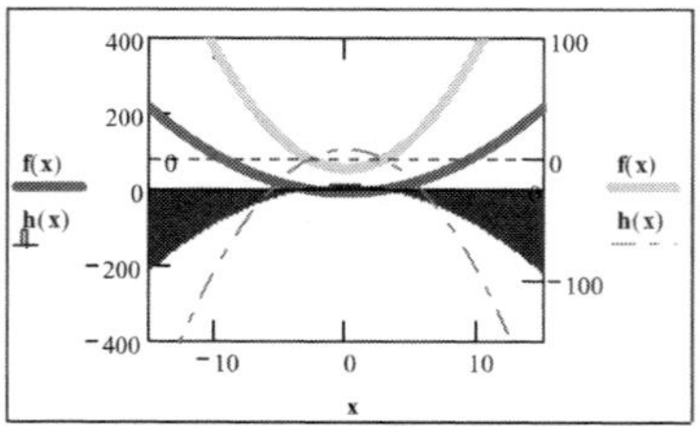

Trace 2 changed to a bar plot. Line weight of trace 1 and trace 3 changed from 1 to 4.

Figure 11.18 Formatting examples

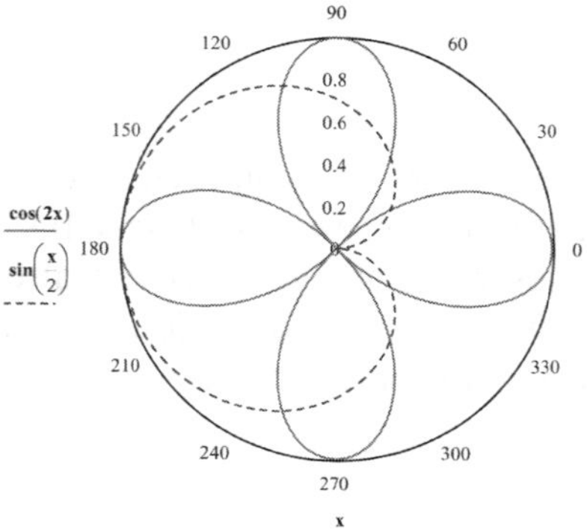

Default QuickPlot of cos(2x) and sin(x/2)

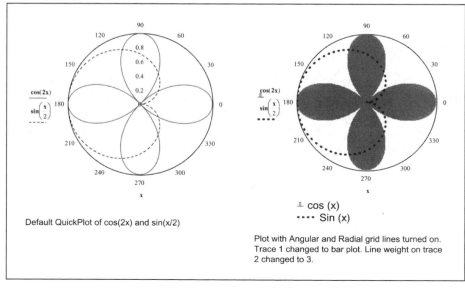

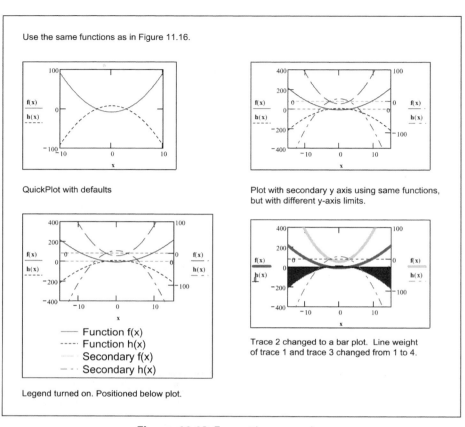

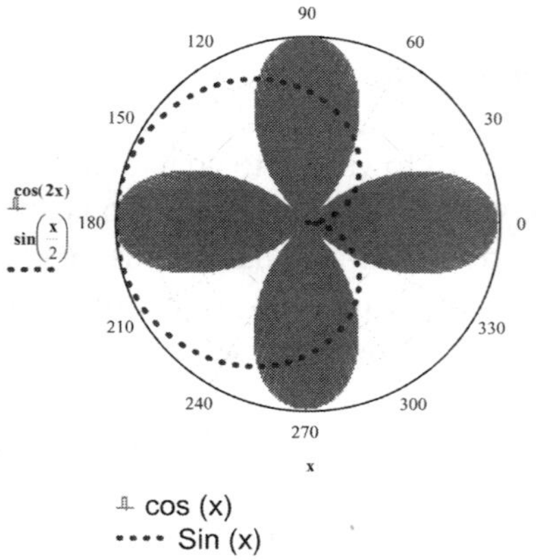

Plot with Angular and Radial grid lines turned on. Trace 1 changed to bar plot. Line weight on trace 2 changed to 3.

Figure 11.19 Formatting examples

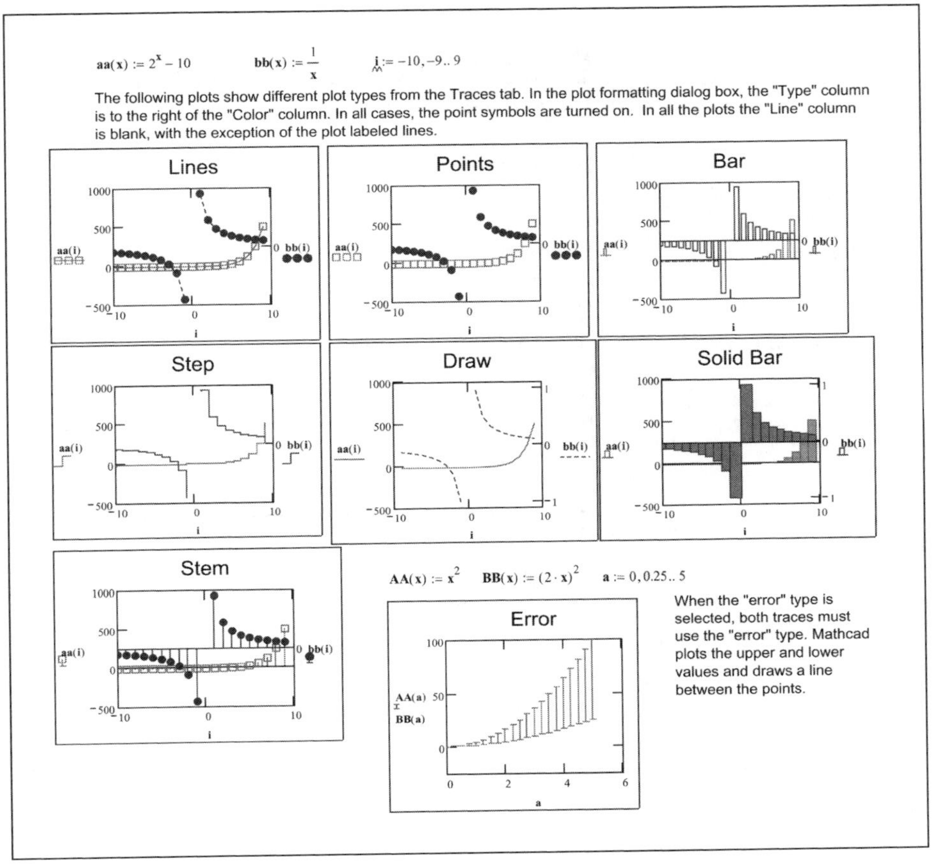

Figure 11.20 Plot types

Labels tab

The Labels tab allows you to apply titles to your plot and to each of the axes. In order for your titles to be visible, the check box associated with each title must be checked.

See Figure 11.21 for an example.

The font used in the plot title and axis labels comes from the "Math Text Font" variable style. To change the font, size, color, or style of your title and labels, change the "Math Text Font" style. To do this, click **Equation** from the **Format** menu. Refer to Chapters 6 and 7 for additional information on customizing styles.

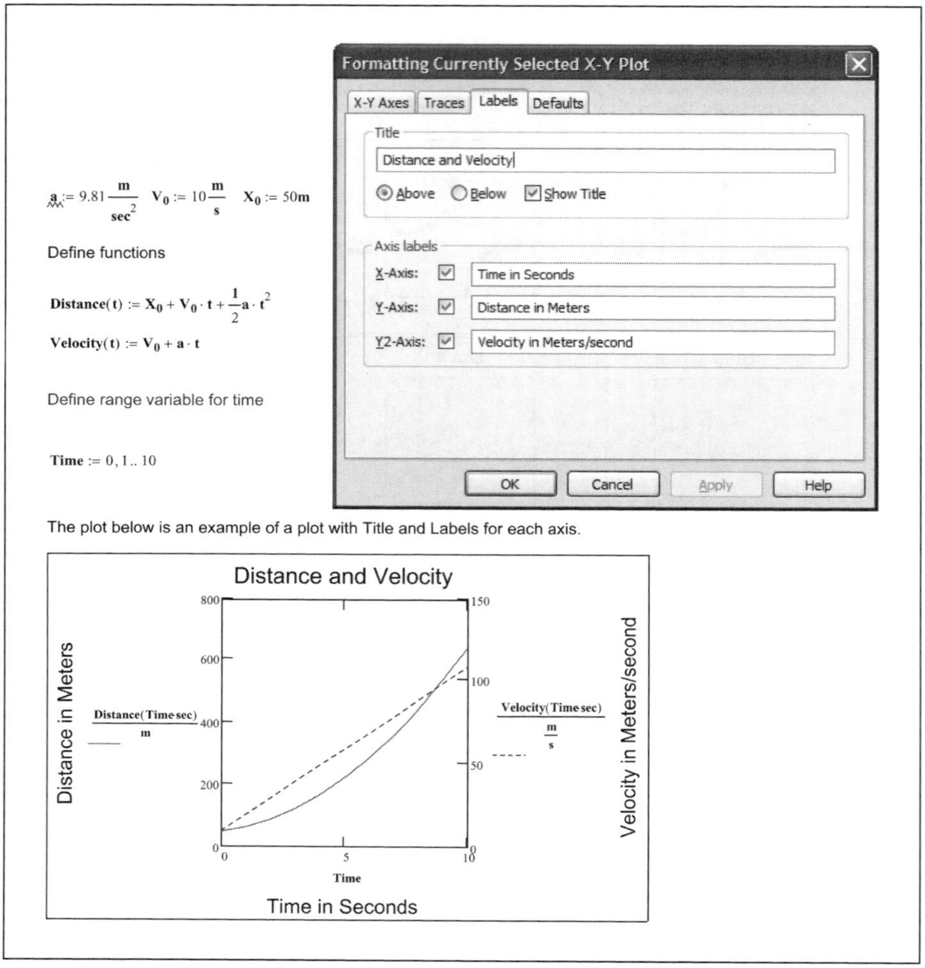

Figure 11.21 Using titles

Defaults tab

This tab allows you to reset your plot to the Mathcad defaults. It also allows you to use the current plot settings as the default settings for the current document.

If you have customized the plot settings and want to reuse the settings for future documents, then save the document as a template. You can even save the plot settings to your customized normal.xmct file so that they will be available for all new documents. See Chapter 7 for additional information about templates.

Zooming

There are times when you may want to zoom in on a plot. Perhaps you want to see where a plot crosses the x-axis, or perhaps you want to see where two plots intersect. To zoom in on a plot, right-click within the plot and select **Zoom**. You may also click on the plot to select it, and then click on the **Zoom** icon from the Graph toolbar. You may also select **Zoom** from the **Graph** option on the **Format** menu. This opens a Zoom dialog box. See Figure 11.22. Once the dialog box is open, click your mouse at one corner of the plot region you want to zoom. Then drag the mouse to include the area you want to zoom. A dashed selection outline will show you what area of the plot will be enlarged. The coordinates of the selection outline will also appear in the Zoom dialog box. Once the selection outline encloses the area of the plot you want to zoom, let go off the mouse button. You can drag the selection outline to fine-tune its location on the plot. Once you are satisfied with the location of the selection outline, click the plus icon in the Zoom dialog box. This temporarily sets the axis limits to the coordinates specified in the Zoom dialog box. You can now zoom in again using the same procedure, or you can zoom back out by clicking the minus icon button

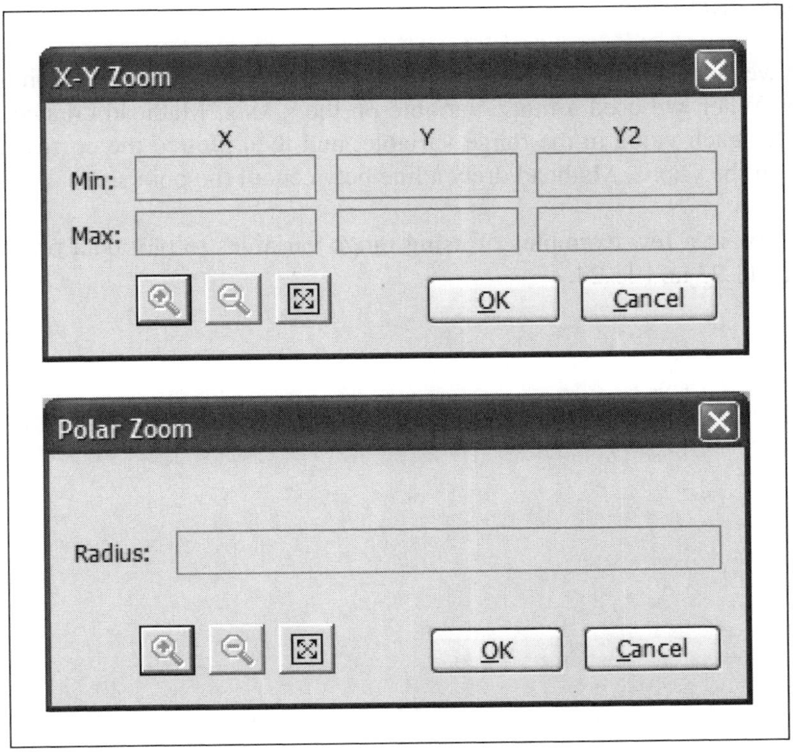

Figure 11.22 Zoom dialog boxes

in the Zoom dialog box. If you have zoomed in several times, you can step out by repeatedly clicking the minus icon, or you can zoom out to the original view by clicking the right icon button in the Zoom dialog box. If you want to make the zoomed-in region the permanent axis limits, then click OK.

Tip! If your Polar plot is not zooming, make sure that the lower plot limit is set to zero.

Another easy way to zoom a plot is to make the plot larger. You can do this by clicking and dragging the bottom right corner grip.

Plotting data points

Up until now, we have focused on graphing functions and expressions. These plots are easily represented with lines. Mathcad also allows you to plot data points. These data points can be created by using range variables, or they can be from a vector or matrix.

Range variables

When we discussed using range variables earlier, we were actually graphing data points. When we used a range variable on the x-axis, Mathcad created a data point for each value in the range variable, and then plotted the corresponding value on the y-axis. Mathcad drew a line between all the points.

Let's look at a few examples of using range variables to plot data points. See Figures 11.23 and 11.24.

$RV_1 := -20, -19 .. 20$ $\quad$ $RV_2 := -20, -10 .. 20$

Range variables actually create data points that are connected by straight lines. The plots used earlier in the chapter did not have the data points displayed.

To turn on the data points, double click on the plot, and select the Traces tab. In the Symbol column, select a symbol from the drop-down menu.

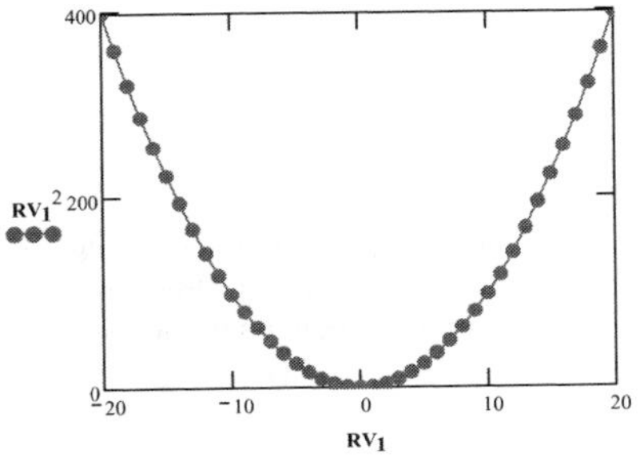

This plot uses a range variable with an increment of 1. Notice how smooth the graph is.

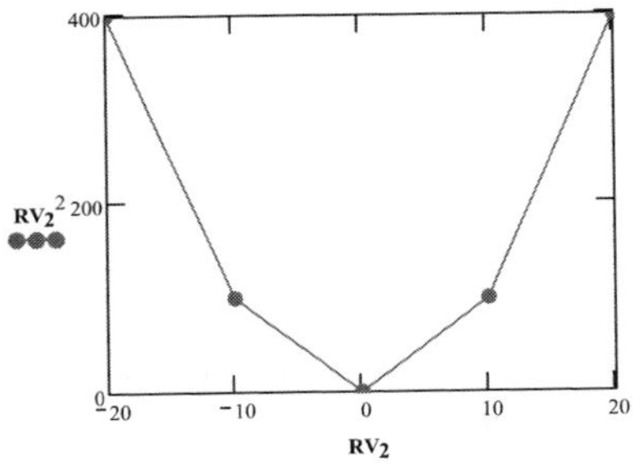

This plot uses a range variable with an increment of 10. Notice how it is not as smooth as the above plot.

Figure 11.23 Plotting data points

$RV_3 := \frac{-\pi}{2}, \frac{-\pi}{2} + \frac{\pi}{24} .. \frac{\pi}{2}$ $\quad$ $RV_4 := \frac{-\pi}{2}, \frac{-\pi}{2} + \frac{\pi}{8} .. \frac{\pi}{2}$

Range variables actually create data points that are connected by straight lines.

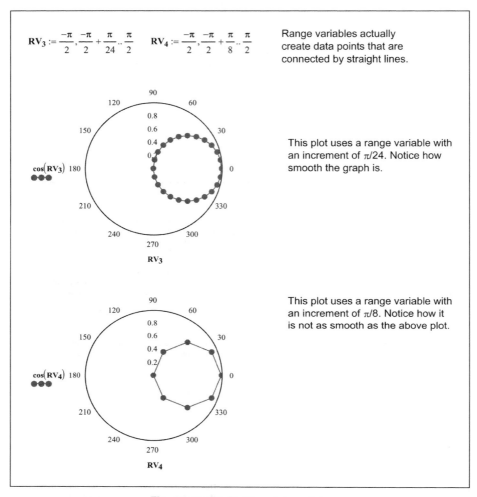

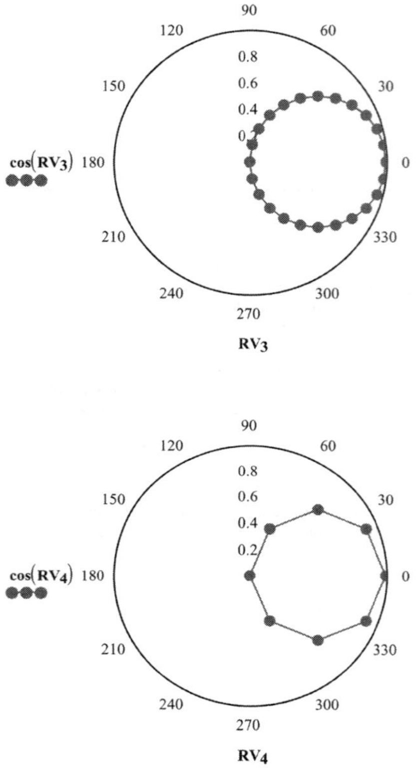

This plot uses a range variable with an increment of π/24. Notice how smooth the graph is.

This plot uses a range variable with an increment of π/8. Notice how it is not as smooth as the above plot.

Figure 11.24 Plotting data points

Data vectors

See Figure 11.25 for an example of plotting a vector of data points.

See Figure 11.26 for an example of plotting matrix data points.

$$\mathbf{DataVector} := \begin{pmatrix} 0.9 \\ 1.9 \\ 3.1 \\ 4.2 \\ 5.1 \\ 5.9 \\ 7.0 \\ 7.9 \\ 9.1 \\ 10.0 \end{pmatrix}$$

To plot a data vector, create a range variable to provide the x-axis value, and to tell Mathcad which data points to plot. The range variable must be in integer increments.

Use the *last*(v) function, which returns the element number of the last element of the vector v.

$\mathbf{last(DataVector)} = 10.00$

$\mathbf{i} := 1, 2 .. \mathbf{last(DataVector)}$

With this range variable, we are telling Mathcad to plot the first through the last elements of the vector "DataVector".

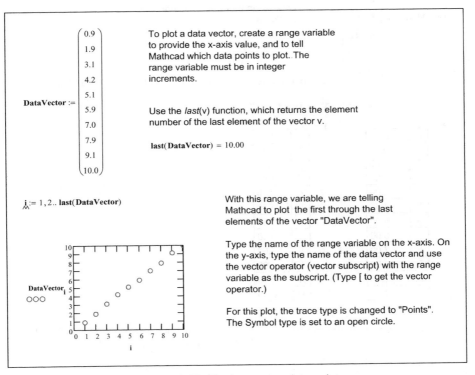

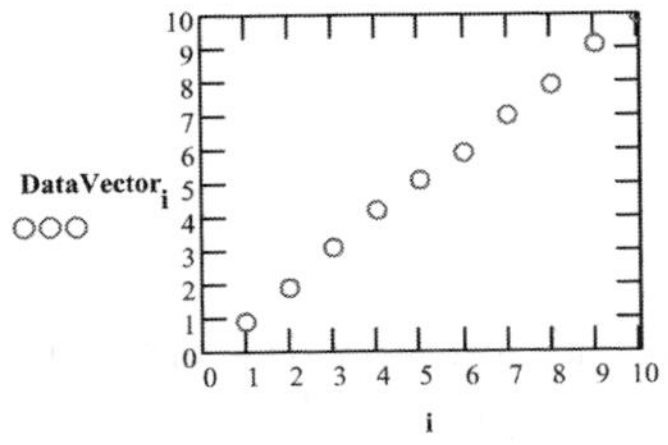

Type the name of the range variable on the x-axis. On the y-axis, type the name of the data vector and use the vector operator (vector subscript) with the range variable as the subscript. (Type [to get the vector operator.)

For this plot, the trace type is changed to "Points". The Symbol type is set to an open circle.

Figure 11.25 Plotting vector data points

$$\mathbf{DataMatrix} := \begin{pmatrix} 0 & 0.1 \\ 1 & 0.9 \\ 2 & 1.9 \\ 3 & 3.1 \\ 4 & 4.2 \\ 5 & 5.1 \\ 6 & 4.3 \\ 7 & 4.0 \\ 10 & 2.9 \\ 11 & 2.1 \\ 12 & 0.8 \end{pmatrix}$$

To plot data from a matrix, assign each column in the matrix to a variable. To do this, use the matrix column operator "M<>". The column operator returns a vector of the matrix column numbered between the brackets. This operator is located on the Matrix toolbar, or press CTRL+6.

$\mathbf{X} := \mathbf{DataMatrix}^{\langle 1 \rangle}$

Note: Type CTRL+6 to get the brackets.

$\mathbf{Y} := \mathbf{DataMatrix}^{\langle 2 \rangle}$

X =

	1
1	0.00
2	1.00
3	2.00
4	3.00
5	4.00
6	5.00
7	6.00
8	7.00
9	10.00
10	11.00
11	12.00

Y =

	1
1	0.10
2	0.90
3	1.90
4	3.10
5	4.20
6	5.10
7	4.30
8	4.00
9	2.90
10	2.10
11	0.80

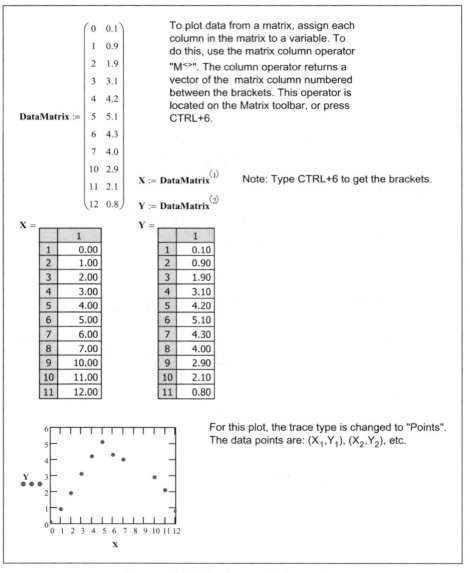

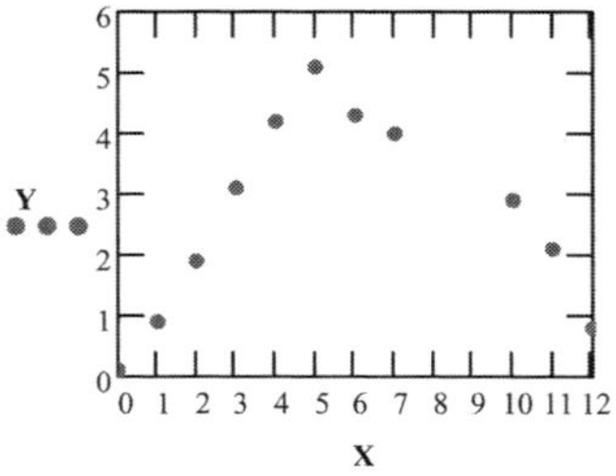

For this plot, the trace type is changed to "Points". The data points are: (X_1,Y_1), (X_2,Y_2), etc.

Figure 11.26 Plotting matrix data points

Numeric display of plotted points (Trace)

You can get numeric display of plotted points by using the Trace dialog box. To open the Trace dialog box, right-click within the plot and choose **Trace**.

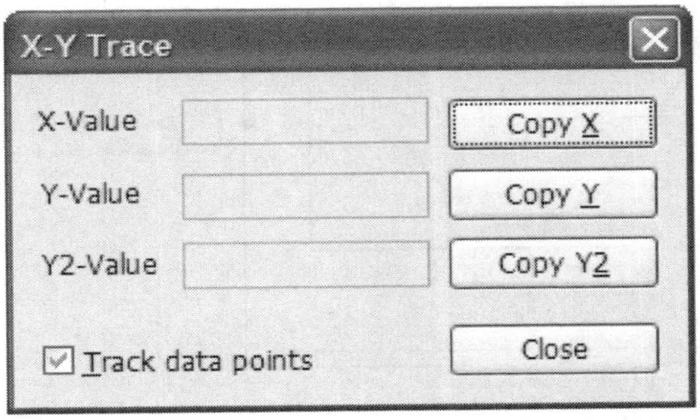

Place a check in the Track data points box. Click and drag your mouse along the trace whose coordinates you want to see. You should see a dotted crosshair move from top point along the trace. The coordinates of each data point will be displayed in the dialog box. If you release the mouse button, you can use the left and right arrows to move to the previous and next data points.

Using plots for finding solutions to problems

One great use of plots is to find the intersection of two functions. In Chapters 14 and 15, we will be discussing the Mathcad solving functions. You can use a plot to get a quick guess for the required input in the ***find*** function. Using the trace feature above, you can also get a quick approximation of the solution. See Figure 11.27 for an example.

More about this topic will be covered in Chapter 14.

$y_1(x) := x^2 + 2$ $\qquad$ $y_2(x) := -x + 15$

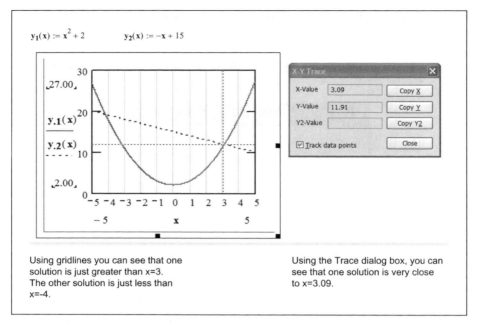

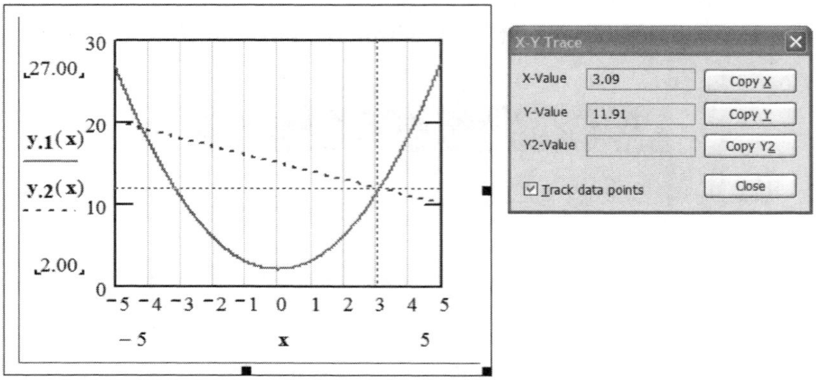

Using gridlines you can see that one solution is just greater than x=3. The other solution is just less than x=-4.

Using the Trace dialog box, you can see that one solution is very close to x=3.09.

Figure 11.27 Using trace to find approximate solutions

Parametric plotting

A parametric plot is one in which a function or expression is plotted against another function or expression that uses the same independent variable. See Figure 11.28 for an example.

$iii := 1, 2..10$

$ZZ(x) := \frac{1}{x^2}$

$YY(x) := x^2$

$ZZ(iii) =$

	1
1	1.000
2	0.250
3	0.111
4	0.063
5	0.040
6	0.028
7	0.020
8	0.016
9	0.012
10	0.010

$YY(iii) =$

	1
1	1.00
2	4.00
3	9.00
4	16.00
5	25.00
6	36.00
7	49.00
8	64.00
9	81.00
10	100.00

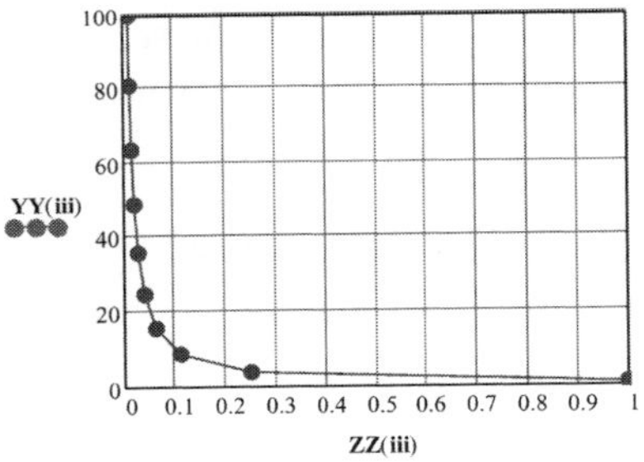

Parametric plot of ZZ(x) on the x-axis and YY(x) on the y-axis.

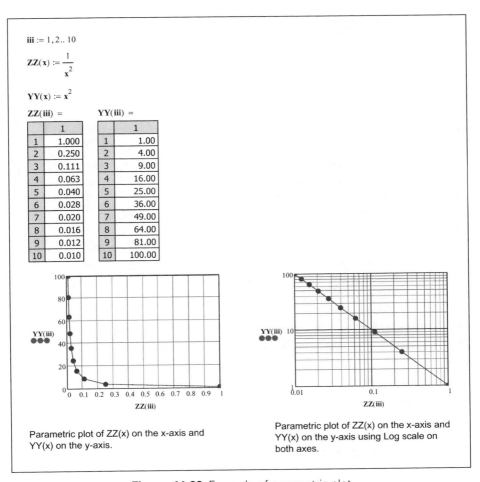

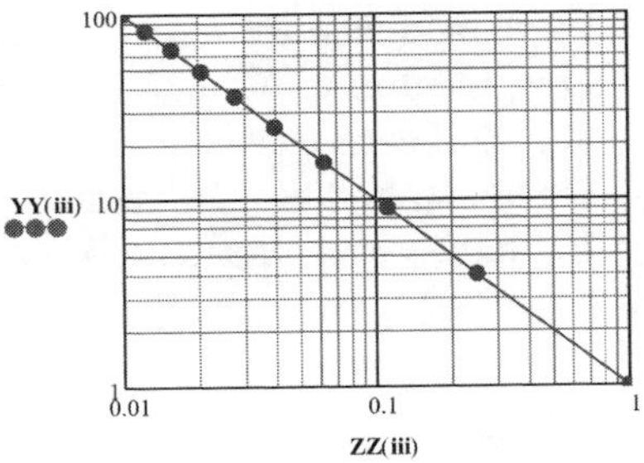

Parametric plot of ZZ(x) on the x-axis and YY(x) on the y-axis using Log scale on both axes.

Figure 11.28 Example of parametric plot

3D plotting

Mathcad can create many types of three-dimensional plots such as a: surface plot, contour plot, 3D bar plot, 3D scatter plot, and vector field plot. These topics are beyond the scope of this book. The Mathcad Help and Mathcad Tutorial are excellent resources to learn more about these topics.

Summary

Plots are useful, easy-to-use tools. They can help visualize equations or functions. They can also be used to help solve equations.

In Chapter 11 we:

- Showed how to create simple X-Y QuickPlots and Polar QuickPlots.
- Showed how to plot functions and expressions.
- Discussed using range variables to set range limits.
- Showed how to plot multiple functions on the same plot.
- Showed how to plot functions on the y-axis and on a secondary y-axis.
- Showed how to plot using units, and how to change the displayed units.
- Explained how to format plots with grid lines, numbers, labels, and titles.
- Explained how to change the trace colors, line type, and line weight.
- Discussed plotting data information from range variables, data vectors, and data matrices.
- Demonstrated how to get numeric readout of plotted traces.
- Explained how plots can be used to help solve systems of equations.

Practice

1. Create separate simple X-Y QuickPlots of the following equations. Use expressions rather than functions on the y-axis.

   [a] x^2
   [b] $x^3+2x^2+3x-10$
   [c] sin(x)

2. Create separate Polar QuickPlots of the following equations. Use expressions rather than functions on the radial axis.

   [a] x/2
   [b] cos(6x)
   [c] tan(x)

3. Create functions for the expressions in Exercises 1 and 2 and plot these functions.
4. Create two range variables for each of the above functions. One range variable should have a small increment; the other should have a large

increment. Plot the above functions using each range variable. Use the Format dialog box and the Traces tab to make each plot look different from the others.

5. The formula to calculate the bending moment at point "x" in a beam (with uniform loading) is $M = (½*w*x)*(L - x)$, where w is in force/length, L is total length of beam and x is distance from one end of the beam. Create a plot with distance x on the x-axis (from zero to L), and moment on the y-axis. Use $w = 2$ N/m and $L = 10$ m. Moment should be displayed as N*m. Provide a title and axis labels.
6. Plot the following data points. Use a range variable for the x-axis. Use a solid box as the symbol. Connect the data points with a dashed line.

1	19.1
2	29.5
3	40.3
4	52.4
5	59.3
6	70.5

7. Plot the following data points. Use a blue solid circle as the symbol. Do not connect the data points.

1.2	2.4
2.3	3.3
4.5	5.3
5.2	6.3
4.5	4.6
5.5	6.4

8. Plot the following equations and use the Trace dialog box to find approximate solutions where the plots intersect. $y_1(x) = 2x^2+3x-10$, $y_2(x) = -x^2+2x+20$.
9. Write a function to describe the vertical motion and a function to describe the horizontal motion of a projectile fired at 700 ft/s with a 35-degree inclination from the horizontal. Each function should be a function of time. Create a parametric plot with the following:

[a] Use a range variable to set the range of the plot. Use a range from 0 to 20 with an increment of 1.

[b] Create a parametric plot with horizontal motion on the x-axis and the vertical motion on the y-axis. Use the range variable for the argument of both functions.

[c] Use units in the functions and the plots.
Hint: remember to multiply the function argument by seconds.

10. Copy the plot from Exercise 8 and plot in terms of meters instead of feet.

12
Simple logic programming

Engineering calculations must have a way to logically reach conclusions based on data calculated. Mathcad programming allows you to write logic programs. Mathcad calls this "programming," but it is essentially a better way to use the *if* function. These "programs" allow Mathcad to choose a result based on specific parameters. Programs can be very complex, and can be used for many things. This chapter focuses solely on simple logic programming. The more complex programs are discussed in Chapter 16—Advanced Programming.

Chapter 12 will:

- Provide several simple Mathcad examples to illustrate the concept of logic programming.
- List the steps necessary to create a logic program.
- Describe the logic Mathcad user to arrive at a conclusion.
- Warn the user about violating the above logic, which could cause Mathcad to make an inaccurate conclusion.
- Reveal new ways of creating logic programs that are not provided in the Mathcad documentation.
- Show how to use a logic program to draw and display conclusions.

Introduction to the programming toolbar

The purpose of this chapter is to get you comfortable with the concept of Mathcad programming. We will use simple examples. You will soon see that you do not need to be a computer programmer to use the Mathcad programming features. You do not need to learn complex programming commands. All the operators you need to use are contained on the Programming toolbar.

To open the Programming toolbar, hover your mouse over **Toolbars** on the **View** menu and click on **Programming**. You may also click on the Programming Toolbar icon on the Math toolbar. The programming toolbar contains the operators you will use when writing a Mathcad program.

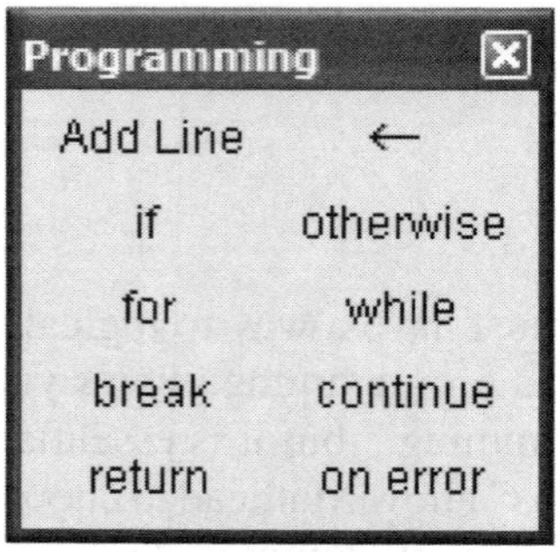

This chapter will focus on the following operators: ***Add Line, if, otherwise, and return***. The remaining operators will be discussed in Chapter 16—Advanced Programming.

One important thing to remember when using Mathcad programming is that all the programming operators must be inserted using the Programming toolbar, or by using keyboard shortcuts. You cannot type "if," "otherwise," or "return." They are operators and must be inserted.

Creating a simple program

You begin a Mathcad program by clicking on "Add Line." This inserts a vertical line with two placeholders. This is called the Programming Operator. We will place a conditional statement in the top placeholder. In the bottom placeholder, we will place a statement about what to do if the conditional statement is false.

Let's look at a two simple examples. See Figures 12.1 and 12.2.

The following example is similar to Figure 10.22, which used the *if* function.

InputLength := 20**ft**

DesignLength := | 25**ft** **if** **InputLength** < 25**ft**
| **InputLength** **otherwise**

DesignLength = 25.00 **ft**

To create this program follow these steps:

1. Open the Programming toolbar.
2. Type the variable name followed by the colon.
3. Click the **Add Line** button on the Programming toolbar.
4. Select the top placeholder and click the **if** button on the Programming toolbar.
5. Select the bottom placeholder and click the **otherwise** button on the Programming toolbar.
6. Fill-in the placeholders.

DesignLength := | ▪
| ▪

DesignLength := | ▪ **if** ▪
| ▪ **otherwise**

Remember. . . When creating a program, typing "if" is not the same as using the **if** operator on the programming toolbar.

Figure 12.1 Simple program

The logic used by the program in Figure 12.1 is, "Do this **if** this statement is true, **otherwise** do this." The ***if*** function could just as easily been used in place of the program. The benefit of using the program operator comes when there are multiple **if** statements.

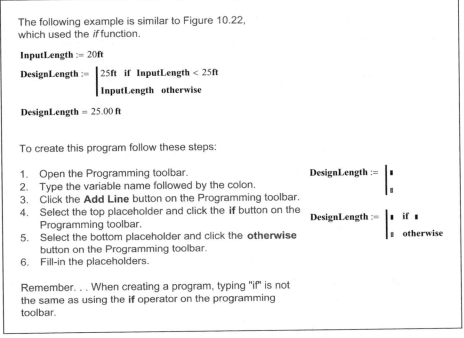

Use three methods to check the program for different lengths.

Refer to Chapter 9 for a discussion of using arrays with equations and functions.

Method 1 -- Use an expression with a range variable and vector to get multiple results.

$$\mathbf{InputLength} := \begin{pmatrix} 5\mathbf{ft} \\ 25\mathbf{ft} \\ 50\mathbf{ft} \end{pmatrix} \qquad \mathbf{i} := 1, 2 .. 3$$

Create and input vector

$$\mathbf{DesignLength_i} := \left| \begin{array}{l} 25\mathbf{ft} \ \ \mathbf{if} \ \ \mathbf{InputLength_i} < 25\mathbf{ft} \\ \mathbf{InputLength_i} \ \ \mathbf{otherwise} \end{array} \right.$$

Define the expression

$$\mathbf{DesignLength} = \begin{pmatrix} 25.00 \\ 25.00 \\ 50.00 \end{pmatrix} \mathbf{ft}$$

OK
OK
OK

Method 2 -- Use a function. Use different arguments in the function.

$$\mathbf{NewLength(x)} := \left| \begin{array}{l} 25\mathbf{ft} \ \ \mathbf{if} \ \ \mathbf{x} < 25\mathbf{ft} \\ \mathbf{x} \ \ \mathbf{otherwise} \end{array} \right.$$

$\mathbf{NewLength(5ft)} = 25.00\,\mathbf{ft}$

$\mathbf{NewLength(25ft)} = 25.00\,\mathbf{ft}$

$\mathbf{NewLength(50ft)} = 50.00\,\mathbf{ft}$

Method 3 -- Use a function with a vector as the argument.

$$\mathbf{ResultLength} := \overrightarrow{\mathbf{NewLength(InputLength)}}$$

$$\mathbf{ResultLength} = \begin{pmatrix} 25.00 \\ 25.00 \\ 50.00 \end{pmatrix} \mathbf{ft}$$

Note that the vectorize operator CTRL+MINUS needs to be added to the function so that Mathcad will recognize the vector with the function.

Figure 12.2 Check the program in Figure 12.1 for different lengths

When using multiple **if** statements in a program, it is important to understand how Mathcad treats the **if** statements. Mathcad evaluates every **if** statement. If a statement is false, the statement is not executed, but Mathcad proceeds to the next line. If the statement is true, the statement is executed, and Mathcad proceeds to the next line. The **otherwise** statement is executed only if all previous **if** statements are false. If there are more than one true **if** statements, Mathcad returns the last true **if** statement. This is important to understand, or you may get incorrect results. Let's look at a few examples. See Figures 12.3 and 12.4.

Assume a previous calculation returned a calculated factor. The minimum value of the factor should be 0.5. The maximum value should be 2.0. Let's write a program as a user-defined function (using the calculated factor as an argument) so that Mathcad will choose a final factor between 0.5 and 2.0.

0.5<Factor<2.0

Let's look at two examples of a program.
This first example creates inaccurate results.

$$\mathbf{Factor}(\mathbf{x}) := \left|\begin{array}{l} 2.0 \;\; \mathbf{if} \;\; \mathbf{x} > 2.0 \\ \mathbf{x} \;\; \mathbf{if} \;\; \mathbf{x} > 0.5 \\ 0.5 \;\; \mathbf{otherwise} \end{array}\right.$$

$\mathbf{F_1} := \mathbf{Factor}(0.25)$ $\quad \mathbf{F_1} = 0.50$ Correct

$\mathbf{F_2} := \mathbf{Factor}(1.0)$ $\quad \mathbf{F_2} = 1.00$ Correct

$\mathbf{F_3} := \mathbf{Factor}(3.0)$ $\quad \mathbf{F_3} = 3.00$ Incorrect (The correct result should be 2.0)

Why did F_3 above example give incorrect results?
It returned incorrect results because Mathcad evaluates every **if** statement, and executes it if it is true. The final true **if** statement is returned. For F_3, the final true statement that is encountered is "if x is greater than 0.5 then return x." Since 3>0.5, Mathcad returned 3.

Rewrite the program so that it creates correct results. Do this by ensuring that the first two statements cannot both be true.

$$\mathbf{RevisedFactor}(\mathbf{x}) := \left|\begin{array}{l} 0.5 \;\; \mathbf{if} \;\; \mathbf{x} < 0.5 \\ 2.0 \;\; \mathbf{if} \;\; \mathbf{x} > 2.0 \\ \mathbf{x} \;\; \mathbf{otherwise} \end{array}\right.$$

$\mathbf{F_4} := \mathbf{RevisedFactor}(0.25)$ $\quad \mathbf{F_4} = 0.50$ Correct

$\mathbf{F_5} := \mathbf{RevisedFactor}(1.0)$ $\quad \mathbf{F_5} = 1.00$ Correct

$\mathbf{F_6} := \mathbf{RevisedFactor}(3.0)$ $\quad \mathbf{F_6} = 2.00$ Correct

Figure 12.3 Using multiple **if** statements

In wood design there is a depth factor that needs to be applied to different depths of wood joists. Let's look at two examples of a conditional program to determine the proper depth factor.
This first example creates incorrect results.

$$\mathbf{Factor_1(d)} := \left| \begin{array}{l} 1.5 \ \ \mathbf{if}\ \ \mathbf{d} \le 3.5\mathbf{in} \\ 1.4 \ \ \mathbf{if}\ \ \mathbf{d} \le 4.5\mathbf{in} \\ 1.3 \ \ \mathbf{if}\ \ \mathbf{d} \le 5.5\mathbf{in} \\ 1.2 \ \ \mathbf{if}\ \ \mathbf{d} \le 7.25\mathbf{in} \\ 1.1 \ \ \mathbf{if}\ \ \mathbf{d} \le 9.25\mathbf{in} \\ 1.0 \ \ \mathbf{if}\ \ \mathbf{d} \le 11.25\mathbf{in} \\ 0.9 \ \ \mathbf{otherwise} \end{array} \right.$$

$\mathbf{Factor_1}(2.5\mathbf{in}) = 1.00$	Incorrect (The result should be 1.5)
$\mathbf{Factor_1}(3.5\mathbf{in}) = 1.00$	Incorrect (The result should be 1.5)
$\mathbf{Factor_1}(4.5\mathbf{in}) = 1.00$	Incorrect (The result should be 1.4)
$\mathbf{Factor_1}(5.5\mathbf{in}) = 1.00$	Incorrect (The result should be 1.3)
$\mathbf{Factor_1}(7.25\mathbf{in}) = 1.00$	Incorrect (The result should be 1.2)
$\mathbf{Factor_1}(9.25\mathbf{in}) = 1.00$	Incorrect (The result should be 1.1)
$\mathbf{Factor_1}(11.25\mathbf{in}) = 1.00$	Correct
$\mathbf{Factor_1}(13.25\mathbf{in}) = 0.90$	Correct

Why did the above example give so many incorrect results?
It returned incorrect results because Mathcad evaluates every **if** statement and executes it if it is true. The final result is the last true statement that is encountered. The last true statement that is encountered is d <=11.25 in. Most of the depths were less than 11.25 in., so 1.0 is the value assigned to Factor_1.

If we reorganize the program, so that false statements are encountered as the depth decreases, we get Correct results.

$$\mathbf{RevisedFactor_2(d)} := \left| \begin{array}{l} 1.0 \ \ \mathbf{if}\ \ \mathbf{d} \le 11.25\mathbf{in} \\ 1.1 \ \ \mathbf{if}\ \ \mathbf{d} \le 9.25\mathbf{in} \\ 1.2 \ \ \mathbf{if}\ \ \mathbf{d} \le 7.25\mathbf{in} \\ 1.3 \ \ \mathbf{if}\ \ \mathbf{d} \le 5.5\mathbf{in} \\ 1.4 \ \ \mathbf{if}\ \ \mathbf{d} \le 4.5\mathbf{in} \\ 1.5 \ \ \mathbf{if}\ \ \mathbf{d} \le 3.5\mathbf{in} \\ 0.9 \ \ \mathbf{otherwise} \end{array} \right.$$

$\mathbf{RevisedFactor_2}(2.5\mathbf{in}) = 1.50$	Correct
$\mathbf{RevisedFactor_2}(3.5\mathbf{in}) = 1.50$	Correct
$\mathbf{RevisedFactor_2}(4.5\mathbf{in}) = 1.40$	Correct
$\mathbf{RevisedFactor_2}(5.5\mathbf{in}) = 1.30$	Correct
$\mathbf{RevisedFactor_2}(7.25\mathbf{in}) = 1.20$	Correct
$\mathbf{RevisedFactor_2}(9.25\mathbf{in}) = 1.10$	Correct
$\mathbf{RevisedFactor_2}(11.25\mathbf{in}) = 1.00$	Correct
$\mathbf{RevisedFactor_2}(13.25\mathbf{in}) = 0.90$	Correct

Figure 12.4 Multiple **if** statements

Return operator

The ***return*** operator is used in conjunction with an **if** statement. When the **if** statement is true, the ***return*** operator tells Mathcad to stop the program and return the value, rather than proceed to the next line in the program. Figure 12.5 shows how to use the ***return*** statement to make the examples from Figure 12.4 more intuitive.

The return operator stops the program. In conjunction with an **if** statement, it can be used to stop the program if the **if** statement is true. If the **if** statement is false, then Mathcad proceeds to the next line.

The first example from Figure 12.4 could be written in the following manner using the return operator:

$$\mathbf{Factor_2(d)} := \left| \begin{array}{l} \mathbf{return}\ 1.5\ \ \mathbf{if}\ \ \mathbf{d} \le 3.5\mathbf{in} \\ \mathbf{return}\ 1.4\ \ \mathbf{if}\ \ \mathbf{d} \le 4.5\mathbf{in} \\ \mathbf{return}\ 1.3\ \ \mathbf{if}\ \ \mathbf{d} \le 5.5\mathbf{in} \\ \mathbf{return}\ 1.2\ \ \mathbf{if}\ \ \mathbf{d} \le 7.25\mathbf{in} \\ \mathbf{return}\ 1.1\ \ \mathbf{if}\ \ \mathbf{d} \le 9.25\mathbf{in} \\ \mathbf{return}\ 1.0\ \ \mathbf{if}\ \ \mathbf{d} \le 11.25\mathbf{in} \\ 0.9\ \ \mathbf{otherwise} \end{array} \right.$$

$\mathbf{Factor_2}(2.5\mathbf{in}) = 1.50$ Correct

$\mathbf{Factor_2}(3.5\mathbf{in}) = 1.50$ Correct

$\mathbf{Factor_2}(4.5\mathbf{in}) = 1.40$ Correct

$\mathbf{Factor_2}(5.5\mathbf{in}) = 1.30$ Correct

$\mathbf{Factor_2}(7.25\mathbf{in}) = 1.20$ Correct

$\mathbf{Factor_2}(9.25\mathbf{in}) = 1.10$ Correct

$\mathbf{Factor_2}(11.25\mathbf{in}) = 1.00$ Correct

$\mathbf{Factor_2}(13.25\mathbf{in}) = 0.90$ Correct

Note: A return could have been placed in front of the last line to make it read more consistent, but it is not required.

Figure 12.5 Return operator

Boolean operators

Boolean operators are very useful when writing logical Mathcad programs. The Boolean operators are located on the Boolean toolbar. To open the Boolean toolbar, hover your mouse over **Toolbars** on the **Format** menu and select **Boolean**. You may also click the  button on the Math toolbar.

To use the Boolean operators you need to use the buttons on the Boolean toolbar or

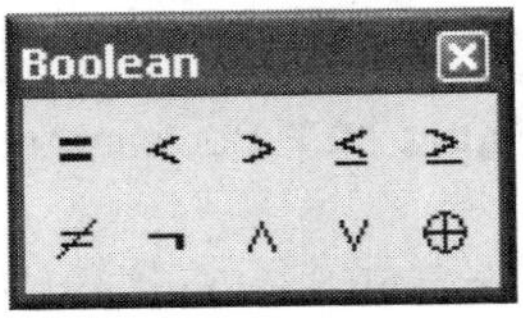

use keyboard shortcuts. The exception is the **Greater Than** and **Less Than** operators, which can be entered from the keyboard.

The first six icons on the toolbar are obvious. The last four are not as obvious.

The ¬ button is Boolean **Not**. This is true if x is zero, and false if x is non-zero. See Figure 12.6 for an example.

Boolean **Not** operator returns true (1) if the value is zero. It returns false (0) if the value is non-zero

$\neg 1 = 0.00 \qquad \neg 0 = 1.00$

Example using the **Not** operator.

$\mathbf{AA}(\mathbf{x}) := \left| \begin{array}{l} \text{"X is zero"} \;\; \textbf{if} \;\; \neg\mathbf{x} \\ \text{"X is not zero"} \;\; \textbf{otherwise} \end{array} \right.$

$\mathbf{AA}(2) = \text{"X is not zero"} \qquad \mathbf{AA}(0) = \text{"X is zero"}$

This is the same result as using **if** x=0.

$\mathbf{AB}(\mathbf{x}) := \left| \begin{array}{l} \text{"X is zero"} \;\; \textbf{if} \;\; \mathbf{x} = 0 \\ \text{"X is not zero"} \;\; \textbf{otherwise} \end{array} \right.$

$\mathbf{AB}(2) = \text{"X is not zero"} \qquad \mathbf{AB}(0) = \text{"X is zero"}$

Figure 12.6 Boolean **Not**

The ∧ button is Boolean **And**. This is true if both statements are true. See Figure 12.7 for an example.

Boolean **And** operator returns true if both statements are true.

$\mathbf{BB}(\mathbf{x},\mathbf{y}) := \left| \begin{array}{l} \text{"This And statement is TRUE"} \;\; \textbf{if} \;\; \mathbf{x} > 0 \wedge \mathbf{y} > 0 \\ \text{"This And statement is false"} \;\; \textbf{otherwise} \end{array} \right.$

$\mathbf{BB}(5,4) = \text{"This And statement is TRUE"} \qquad \mathbf{BB}(5,0) = \text{"This And statement is false"}$

$\mathbf{BB}(0,3) = \text{"This And statement is false"} \qquad \mathbf{BB}(0,0) = \text{"This And statement is false"}$

Figure 12.7 Boolean **And**

The ∨ button is Boolean **Or**. This is true if either statement is true.

Warning: If the first statement is true, then the second statement is not evaluated. Therefore, the second statement may contain an error that is not initially detected because Mathcad stopped evaluating after the first statement. See Figure 12.8 for examples.

Boolean **OR** operator returns true if either statement is true.

$$CC(x,y) := \begin{cases} \text{"This Or statement is TRUE"} & \textbf{if } x > 0 \vee y > 0 \\ \text{"This Or statement is false"} & \textbf{otherwise} \end{cases}$$

CC(5, 4) = "This Or statement is TRUE"   CC(0, 4) = "This Or statement is TRUE"

CC(5, 0) = "This Or statement is TRUE"   CC(0, 0) = "This Or statement is false"

When using the **OR** operator, if the first statement is true, the second statement will not be checked. This could prevent an error from being detected.

$$DD(x,y) := \begin{cases} \text{"This Or statement is TRUE"} & \textbf{if } x > 0 \vee \frac{y}{0} > 0 \\ \text{"This Or statement is false"} & \textbf{otherwise} \end{cases}$$

This divide by zero error is not caught, unless the first statement is false.

DD(5, 4) = "This Or statement is TRUE"   DD(0, 4) = ▮

DD(0, 4) = ▮▮
Divide by zero.

DD(5, 0) = "This Or statement is TRUE"   DD(0, 0) = ▮

Figure 12.8 Boolean **Or**

The ⊕ button is Boolean **Xor** (exclusive Or). This is true if either the first or the second statement is non-zero, but not both. Thus if both statement are non-zero, the result is false. See Figure 12.9 for an example.

Boolean **Xor** operator returns true if one statement is true, but not both.

$$EE(x,y) := \begin{cases} \text{"The xor statement is TRUE"} & \textbf{if } x > 0 \oplus y > 0 \\ \text{"The xor statement is false"} & \textbf{otherwise} \end{cases}$$

EE(5, 6) = "The xor statement is false"   False because one of the statements is not false.

EE(0, 6) = "The xor statement is TRUE"   True because one, but not both statements is true.

EE(5, 0) = "The xor statement is TRUE"   True because one, but not both statements is true.

EE(0, 0) = "The xor statement is false"   False because one of the statements is not false.

Figure 12.9 Boolean **Xor**

Adding lines to a program

It is possible to add nested programs inside a program. Creative use of nested programs sometimes makes it easier to create conditional programs.

If you are inside a program, Mathcad will add a new programming line or a new placeholder when you click the Add Line button. The location of the new line or placeholder depends where the vertical editing line is located and what information is selected. It is much easier to illustrate the behavior than try to describe it. In the following examples, notice what is selected and where the vertical editing line is located when the Add Line button is clicked.

1.

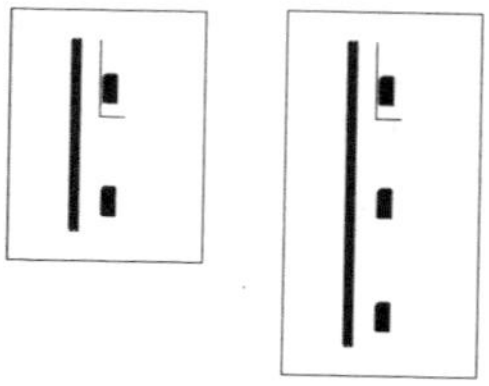

2.

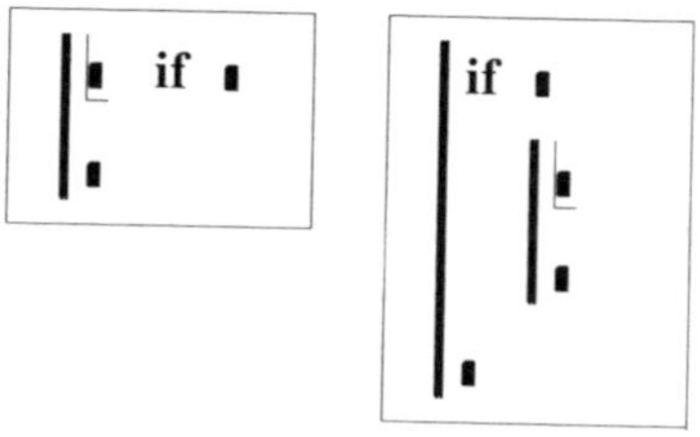

3.

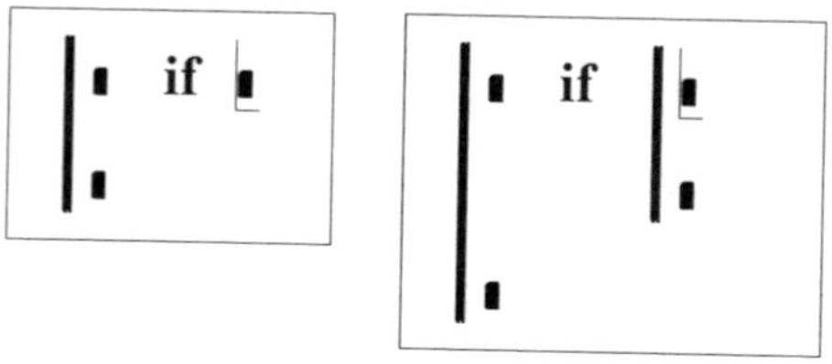

4.

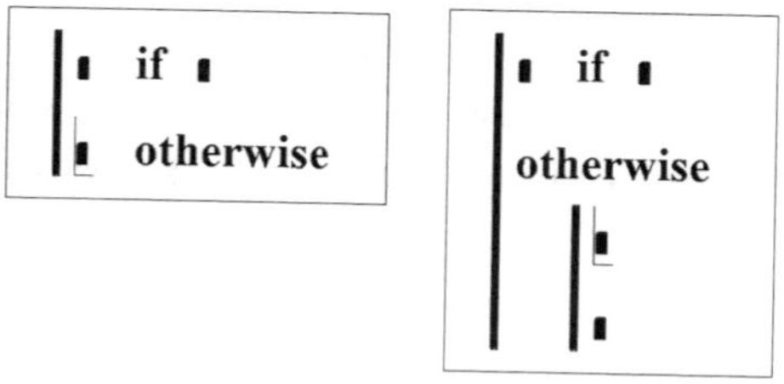

5.

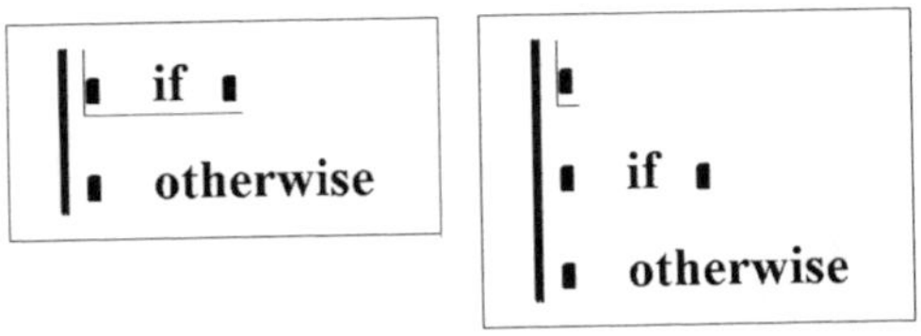

6.

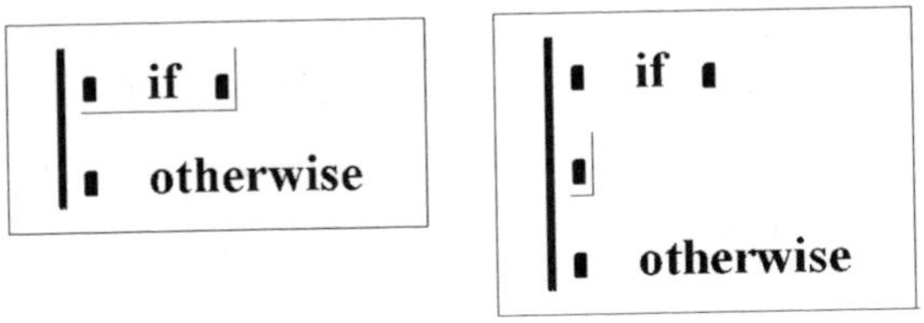

7.

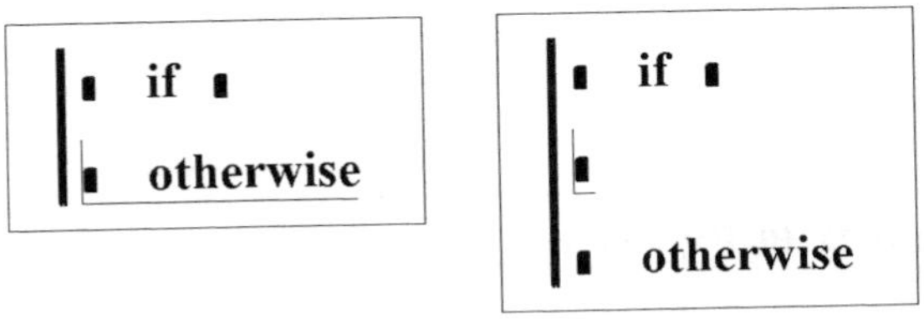

8.

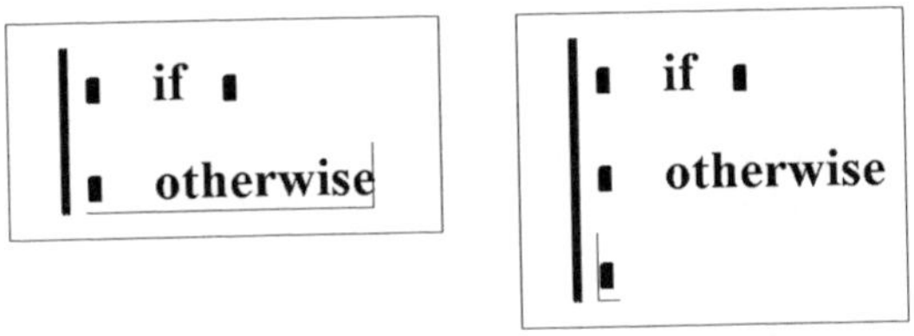

Now, let's illustrate how these different forms of the program work. In Figures 12.10–12.13 we try to arrive at consistent results using different forms of the programming lines. Notice how each program is a little different from the others, but still achieves the same result. In these examples, we use a range variable and vector math.

$i := 1, 2 .. 4$ $\quad x := \begin{pmatrix} 3 \\ 4 \\ -7 \\ -10 \end{pmatrix}$ $\quad y := \begin{pmatrix} 2 \\ -1 \\ 3 \\ -2 \end{pmatrix}$

```
■ if ■
■ if ■
■ otherwise
```

$$Test\_1_i := \left| \begin{array}{l} 100 \;\; \textbf{if} \;\; x_i \geq 0 \wedge y_i \geq 0 \\ 200 \;\; \textbf{if} \;\; x_i < 0 \wedge y_i \geq 0 \\ 300 \;\; \textbf{if} \;\; x_i \geq 0 \wedge y_i < 0 \\ 400 \;\; \textbf{otherwise} \end{array} \right.$$

This program uses the "And" Boolean operator. Both statements must be true in order for the line to be true.

$Test\_1_i =$

100.00
300.00
200.00
400.00

Figure 12.10 Programming Form 1

$x = \begin{pmatrix} 3.00 \\ 4.00 \\ -7.00 \\ -10.00 \end{pmatrix}$ $\quad y = \begin{pmatrix} 2.00 \\ -1.00 \\ 3.00 \\ -2.00 \end{pmatrix}$

```
■ if ■
■ otherwise
```

$$Test\_2_i := \left| \begin{array}{l} \textbf{if} \;\; x_i \geq 0 \\ \quad \left| \begin{array}{l} 100 \;\; \textbf{if} \;\; y_i \geq 0 \\ 200 \;\; \textbf{otherwise} \end{array} \right. \\ \textbf{if} \;\; x_i < 0 \\ \quad \left| \begin{array}{l} 300 \;\; \textbf{if} \;\; y_i \geq 0 \\ 400 \;\; \textbf{otherwise} \end{array} \right. \end{array} \right.$$

This form of the program allows you to have an additional conditional statement branching off of the first conditional statement.

$Test\_2_i =$

100.00
200.00
300.00
400.00

$Test\_1_i =$

100.00
300.00
200.00
400.00

Figure 12.11 Programming Form 2

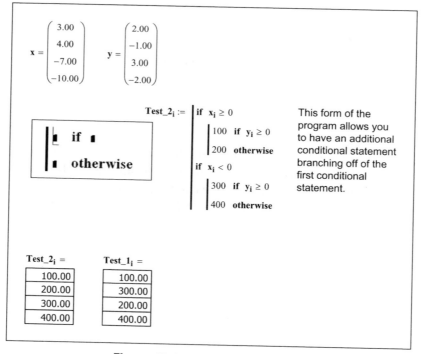

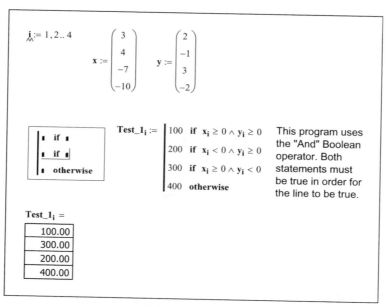

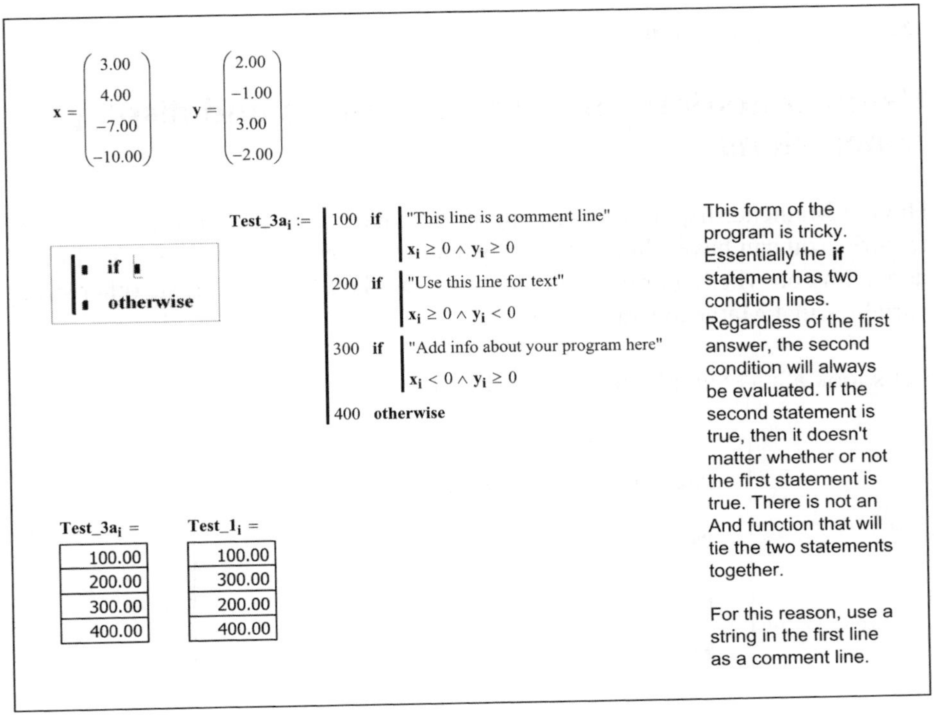

Figure 12.12 Programming Form 3

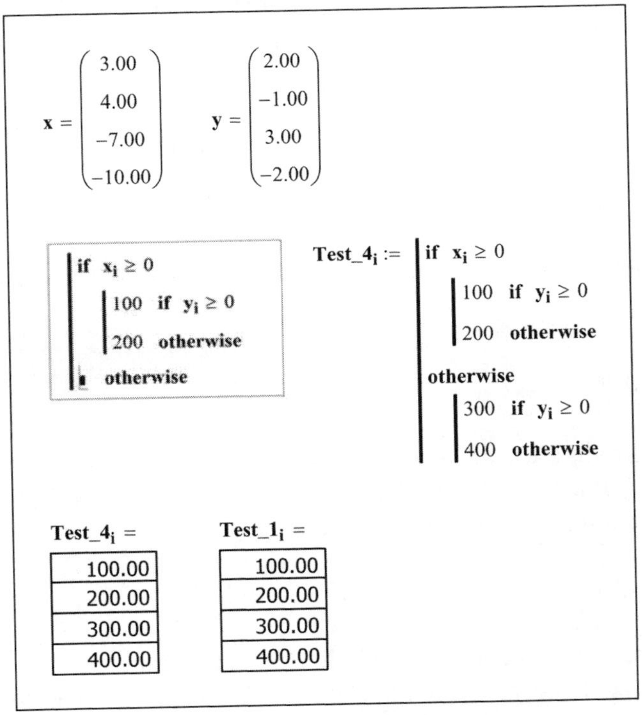

Figure 12.13 Programming Form 4

Using conditional programs to make and display conclusions

Having Mathcad display a statement at the conclusion of a problem is a great feature. You can have Mathcad display statements such as: "Passes" or "Fails," depending on whether or not the calculation worked. You can create other string variables that Mathcad can display.

Let's look at an example. See Figure 12.14.

Create three string variables.

Passes := "Passes - Solution Works"

Fails := "Fails - Retry"

Zero := "Result is Zero"

$i := 1, 2 .. 3$

$$\mathbf{Result} := \begin{pmatrix} -5 \\ 0 \\ 5 \end{pmatrix} \qquad \mathbf{Test}_i := \left| \begin{array}{l} \mathbf{Passes}\ \ \mathbf{if}\ \ \mathbf{Result}_i > 0 \\ \mathbf{Fails}\ \ \mathbf{if}\ \ \mathbf{Result}_i < 0 \\ \mathbf{Zero}\ \ \mathbf{otherwise} \end{array} \right.$$

$\mathbf{Test}_1$ = "Fails - Retry"

$\mathbf{Test}_2$ = "Result is Zero"

$\mathbf{Test}_3$ = "Passes - Solution Works"

Figure 12.14 Using conditional programming to display conclusions

Summary

Logical programs are used in place of the ***if*** function. They are easier to visualize than the ***if*** function, and they are easier to write and check. Chapter 16 will discuss programming in much more detail.

In Chapter 12 we:

- Explained how to create and add lines to a program.
- Emphasized the need to use the Programming tool bar to insert the programming operators. Typing the operators will not work.

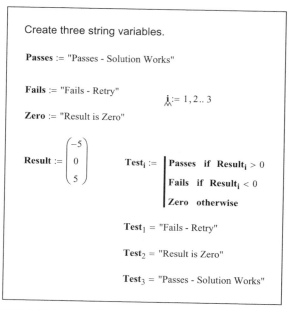

- Warned about checking the logic in the program so that a subsequent true statement does not change the result.
- Learned about Boolean operators.
- Showed how to insert the if operator into different locations within the program.
- Demonstrated how to use logic programs to draw and display conclusions.

Practice

1. Use the ***if*** function for the following logic:

    [a] Create variable x = 3.
    [b] If x = 3 the variable "Result" = 40.
    [c] If x is not equal to 3 then "Result" = 100.

2. Create a Mathcad program to achieve the same result as above.
3. Rewrite the program to place the if operator in a different location.
4. Write three Mathcad programs that use the Boolean operator **And**.
5. Write three Mathcad programs that use the Boolean operator **Or**.
6. Create and solve five problems from your field of study. At the conclusion of each problem, write a logical program to have Mathcad display a concluding statement depending on the result of the problem. Change input variables to ensure that the display is accurate for all conditions.

13 Useful information—Part II

This chapter is a collection of topics and information related to the chapters in Part II. There will not be a summary or practice exercises.

Vectors and matrices

Converting a range variable to a vector

Chapter 9 discusses the differences between range variables and vectors. It may be helpful to review the differences prior to reading the following procedures. The following is based on an article in the February 2002 issue of the Mathcad Advisor Newsletter.

There may be times when you want to convert a range variable to a vector so that you can access individual elements of the range variable. There are three methods available.

Method 1

This method manipulates the values of the range variable to create integer array subscripts beginning with ORIGIN.

Let's begin with a range variable i: = −3, −2.75.. 3. All the examples in this book use the ORIGIN set to 1, so the first element of the vector "V" will be V_1.

The goal in this method is to have the value of V_1 be equal to the first value of the range variable "i" (−3). The second element of the vector "V" will be V_2 and will be the second value of the range variable "i" (−2.75).

The array subscripts need to be integers beginning at ORIGIN, so we need to derive a relationship for the variables "a" and "b" such that $a*(-3)+b=1$ and $a*(-2.75)+b=2$, etc. Solving these two equations yields $a=4$ and $b=13$. Thus the first array subscript for the range "i" is $4*(-3)+13=1$. The general equation to derive the variable "a" is 1/I, where I is the increment of the range variable. The general equation to derive the variable "b" is ORIGIN-(R/I), where R is the first value of the range variable. See Figure 13.1 for an example.

$i := -3, -2.75 .. 3$

Create a range variable "i" beginning at -3, with an increment of 0.25, and ending with 3.

$V_{4 \cdot i+13} := i$

Next, create a vector such that:
V_1=1st value of i
V_2=2nd value of i
V_3=3rd value of i
etc, until the last value of i is reached.

See the discussion below.

$4 \cdot i + 13 =$	V =	1
1.00	1	-3.00
2.00	2	-2.75
3.00	3	-2.50
4.00	4	-2.25
5.00	5	-2.00
6.00	6	-1.75
7.00	7	-1.50
8.00	8	-1.25
9.00	9	-1.00
10.00	10	-0.75
11.00	11	-0.50
12.00	12	-0.25
13.00	13	0.00
14.00	14	0.25
15.00	15	0.50
16.00	16	0.75
17.00	17	1.00
18.00	18	1.25
19.00	19	1.50
20.00	20	1.75
21.00	21	2.00
22.00	22	2.25
23.00	23	2.50
24.00	24	2.75
25.00	25	3.00

Let R = the beginning value of the range variable "i," R = -3.
Let I = the increment of the range variable "i," i = 0.25.
We need to derive an equation so that for every value in the range variable "i," we arrive at an integer beginning with ORIGIN. In this worksheet ORIGIN is set to 1.
The general equation for the first value is a*R+b=ORIGIN.
The general equation for the second value is a*(R+I)+b=ORIGIN+1.
The general solution to these equations is a=1/I and b=ORIGIN-R/I.
For the range variable "i," a=1/0.25=4.0 and b=1-(-3.0/0.25)=1+12=13.
Now create a vector "V" with the first element equal to the first value of the range variable "i." The equation for the subscript operator is V_{4*i+13}.

rows(V) = 25.00 There are 25 rows in the vector "V"

$V_1 = -3.00$ $V_{25} = 3.00$

Figure 13.1 Converting a range variable into a vector—method 1

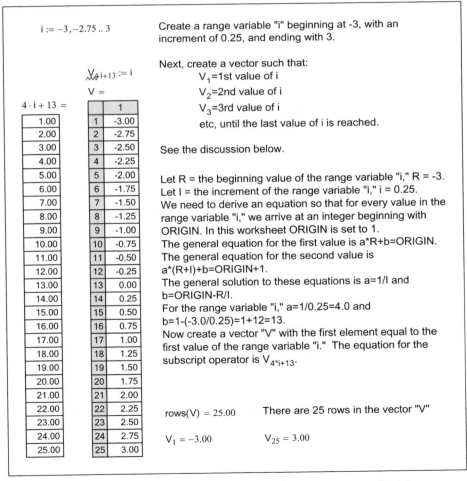

Method 2

This method begins with a new range variable starting at ORIGIN and incrementing by one to the number of elements in the original range variable. The values assigned to the vector are manipulated so that they equal the values of the original range variable.

The first step is to determine a formula to calculate the range variable values. This method creates a new range variable "j" beginning at ORIGIN and incrementing by one. We will use a similar formula as above: a*ORIGIN+b = R and a*(ORIGIN+1)+b = R+I. Solving for these two equations yields a = I and b = R−ORIGIN*I. For the same range variable "i" as in method 1, a = 0.25 and b = −3.25. Now create a vector "W" whose value use the equation: 0.25*j−3.25. See Figure 13.2.

We want to convert the range variable "i" from Figure 13.1. This range variable begins at R = -3, has an increment I=0.25, and ends at 3.0. This range variable has 25 elements

$j := 1 .. 25$

Create a new range variable "j" beginning at 1, with an increment of 1.0, and ending with 25.

$W_j := 0.25 \cdot j - 3.25$

W =

	1
1	-3.00
2	-2.75
3	-2.50
4	-2.25
5	-2.00
6	-1.75
7	-1.50
8	-1.25
9	-1.00
10	-0.75
11	-0.50
12	-0.25
13	0.00
14	0.25
15	0.50
16	0.75
17	1.00
18	1.25
19	1.50
20	1.75
21	2.00
22	2.25
23	2.50
24	2.75
25	3.00

Create a new vector "W," whose first element W_1 is equal to the first value of the range variable "i" (-3.0).

Let R = the beginning value of the range variable "i," R = -3.
Let I = the increment of the range variable "i," i = 0.25.

We need to derive an equation so that for values 1, 2, 3, etc the results are R, R+I,R+2*I etc.

The general form for this equation is a*j+b, where j is a range variable from 1 to 25.

The general equation for the first value is a*(ORIGIN)+b=R.
The general equation for the second value is a*(ORIGIN+1)+b=R+I.
The general solution to these equations is a=I and b=R-(ORIGIN*I).

For the range variable "i," a=0.25 and b=-3-(1*0.25)=-3-0.25=-3.25.
Now create a vector "W" with the first element equal to the first value of the range variable "i."

W_j=0.25*j-3.25

rows(W) = 25.00

Figure 13.2 Converting a range into a vector—method 2

Method 3

The third method creates a user-defined function called Range2Vec. It uses some programming features that have not yet been introduced. This method is illustrated in Figure 16.8.

Sorting arrays

The following are sorting functions that can be used with vectors or matrices:

- ***sort***(v): Returns a vector with the values from "v" sorted in ascending order.
- ***reverse***(A): Reverses the order of elements in a vector, or the order of rows in a matrix A.
- ***csort***(A,n): Returns an array formed by rearranging rows of A until column n is in ascending order.
- ***rsort***(A,n): Returns an array formed by rearranging columns of A until row n is in ascending order.
- ***reverse***(sort(v)): Sorts in descending order.

Refer to Mathcad Help for additional information. The September 2004 Mathcad Advisor Newsletter also has an excellent article on sorting functions.

Functions

Namespace operator

The namespace operator was introduced in Mathcad version 12.

If you redefine a unit or a built-in variable, the new variable definition is used by Mathcad and the old variable definition is no longer available. The namespace operator allows you to reference the original value.

To use the namespace operator type the name of the variable and then type **CTRL+SHIFT+N**. This places bracketed subscripts below the variable name. There are four module names that can be used within the brackets.

- mc: For any built-in Mathcad function or built-in dimensionless constant such as e or π.
- unit: For any built-in Mathcad unit or dimensioned constant.
- doc: Referring to the most recent previous definition in the document.
- user: For any function in a UserDLL (Dynamic Link Library).

For example, in Figures 13.1 and 13.2 the unit V was redefined as a vector. Using the namespace operator it is possible to still use the unit V = volt. If any unit is redefined, use the namespace operator to get the original Mathcad definition. You can also use the namespace operator in a built-in Mathcad function which is redefined. See Figure 13.3 for examples.

Assign a voltage:

$\text{Voltage}_1 := 3V$ — We think we are assigning a voltage to a number

This gives an incorrect result because "V" was redefined in Figure 13.1 to be a vector.

$\text{Voltage}_1 =$

	1
1	-9.00
2	-8.25
3	-7.50
4	-6.75
5	-6.00
6	-5.25
7	-4.50
8	-3.75
9	-3.00
10	-2.25
11	-1.50
12	-0.75
13	0.00
14	0.75
15	1.50
16	2.25

In order to assign a voltage, we must now use the namespace operator.

$\text{Voltage}_2 := 3V_{[\text{unit}]}$

$\text{Voltage}_2 = 3.00\ \text{V}$

$m := 50\text{kg}$ — If the mass of an object is defined as a variable "m", it redefines the built-in variable "m" for meters.

$\text{Length}_1 := 50\text{m}$ $\quad \text{Length}_1 = 2500.00\ \text{kg}$ — The variable "Length" is now assigned a mass.

$\text{Length}_2 := 50\text{m}_{[\text{unit}]}$ $\quad \text{Length}_2 = 50.00\ \text{m}$ — The namespace operator can be used to tell Mathcad to use meter if the built-in unit "m" has been redefined.

<u>Using the namespace operator for a built-in Mathcad function.</u>

$\text{sin}(x) := 4 \cdot x$ — The function sin is redefined

$\text{sin}\left(\frac{\pi}{4}\right) = 3.14$ — The original Mathcad function is no longer available.

$\text{sin}_{[\text{mc}]}\left(\frac{\pi}{4}\right) = 0.71$ — Using namespace operator CTRL+SHIFT+N, you can use the original Mathcad function for sin.

$\text{sin}(x) := \text{sin}_{[\text{mc}]}(x)$ — In order to reassign the function sin to its original function, use the namespace operator with the mc designation.

$\text{sin}\left(\frac{\pi}{4}\right) = 0.71$ — The original sin function is now reassigned.

Figure 13.3 Namespace operator CTRL+SHIFT+N

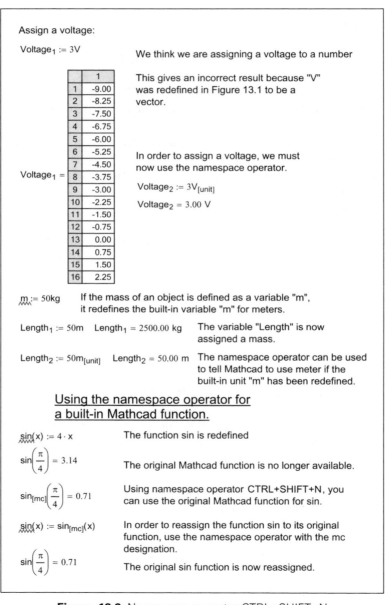

Even though this operator is available, it is still best to follow the practice of not overwriting built-in Mathcad units, functions, or constants.

Error *function*

The ***error()*** function allows you to create your own custom error messages. This is useful when you are writing user-defined functions, or if you are creating Mathcad programs. See Figure 13.4 for examples.

Using the Error function within a function

$$\text{ErrorTest}(a) := \text{if}\left(a > 0, \frac{1}{a}, \text{error}(\text{"Argument must be greater than zero"})\right)$$

$\text{ErrorTest}(2) = 0.50$

A custom error message appears when you use a negative number as an argument

$\text{ErrorTest}(3) = 0.33$

$\text{ErrorTest}(-3) = ▮$

ErrorTest(−3) = ▮▮

Argument must be greater than zero

Using the Error function within a program

$$\text{ErrorTest\_2}(x) := \begin{array}{|l} \text{error}(\text{"You must enter a positive number"}) \;\; \text{if} \;\; x \le 0 \\ \dfrac{1}{\sqrt{x}} \;\; \text{otherwise} \end{array}$$

$\text{ErrorTest\_2}(4) = 0.50$

A custom error message appears when you use a negative number as an argument

$\text{ErrorTest\_2}(9) = 0.33$

$\text{ErrorTest\_2}(-4) = ▮$

ErrorTest_2(−4) = ▮▮

You must enter a positive number

Figure 13.4 Error function

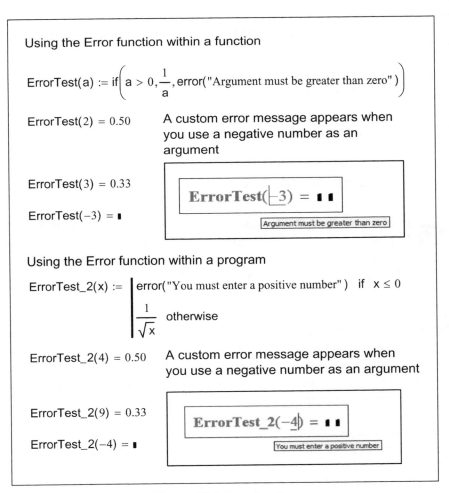

String functions

Mathcad has many string functions that can add to or manipulate strings. These will not be described in any detail, but are presented to inform you of Mathcad's ability to work with strings. For further information, refer to Mathcad Help. The following can be done with strings:

- Add strings together to form one string using concat().
- Return only a portion of a string using substr().
- Tell where a certain phrase begins in a string using search().
- Return how many characters are in a string using strlen().
- Convert a scalar into a string using num2str().
- Convert a number to a string using str2num().
- Convert a string to a vector of ANSI codes using str2vec().
- Convert a vector of integer ANSI codes to a string using vec2str().
- Ask Mathcad if a variable is a string using IsString().

The November 2002 Mathcad Advisor Newsletter has an excellent article on the use of the string functions.

Picture functions and image processing

You can insert graphic images into Mathcad. Mathcad supports the following graphic types: BMP, GIF, JPG, PCX, and TGA.

> **Tip!** Mathcad currently does not support some digital camera images. If Mathcad does not load a digital camera image, open the file in an editing program and resave the file as a JPEG file in JFIF format. Mathsoft is aware of the problem, and is working on a solution.

To insert a picture into Mathcad, click **Picture** from the **Insert** menu, or type `CTRL+T`. This inserts a square region with a placeholder in the lower left corner. Enter a string containing the path and file name of the picture you want to insert. Unfortunately, there is not a browse command with this feature. The string with the path and filename can be assigned to a variable. This variable can then be used in the picture placeholder.

Once the picture is inserted, if you click on the picture, Mathcad opens the Picture Toolbar. From this toolbar you can modify the orientation, brightness, contrast, magnification, and grayscale/color mapping. You can hide the path and filename by right-clicking on the image and selecting "Hide Arguments" from the pop-up menu.

There are many other Mathcad functions that can be used with picture images. Refer to Mathcad help for additional information.

Curve fitting, regression, and data analysis

Mathcad has numerous curve fitting and data analysis functions. It has functions for linear regression, generalized regression, polynomial regression, and specialized regression. There is even a Data Analysis Extension Pack that can be purchased. This adds additional capacity to analyze data. The topic of data analysis can fill chapters. These topics are beyond the scope of this book. For additional information, you are referred to: Mathcad Help, QuickSheets, and Tutorial. There are also many resources available at www.mathcad.com such as the Mathcad Advisory Newsletter.

Complex numbers, polar coordinates, and mapping functions

Complex numbers

Mathcad recognizes either i or j to represent the imaginary portion of a complex number. When entering complex numbers, you must use a number in front of the i or j. For example type 1+1i or 1+1j, do not type 1+i or 1+j. Also, do not type 1+1*i or 1+1*j. If you type either i or j by itself, or if you use a multiplication, then Mathcad will look for a variable i or j. Once you type 1i or 1j, the number 1 will disappear and only show an i or j. You can choose to display complex results with either i or j. This is controlled by the Display Options tab in the Result Format dialog box, found on the **Format** menu.

Mathcad has several functions for working with complex numbers. The easy to remember ones are: ***Re(Z)*** returns the real part of Z and ***Im(Z)*** returns the imaginary part of Z.

Mapping functions

Mathcad has several mapping functions that are used in 2D and 3D plotting. These functions convert from rectangular coordinates to polar coordinates [***xy2pol***(x,y)], spherical coordinates [***xyz2sph***(x,y,z)], and to cylindrical coordinates [***xyz2cyl***(x,y,z)]. These functions need unitless numbers, and the results are returned as a vector with unitless radius and angles in radians. The inverse to these functions are ***pol2xy***(r,theta), ***sph2xy***(r,theta,phi), and ***cyl2xy***(r,q,f).

Polar notation

It is possible to use complex numbers to convert between rectangular and polar coordinates, and to add rectangular coordinates. The following procedure was featured in the February 2004 Mathcad Advisor Newsletter.

Let the real part of the complex number represent the x-coordinate and the imaginary part represent the y-coordinate. You can now add or subtract a series of x- and y-coordinates by adding and subtracting the complex numbers.

Now let's see how to convert to polar coordinates. Let the complex number be represented by "z." The radius for polar coordinates is the absolute value of "z." The angle is calculated using the function ***arg***(z). The angle is measured from π to $-\pi$. The result can also be displayed in degrees, if you add deg to the unit placeholder. You can create a user-defined function to convert from polar coordinates back to polar notation. See Figure 13.5 for the procedure.

Let the x-coordinate be represented by the real part and the y-coordinate be represented by the imaginary part.

$PolarExample_1 := 3 + 3i$

$Radius_1 := |PolarExample_1| \qquad Radius_1 = 4.24$

$Angle_1 := arg(PolarExample_1) \qquad Angle_1 = 0.79 \qquad Angle_1 = 45.00\,deg$

$PolarExample_2 := -3 - 3i$

$Radius_2 := |PolarExample_2| \qquad Radius_2 = 4.24$

$Angle_2 := arg(PolarExample_2) \qquad Angle_2 = -2.36 \qquad Angle_2 = -135.00\,deg$

This polar notation allows you to add x and y coordinates and then get a final radius and angle.

$PolarExample_3 := (3 + 3i) + (4 + 12i) + (3 - 5i) \qquad PolarExample_3 = 10.00 + 10.00i$

$Radius_3 := |PolarExample_3| \qquad Radius_3 = 14.14$

$Angle_3 := arg(PolarExample_3) \qquad Angle_3 = 0.79 \qquad Angle_3 = 45.00\,deg$

Create a user defined function to convert from polar coordinates to a complex number.

$i := i$ Reset i from the range variable used in Figure 13.1

$P2i(mag, angle) := |mag| \cdot (\cos(angle \cdot deg) + \sin(angle \cdot deg)\, i)$

$P2i(2, 45) = 1.41 + 1.41i$

$P2i(10\sqrt{2}, -135) = -10.00 - 10.00i$

Assign a variable so that results can be reused.

$PolarExample_4 := P2i(20, 145) \qquad PolarExample_4 = -16.38 + 11.47i$

Figure 13.5 Polar notation

Angle functions

It may be useful to compare the results of Mathcad's various angle functions: ***angle***(x,y), ***atan***(y/x), ***atan2***(x,y), and ***arg***(x+yi). These functions return the angle from the x-axis to a line going through the origin and the point (x,y).

angle(x,y)**:** Returns the angle in radians between 0 and 2π, excluding 2π.
atan(y/x)**:** Returns the angle in radians between $-\pi/2$ and $\pi/2$.
atan2(x,y)**:** Returns the angle in radians between $-\pi$ and π, excluding $-\pi$.
arg(x+yi)**:** Returns the angle in radians between $-\pi$ and π, excluding $-\pi$. The only difference between this function and ***atan2***(x,y) is the way the arguments are input.

Plots

Plotting a range over a log scale

If you are plotting a series of points using a log scale, you will want to have your points closer together the closer you get to zero, and further apart the further you get away from zero. For example, if you are plotting from 0.001 to 10,000, you need to have a very small increment in order to see the points near the beginning of the plot. However, if you select a small increment for a range variable, then you will be plotting millions of unnecessary points as you move toward 10,000.

The solution to this is to use a variable plotting range. The following procedure is discussed in the December 2002 Mathcad Advisor Newsletter.

Let's define three variables: low, high, and interval for the first plot value the last plot value and the number of intervals to use. See Figures 13.6 and 13.7 for the example.

Most plot ranges occur over a uniform increment. This example uses a variable increment.

Low := 0.001 High := 10000 Intervals := 20

$$\text{Step} := \frac{\log\left(\frac{\text{High}}{\text{Low}}\right)}{\text{Intervals}} \quad \text{Step} = 0.3500 \quad \frac{\text{High}}{\text{Low}} = 1.00 \times 10^{7} \quad \log\left(\frac{\text{High}}{\text{Low}}\right) = 7.00000$$

r := 1, 2 .. Intervals + 1 Range variable (for this example ORIGIN is set to 1)

Create a vector of values to be plotted. Notice that as "r" gets larger, so does the distance between each point.

$X_r := \text{Low} \cdot 10^{(r-1)\cdot \text{Step}}$

X =

	1
1	0.00100
2	0.00224
3	0.00501
4	0.01122
5	0.02512
6	0.05623
7	0.12589
8	0.28184
9	0.63096
10	1.41254
11	3.16228
12	7.07946
13	15.84893
14	35.48134
15	79.43282
16	177.82794
17	398.10717
18	891.25094
19	1995.26231
20	4466.83592

Figure 13.6 Plotting using a variable range

Continued from Figure 13.6

Function(a) := a

$$s := 0.001, \frac{10000 - 0.001}{20} .. 10000$$

Uniform Range

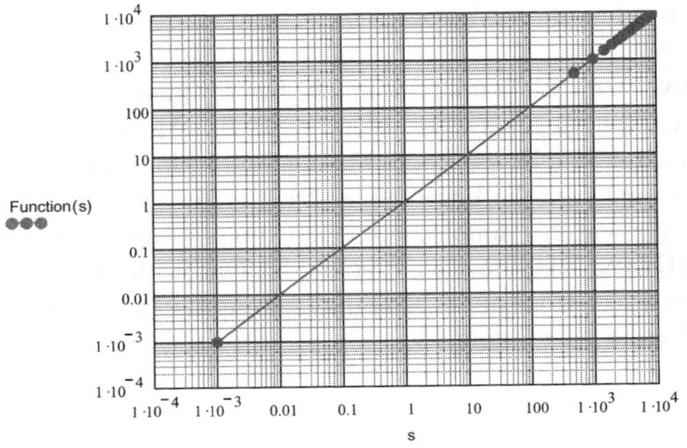

20 Uniform Increments

s =

	1
1	0.0
2	500.0
3	1000.0
4	1500.0
5	2000.0
6	2500.0
7	3000.0
8	3500.0
9	4000.0
10	4500.0
11	5000.0
12	5500.0
13	6000.0
14	6500.0
15	7000.0
16	7500.0

Variable Range

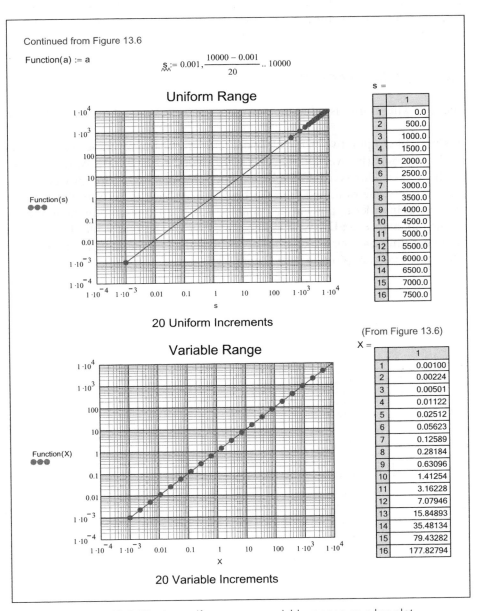

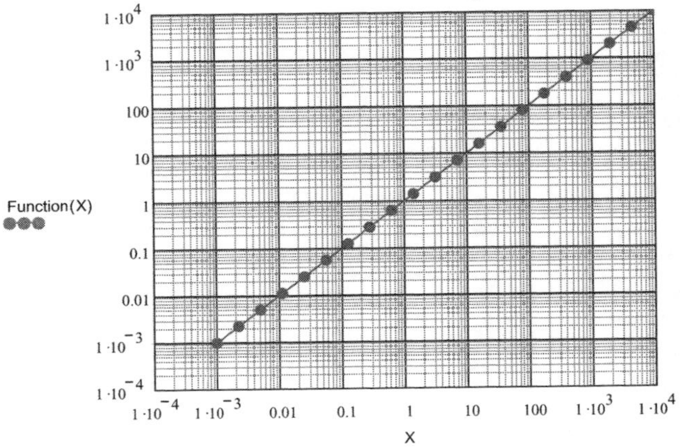

20 Variable Increments

(From Figure 13.6)

X =

	1
1	0.00100
2	0.00224
3	0.00501
4	0.01122
5	0.02512
6	0.05623
7	0.12589
8	0.28184
9	0.63096
10	1.41254
11	3.16228
12	7.07946
13	15.84893
14	35.48134
15	79.43282
16	177.82794

Figure 13.7 Plotting uniform versus variable ranges on a log plot

Plotting conics

The January 2002 issue of the Mathcad Advisor Newsletter has an excellent discussion on the graphing of circles, ellipses, and hyperbolas. Please refer to this article for information on this topic.

Plotting a family of curves

The following is discussed in the November 2001 Mathcad Advisor Newsletter.

It is possible to plot two parameters on a 2D graph. For example, you can plot F(a,b): = a^2*cos(b) over the ranges a = 0,1..4 and b = 0,0.1..15, with b plotted on the x-axis. This is actually a single trace. Mathcad will plot all values of b for a single value of a, and then the line doubles back to zero to plot the next value of a. In order to prevent this double-back line from plotting, double-click on the plot and select **Draw** from the trace type on the Traces tab. Be sure that the upper limit on your x-axis does not exceed the maximum value in the range over which you are plotting. See Figure 13.8.

Family(a, b) := $a^2 \cdot \cos(b)$ a := 0, 1 .. 4   b := 0, 0.1 .. 15

Since this is actually only one trace, the lines double back for each value of "a" plotted.

Trace type set to "Line"

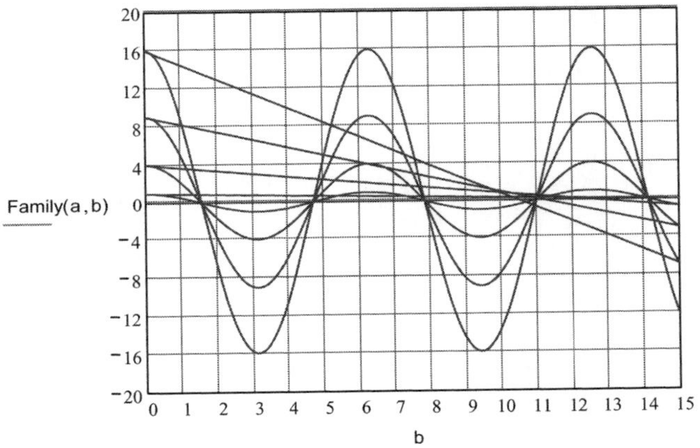

In order to prevent the extra lines, select "Draw" from the trace type.

Trace type set to "Draw"

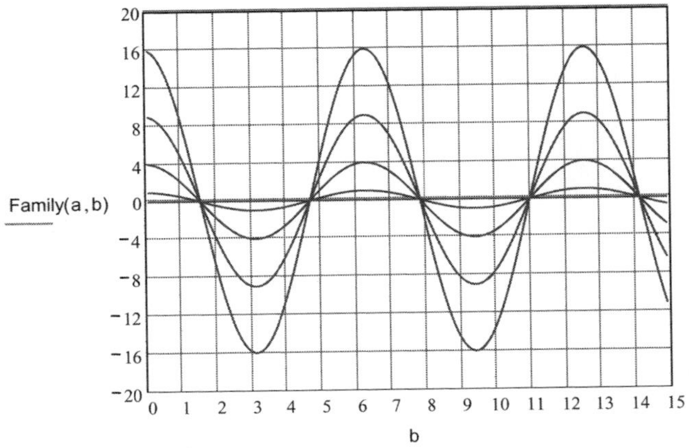

Figure 13.8 Plotting a family of curves

Part III

Power Tools for your Mathcad Toolbox

You have just filled your Mathcad toolbox with many simple, yet useful tools. Part III will now add some very powerful Mathcad tools.

The purpose of Part II was to introduce features and functions that were not too confusing. It also gave simple examples. The features and functions introduced in Part III tend to be more complex and require more complex examples. We postponed the discussion of these topics until after you had a thorough understanding of Mathcad basics.

Part III discusses the powerful topics of symbolic calculations, root finding, solve blocks, and advanced Mathcad programming.

14 Introduction to symbolic calculations

Symbolic calculations returns algebraic results rather than numeric results. Mathcad has a sophisticated symbolic processor that can solve very complex problems algebraically. The symbolic processor is different from the numeric processor.

The intent of this chapter is to wet your appetite for symbolic calculations. We will only cover a few of the topics in symbolic calculations. The topics we discuss are useful to help solve some basic engineering problems. We will not get into solving complex mathematical equations.

As always, the Mathcad Help and Tutorials provide excellent examples of topics not covered in this chapter.

Chapter 14 will:

- Introduce symbolic calculations.
- Begin by showing how to solve a polynomial expression.
- Show how to solve polynomial expressions for algebraic and numeric solutions.
- Demonstrate how to get quick static symbolic results by using the Symbolics menu commands.
- Discuss live symbolics and use of the symbolic equal sign.
- Tell how to get numeric rather than algebraic results.
- Show how to use the "explicit" keyword to display the values of variables used in your expressions.
- Demonstrate how to string a series of symbolic keywords.

Getting started with symbolic calculations

Let's get started by showing how simple it is to solve for the roots of a polynomial expression. Figure 14.1 shows how to solve for symbolic results with variables as coefficients. You can also solve equations with numbers as coefficients. This is shown in Figure 14.2.

To solve for the following expression, select either one of the instances of variable "x" in the below equation. From the **Symbolics** menu, hover your mouse over **Variable**, and click **Solve**.

$$a \cdot x^2 + b \cdot x + c$$

$$\begin{bmatrix} \frac{1}{2 \cdot a} \cdot \left[(-b) + \left(b^2 - 4 \cdot a \cdot c\right)^{\frac{1}{2}}\right] \\ \frac{1}{2 \cdot a} \cdot \left[(-b) - \left(b^2 - 4 \cdot a \cdot c\right)^{\frac{1}{2}}\right] \end{bmatrix}$$

Mathcad returns both algebraic solutions to the expression. You will recognize the solutions as both solutions to the quadratic equation.

The below examples use the same procedure, but each example selects a different variable to solve for.

Solving for "a"	Solving for "b"	Solving for "c"
$a \cdot x^2 + b \cdot x + c$	$a \cdot x^2 + b \cdot x + c$	$a \cdot x^2 + b \cdot x + c$
$\frac{-(b \cdot x + c)}{x^2}$	$\frac{-\left(a \cdot x^2 + c\right)}{x}$	$(-a) \cdot x^2 - b \cdot x$

Figure 14.1 Symbolically solving for a variable

Use the same procedure as described in Figure 14.1. The solutions can be real or complex. The symbolic processor does not default to decimal answers.

$$x^2 + 5 \cdot x + 6 \qquad x^2 + x + 3 \qquad 6 \cdot x + 2 \qquad x^3 + 1$$

$$\begin{pmatrix} -3 \\ -2 \end{pmatrix} \qquad \begin{pmatrix} \frac{-1}{2} + \frac{1}{2} \cdot i \cdot 11^{\frac{1}{2}} \\ \frac{-1}{2} - \frac{1}{2} \cdot i \cdot 11^{\frac{1}{2}} \end{pmatrix} \qquad \frac{-1}{3} \qquad \begin{pmatrix} -1 \\ \frac{1}{2} - \frac{1}{2} \cdot i \cdot 3^{\frac{1}{2}} \\ \frac{1}{2} + \frac{1}{2} \cdot i \cdot 3^{\frac{1}{2}} \end{pmatrix}$$

Figure 14.2 Using the symbolic processor to solve for numeric results

The results from using the symbolic solve with variable coefficients can become very complex as can be seen in Figure 14.3.

$$x^3 + x + 1$$

$$\begin{bmatrix} \frac{-1}{6} \cdot \left(108 + 12 \cdot 93^{\frac{1}{2}}\right)^{\frac{1}{3}} + \frac{2}{\left(108 + 12 \cdot 93^{\frac{1}{2}}\right)^{\frac{1}{3}}} \\ \frac{1}{12} \cdot \left(108 + 12 \cdot 93^{\frac{1}{2}}\right)^{\frac{1}{3}} - \frac{1}{\left(108 + 12 \cdot 93^{\frac{1}{2}}\right)^{\frac{1}{3}}} + \frac{1}{2} \cdot i \cdot 3^{\frac{1}{2}} \cdot \left[\frac{-1}{6} \cdot \left(108 + 12 \cdot 93^{\frac{1}{2}}\right)^{\frac{1}{3}} - \frac{2}{\left(108 + 12 \cdot 93^{\frac{1}{2}}\right)^{\frac{1}{3}}}\right] \\ \frac{1}{12} \cdot \left(108 + 12 \cdot 93^{\frac{1}{2}}\right)^{\frac{1}{3}} - \frac{1}{\left(108 + 12 \cdot 93^{\frac{1}{2}}\right)^{\frac{1}{3}}} - \frac{1}{2} \cdot i \cdot 3^{\frac{1}{2}} \cdot \left[\frac{-1}{6} \cdot \left(108 + 12 \cdot 93^{\frac{1}{2}}\right)^{\frac{1}{3}} - \frac{2}{\left(108 + 12 \cdot 93^{\frac{1}{2}}\right)^{\frac{1}{3}}}\right] \end{bmatrix}$$

Figure 14.3 Solving for a variable

These examples show how easily you can solve for any variable in your equation. There is a drawback to this approach however. As you can see, the results are static. If you change the equation, the result does not change. This can cause errors in your calculations. It would be much nicer to have Mathcad provide a dynamic result.

Mathcad does provide this capability! It is called the symbolic equal sign or live symbolics. You can get similar results as we did above, but with the advantage of being updated automatically whenever there is a change to the equations.

The symbolic equal sign is an arrow pointing to the right $\rightarrow$. The symbolic equal sign is located on the Symbolic toolbar. You can open this toolbar from the **View** menu, or you can click the  icon on the Math toolbar. In order to solve an equation with the symbolic equal sign, you need to include a keyword with it. The keyword tells Mathcad what you want to do symbolically. In our case, the keyword is "solve." The keyword is located in a placeholder located just before the symbolic equal sign ▪→ . To insert the symbolic equal sign with the keyword placeholder, click the icon on the Symbolic toolbar or type **CTRL+SHIFT+PERIOD**. You also need to tell Mathcad which variable to solve for. To do this, type a comma after the keyword, followed by the name of the variable to solve for.

Let's look at how you can use the symbolic equal sign to solve for the same equations as in Figures 14.1–14.3. See Figures 14.4–14.6.

These examples are the same as used in Figure 14.1, but using live symbolics. If you change the function then the symbolic result will also change.

Follow these steps to solve using the symbolic equal sign:
1. Type the equation
2. Click the symbolic equal sign with a placeholder, or type CTRL+SHIFT+PERIOD.
3. Click in the placeholder
4. Type "solve,x"
5. Click outside the region

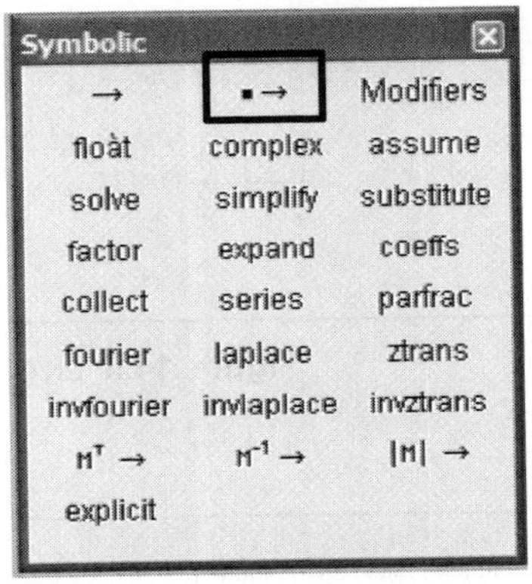

$$\mathbf{a}\cdot\mathbf{x}^2+\mathbf{b}\cdot\mathbf{x}+\mathbf{c}\ \textbf{solve},\mathbf{x}\ \rightarrow \begin{bmatrix} \dfrac{1}{2\cdot\mathbf{a}}\cdot\left[(-\mathbf{b})+\left(\mathbf{b}^2-4\cdot\mathbf{a}\cdot\mathbf{c}\right)^{\frac{1}{2}}\right] \\ \dfrac{1}{2\cdot\mathbf{a}}\cdot\left[(-\mathbf{b})-\left(\mathbf{b}^2-4\cdot\mathbf{a}\cdot\mathbf{c}\right)^{\frac{1}{2}}\right] \end{bmatrix}$$

Solving for "a"

$$\mathbf{a}\cdot\mathbf{x}^2+\mathbf{b}\cdot\mathbf{x}+\mathbf{c}\ \textbf{solve},\mathbf{a}\ \rightarrow \frac{-(\mathbf{b}\cdot\mathbf{x}+\mathbf{c})}{\mathbf{x}^2}$$

Solving for "b"

$$\mathbf{a}\cdot\mathbf{x}^2+\mathbf{b}\cdot\mathbf{x}+\mathbf{c}\ \textbf{solve},\mathbf{b}\ \rightarrow \frac{-\left(\mathbf{a}\cdot\mathbf{x}^2+\mathbf{c}\right)}{\mathbf{x}}$$

Solving for "c"

$$\mathbf{a}\cdot\mathbf{x}^2+\mathbf{b}\cdot\mathbf{x}+\mathbf{c}\ \textbf{solve},\mathbf{c}\ \rightarrow (-\mathbf{a})\cdot\mathbf{x}^2-\mathbf{b}\cdot\mathbf{x}$$

Figure 14.4 Live symbolics

These equations are the same as in Figure 14.2, but they are now dynamic. If you change the equation, then the result will also change.

$$x^2 + 5x + 6 \textbf{ solve}, x \rightarrow \begin{pmatrix} -3 \\ -2 \end{pmatrix} \qquad 6x + 2 \textbf{ solve}, x \rightarrow \frac{-1}{3}$$

$$x^2 + x + 3 \textbf{ solve}, x \rightarrow \begin{pmatrix} \frac{-1}{2} + \frac{1}{2} \cdot i \cdot 11^{\frac{1}{2}} \\ \frac{-1}{2} - \frac{1}{2} \cdot i \cdot 11^{\frac{1}{2}} \end{pmatrix} \qquad x^3 + 1 \textbf{ solve}, x \rightarrow \begin{pmatrix} -1 \\ \frac{1}{2} - \frac{1}{2} \cdot i \cdot 3^{\frac{1}{2}} \\ \frac{1}{2} + \frac{1}{2} \cdot i \cdot 3^{\frac{1}{2}} \end{pmatrix}$$

Figure 14.5 Live symbolics

This is the same equation as in Figure 14.3, except that it is dynamic.
If you change the equation, then the result will also change.

$$x^3 + x + 1 \textbf{ solve}, x \rightarrow \begin{bmatrix} \frac{-1}{6} \cdot \left(108 + 12 \cdot 93^{\frac{1}{2}}\right)^{\frac{1}{3}} + \frac{2}{\left(108 + 12 \cdot 93^{\frac{1}{2}}\right)^{\frac{1}{3}}} \\ \frac{1}{12} \cdot \left(108 + 12 \cdot 93^{\frac{1}{2}}\right)^{\frac{1}{3}} - \frac{1}{\left(108 + 12 \cdot 93^{\frac{1}{2}}\right)^{\frac{1}{3}}} + \frac{1}{2} \cdot i \cdot 3^{\frac{1}{2}} \cdot \left[\frac{-1}{6} \cdot \left(108 + 12 \cdot 93^{\frac{1}{2}}\right)^{\frac{1}{3}} - \frac{2}{\left(108 + 12 \cdot 93^{\frac{1}{2}}\right)^{\frac{1}{3}}}\right] \\ \frac{1}{12} \cdot \left(108 + 12 \cdot 93^{\frac{1}{2}}\right)^{\frac{1}{3}} - \frac{1}{\left(108 + 12 \cdot 93^{\frac{1}{2}}\right)^{\frac{1}{3}}} - \frac{1}{2} \cdot i \cdot 3^{\frac{1}{2}} \cdot \left[\frac{-1}{6} \cdot \left(108 + 12 \cdot 93^{\frac{1}{2}}\right)^{\frac{1}{3}} - \frac{2}{\left(108 + 12 \cdot 93^{\frac{1}{2}}\right)^{\frac{1}{3}}}\right] \end{bmatrix}$$

Figure 14.6 Live symbolics

The previous examples only used equations. You may also create user-defined functions using symbolic math. You do this by naming the user-defined function and listing its arguments. You then enter the equation, the keyword "solve," and the variable you want to solve for. You can now add numeric arguments in the function and the function will return numeric results. See Figure 14.7.

You can assign a user-defined function in conjunction with symbolic calculations.

The function below solves for both roots of a quadratic equation.

$$\mathbf{Function_1(a,b,c) := a \cdot x^2 + b \cdot x + c\ solve, x} \rightarrow \begin{bmatrix} \frac{1}{2 \cdot a} \cdot \left[(-b) + \left(b^2 - 4 \cdot a \cdot c\right)^{\frac{1}{2}}\right] \\ \frac{1}{2 \cdot a} \cdot \left[(-b) - \left(b^2 - 4 \cdot a \cdot c\right)^{\frac{1}{2}}\right] \end{bmatrix}$$

$$\mathbf{Function_1}(1,5,6) = \begin{pmatrix} -2.00 \\ -3.00 \end{pmatrix}$$

$$\mathbf{Function_1}(2,3,4) = \begin{pmatrix} -0.75 + 1.20i \\ -0.75 - 1.20i \end{pmatrix}$$

You can also create functions from the other equations used in Figure 14.4.

$$\mathbf{Function_2(b,c,x) := a \cdot x^2 + b \cdot x + c\ solve, a} \rightarrow \frac{-(b \cdot x + c)}{x^2}$$

$$\mathbf{Function_2}(5,6,1) = -11.00$$

$$\mathbf{Function_3(a,c,x) := a \cdot x^2 + b \cdot x + c\ solve, b} \rightarrow \frac{-\left(a \cdot x^2 + c\right)}{x}$$

$$\mathbf{Function_3}(1,6,1) = -7.00$$

$$\mathbf{Function_4(a,b,x) := a \cdot x^2 + b \cdot x + c\ solve, c} \rightarrow (-a) \cdot x^2 - b \cdot x$$

$$\mathbf{Function_4}(1,5,1) = -6.00$$

Figure 14.7 Creating user-defined functions using symbolic math

Evaluate

Whenever you use the symbolic equal sign without any keywords, Mathcad evaluates the expression to the left of the symbolic equal sign. If the expression cannot be simplified further, Mathcad returns the original expression. If variables are not defined prior to using the symbolic equal sign, Mathcad just returns the same expression. When using the symbolic equal sign with numeric input, Mathcad does not express numbers in decimal format. It uses fractions instead of decimals. Let's look at a few examples comparing the symbolic equal sign with the numeric equal sign. See Figures 14.8 and 14.9.

Type CTRL+PERIOD to get the symbolic equal sign.

Symbolic

When using the symbolic equal sign, if A_1 and A_2 are not defined, then the symbolic equal sign returns the original expressions.

$A_1 + A_2 \rightarrow A_1 + A_2$

$A_1 \cdot A_2 \rightarrow A_1 \cdot A_2$

$A_1 := 2ft \quad A_2 := 4sec$

When using the symbolic equal sign, if the units of the equation are not consistent, then Mathcad simplifies the equation as much as possible, but does not return an error message. Units are not converted to the default unit system.

$A_1 + A_2 \rightarrow 2 \cdot ft + 4 \cdot sec$

$A_1 \cdot A_2 \rightarrow 8 \cdot ft \cdot sec$

Numeric

When using the numeric equal sign, if A_1 and A_2 are not defined, Mathcad gives an error stating that the variable is not defined.

$A_1 + A_2 = \blacksquare$

$A_1 \cdot A_2 = \blacksquare$

$A_1 \cdot A_2 = \blacksquare\blacksquare$

This variable is undefined.

When using the numeric equal sign, Mathcad returns an error if it tries to add inconsistent units. The multiplication returns units in the default unit system.

$A_1 + A_2 = \blacksquare$

$A_1 \cdot A_2 = 2.44\ m \cdot s$

$A_1 + A_2 = \blacksquare\blacksquare$

This value has units: Time, but must have units: Length.

Figure 14.8 Comparison of how the symbolic equal sign works compared to the numeric equal sign

Float

In the previous examples, Mathcad returned the symbolic results as fractions without reducing the numbers to decimals. This is because Mathcad's symbolic processor does not reduce terms. In Figure 14.9, we showed how to have Mathcad display the numeric results by using the numeric equal sign at the end of the expression. There is another way to have the symbolic processor return numeric results. You do this by using the keyword "float." This is called floating point calculation. Using this keyword skips the display of the symbolic result. When using floating point calculations, Mathcad uses a default level of precision equal to 20. You can control the level of precision by typing a comma following the "float" keyword and typing an integer, which defines the desired precision. When you specify a precision, Mathcad returns a result based on that level of precision. It does not calculate to a higher level of precision and then display the result to the requested number of decimal places. The precision can be between 1 and 250 for live symbolics. If you use the Evaluate>Floating Point

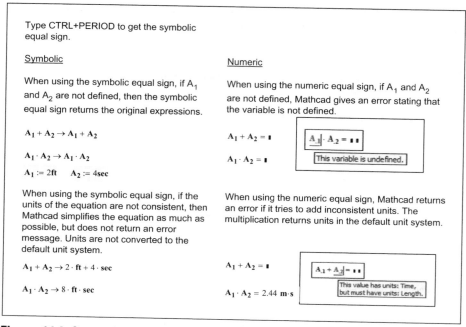

Symbolic evaluation		Numeric evaluation	
$\cos(x_1 + x_2) \to \cos(x_1 + x_2)$	If x_1 and x_2 are undefined, returns the original expression.	$\cos(\;\; + x_2) = \blacksquare$	If x_1 and x_2 are undefined, gives error for undefined variable
$x_1 := \frac{\pi}{4} \quad x_2 := \frac{4\pi}{3}$	$x_1 \to \frac{1}{4} \cdot \pi$	$x_1 = 0.79$	
$\cos(x_1) \to \frac{1}{2} \cdot 2^{\frac{1}{2}}$	Symbolic results do not use decimals. They are expressed as fractions.	$\cos(x_1) = 0.71$	Numeric results
$\cos(x_2) \to \frac{-1}{2}$		$\cos(x_2) = -0.50$	
$\cos(x_1 + x_2) \to \frac{1}{4} \cdot 6^{\frac{1}{2}} \cdot \left(1 - \frac{1}{3} \cdot 3^{\frac{1}{2}}\right)$		$\cos(x_1 + x_2) = 0.26$	
		Check $\frac{1}{4} \cdot \sqrt{6} \cdot \left(1 - \frac{1}{3} \cdot \sqrt{3}\right) = 0.26$	
$x_1 + x_2 \to \frac{19}{12} \cdot \pi$		$x_1 + x_2 = 4.97$	

You can cause Mathcad to return a numeric evaluation of a live symbolics by adding a numeric equal sign at the end of the expression.

$$\cos(x_1 + x_2) \to \frac{1}{4} \cdot 6^{\frac{1}{2}} \cdot \left(1 - \frac{1}{3} \cdot 3^{\frac{1}{2}}\right) = 0.26$$

$$x_1 + x_2 \to \frac{19}{12} \cdot \pi = 4.97$$

$$x_1 \to \frac{1}{4} \cdot \pi = 0.79$$

Figure 14.9 Additional examples of symbolic evaluation compared to numeric evaluation

command from the **Symbolics** menu, you can specify a precision as high as 4000. See Figure 14.10.

> **Tip!** I suggest keeping the precision at 20. The accuracy of the symbolic processor goes down when you choose a lower precision. If you want a precision less than 20, never use less than 5.

$y_1 := \frac{\pi}{8}$ $\quad$ $y_2 := \frac{\pi}{4}$ $\quad$ $y_3 := \pi$

You can force numeric evaluation of live symbolic calculations by using the "float" keyword. The default precision is 20.

Using standard evaluation with numeric equal sign | Using keyword "float"

$y_1 \rightarrow \frac{1}{8} \cdot \pi = 0.392699$ $\quad$ $y_1\ \textbf{float} \rightarrow .39269908169872415481$

$y_2 \rightarrow \frac{1}{4} \cdot \pi = 0.785398$ $\quad$ $y_2\ \textbf{float} \rightarrow .78539816339744830963$

$y_1 + y_2 \rightarrow \frac{3}{8} \cdot \pi = 1.178097$ $\quad$ $y_1 + y_2\ \textbf{float} \rightarrow 1.1780972450961724644$

Use a comma and a number after the keyword to control the desired precision.

$y_1\ \textbf{float}, 5 \rightarrow .39270$ $\quad$ $y_2\ \textbf{float}, 5 \rightarrow .78540$ $\quad$ $y_1 + y_2\ \textbf{float}, 5 \rightarrow 1.1781$

$\cos(y_1 + y_2) \rightarrow \frac{1}{2} \cdot \left(2 - 2^{\frac{1}{2}}\right)^{\frac{1}{2}} = 0.38268$ $\quad$ $\cos(y_1 + y_2)\ \textbf{float}, 4 \rightarrow .3828$

The floating point calculations only provide numbers to the level of precision you specify. Note the lower accuracy when the precision is set low. It is best to keep the default precision. Compare the numbers below with the numeric results. The precision can be as high as 250 for live symbolics.

$AA := y_1\ \textbf{float}, 4 \rightarrow .3928$ $\quad$ $BB := y_2\ \textbf{float}, 2 \rightarrow .78$ $\quad$ $CC := \cos(y_1 + y_2)\ \textbf{float}, 6 \rightarrow .382685$

$AA = 0.392800$ $\quad$ $BB = 0.780000$ $\quad$ $CC = 0.38268500$

$y_1 = 0.39269908$ $\quad$ $y_2 = 0.78539816$ $\quad$ $\cos(y_1 + y_2) = 0.38268343$

$y_3 = 3.14159265358979300$ Numeric evaluation

$y_3\ \textbf{float}, 80 \rightarrow 3.1415926535897932384626433832795028841971693993751058209749445923078164062862090$

$e\ \textbf{float}, 80 \rightarrow 2.7182818284590452353602874713526624977572470936999595749669676277240766303535476$

Figure 14.10 Using floating point calculations

You can also force Mathcad to use floating point calculations by including a decimal point in the input numbers. See Figure 14.11.

You can force Mathcad to use floating point calculations by including a decimal point in your numbers.

$e^3 \rightarrow e^3 = 20.08553692318767000$ Normal live symbolic result with numeric equal sign.

$e^3 \textbf{ float} \rightarrow 20.085536923187667741$ Live symbolic with floating point calculation and default level of precision = 20.

$e^{3.0} \rightarrow 20.085536923187667741$ Forced floating point calculation (uses default precision of 20) by using decimal number.

$e^3 = 20.08553692318766400$ Numeric calculation with number of decimal places set to 17. Note slight differences between symbolic and numeric calculation.

See below for additional examples of forced floating point calculations:

$\ln(2) \rightarrow \ln(2)$ $\cos(1) \rightarrow \cos(1)$ $5\frac{3}{4} \rightarrow 5\frac{3}{4}$

$\ln(2.0) \rightarrow .69314718055994530942$ $\cos(1.0) \rightarrow .54030230586813971740$ $5\frac{3.0}{4} \rightarrow 5.7500000000000000000$

(Note: Use the mixed number icon on the Calculator toolbar get the above mixed fraction.)

Note that once a forced floating point symbolic evaluation is made (by using a decimal), you cannot use a precision greater than 20.

$\ln(3) \rightarrow \ln(3)$

$\ln(3.0) \rightarrow 1.0986122886681096914 \textbf{ float}, 40 \rightarrow 1.0986122886681096914$

$\ln(3) \textbf{ float}, 40 \rightarrow 1.098612288668109691395245236922525704647$

The forced floating point calculation set the precision at 20. You cannot then use a greater precision.

Figure 14.11 Forcing floating point calculations

Expand, simplify, and factor

You can also use the symbolic menu to expand, simplify, and factor equations.

When Mathcad expands an equation, it multiplies the various elements and expands the exponents. Let's look at how Mathcad can expand algebraic equations. See Figures 14.12 and 14.13.

To expand the following expression, select the entire equation.
Click on **Expand** from the **Symbolics** menu.

$(x+1)^3 + (x+3)^2 + x - 3$

$x^3 + 4 \cdot x^2 + 10 \cdot x + 7$

You can also expand just a portion of the equation. To do this, just select the portion you want to expand and click **Expand** from the **Symbolics** menu.

$(x+1)^3 + (x+3)^2 + x - 3$

$x^3 + 3 \cdot x^2 + 3 \cdot x + 1 + (x+3)^2 + x - 3$ For this case, just the $(x+1)^3$ was selected before expanding.

$(x+1)^3 + \left[(x+3)^2\right] + x - 3$

$(x+1)^3 + \left(x^2 + 6 \cdot x + 9\right) + x - 3$ For this case, just the $(x+3)^2$ was selected before expanding.

Use the symbolic equal sign to expand the above equations. Use the keyword "expand". The entire equation is selected when you use the symbolic equal sign.

$(x+1)^3 + (x+3)^2 + x - 3 \text{ expand } \rightarrow x^3 + 4 \cdot x^2 + 10 \cdot x + 7$

Figure 14.12 Expanding expressions

Below are more examples of expanding algebraic equations using **Expand** from the Symbolics menu.

$(\cot(x) + 1)^2 + (\tan(x) + 2)^2$

$\cot(x)^2 + 2 \cdot \cot(x) + 5 + \tan(x)^2 + 4 \cdot \tan(x)$

$(a \cdot x - b \cdot y) \cdot (c \cdot z + d \cdot w)$

$c \cdot z \cdot a \cdot x + a \cdot x \cdot d \cdot w - c \cdot z \cdot b \cdot y - b \cdot y \cdot d \cdot w$

$\frac{1}{2} + \frac{1}{3} + \frac{1}{4}$

$\frac{13}{12}$

Use the symbolic equal sign to expand the above equations.

$(\cot(x) + 1)^2 + (\tan(x) + 2)^2 \text{ expand } \rightarrow \cot(x)^2 + 2 \cdot \cot(x) + 5 + \tan(x)^2 + 4 \cdot \tan(x)$

$(a \cdot x - b \cdot y) \cdot (c \cdot z + d \cdot w) \text{ expand } \rightarrow a \cdot x \cdot c \cdot z + a \cdot x \cdot d \cdot w - b \cdot y \cdot c \cdot z - b \cdot y \cdot d \cdot w$

$\frac{1}{2} + \frac{1}{3} + \frac{1}{4} \text{ expand } \rightarrow \frac{13}{12} = 1.08$

Figure 14.13 Expanding expressions

When Mathcad simplifies an equation, it tries to reduce the equation to the simplest form. Figure 14.14 shows the results of some simplifications.

Simplify returns a simplified expression, which cancels common factor, and returns trigonometric identities. Select the entire equation and click **Simplify** from the **Symbolics** menu.

$$(\cot(x)+1)^2+(\tan(x)+2)^2 \qquad \frac{x^2+5x+6}{x+3} \qquad 1-\sin(x)^2$$

$$\frac{3\cdot\cos(x)^4+2\cdot\cos(x)^3\cdot\sin(x)-4\cdot\sin(x)\cdot\cos(x)-3\cdot\cos(x)^2-1}{\cos(x)^2\cdot\left[(-1)+\cos(x)^2\right]} \qquad x+2 \qquad \cos(x)^2$$

Use the symbolic equal sign to simplify the above equations.

$$(\cot(x)+1)^2+(\tan(x)+2)^2 \text{ simplify } \rightarrow \frac{3\cdot\cos(x)^4+2\cdot\cos(x)^3\cdot\sin(x)-4\cdot\sin(x)\cdot\cos(x)-3\cdot\cos(x)^2-1}{\cos(x)^2\cdot\left[(-1)+\cos(x)^2\right]}$$

$$\frac{x^2+5x+6}{x+3} \text{ simplify } \rightarrow x+2$$

$$1-\sin(x)^2 \text{ simplify } \rightarrow \cos(x)^2$$

Figure 14.14 Simplifying expressions

When you factor an expression, Mathcad breaks the equation into all of its parts. Figure 14.15 gives examples of what happens when you use "factor."

The best way to learn what each of these do is to experiment and practice.

Factoring returns the entire expression as a product of all its parts. It is the opposite of expanding. Select the entire equation and click **Factor** from the **Symbolics** menu

$21 \quad 462 \quad x^2 - 5x + 6 \quad x^3 - x^2 - 17 \cdot x - 15 \quad x^4 - 4 \cdot x^3 - 7 \cdot x^2 + 22 \cdot x + 24$

$3 \cdot 7 \quad 2 \cdot 3 \cdot 7 \cdot 11 \quad (x - 2) \cdot (x - 3) \quad (x + 1) \cdot (x - 5) \cdot (x + 3) \quad (x + 1) \cdot (x + 2) \cdot (x - 3) \cdot (x - 4)$

Use the symbolic equal sign to factor the above equations.

$21 \text{ factor } \rightarrow 3 \cdot 7$

$462 \text{ factor } \rightarrow 2 \cdot 3 \cdot 7 \cdot 11$

$x^2 + 5x + 6 \text{ factor } \rightarrow (x + 3) \cdot (x + 2)$

$x^3 - x^2 - 17 \cdot x - 15 \text{ factor } \rightarrow (x - 5) \cdot (x + 3) \cdot (x + 1)$

$x^4 - 4 \cdot x^3 - 7 \cdot x^2 + 22 \cdot x + 24 \text{ factor } \rightarrow (x - 3) \cdot (x - 4) \cdot (x + 2) \cdot (x + 1)$

Figure 14.15 Factoring equations

Explicit

Mathcad 13 added the keyword "explicit." This is a very exciting feature because it allows you to display the values of the variables used in your expressions. To use this feature, type "explicit" as a keyword followed by a comma and a list of the variables you want to display. See Figures 14.16–14.19.

The explicit keyword allows you to show intermediate results in your calculations.

Given the following input variable definitions:

$C_1 := 5.1N$ $C_2 := 4.5N$ $D_1 := 3.2m$ $D_2 := 4.8m$

A normal equation would look like this:

$M_1 := C_1 \cdot D_1 + C_2 \cdot D_2$ $M_1 = 37.92\,N \cdot m$

Using explicit calculations, you are able to show the values that Mathcad uses to calculate the results:

$M_2 := C_1 \cdot D_1 + C_2 \cdot D_2 \text{ explicit}, C_1, D_1, C_2, D_2 \rightarrow 5.1 \cdot N \cdot 3.2 \cdot m + 4.5 \cdot N \cdot 4.8 \cdot m \rightarrow 37.92 \cdot N \cdot m$

$M_2 = 37.92\,N \cdot m$

$LineLoad_1 := \frac{C_1}{D_1} + \frac{C_2}{D_2}$ $LineLoad_1 = 2.53 \frac{kg}{s^2}$ $LineLoad_1 = 2.53 \frac{N}{m}$

$LineLoad_2 := \frac{C_1}{D_1} + \frac{C_2}{D_2} \text{ explicit}, C_1, C_2, D_1, D_2 \rightarrow \frac{5.1 \cdot N}{3.2 \cdot m} + \frac{4.5 \cdot N}{4.8 \cdot m} \rightarrow 2.5312500000000000000 \cdot \frac{N}{m}$

$LineLoad_2 = 2.53 \frac{N}{m}$

Figure 14.16 Using the keyword "explicit"

$Time := 3sec$ $Velocity := 5\frac{m}{sec}$ $InitialDist := 5.2m$

A normal equation would look like this

$Distance_1 := InitialDist + Time \cdot Velocity$ $Distance_1 = 20.20\ m$

In the above example, if the definitions for InitialDist, Time, and Velocity were on a different page, you would not know what values Mathcad is using for the three variables. With the keyword "explicit," you can have Mathcad display the input variables.

$Distance_2 := InitialDist + Time \cdot Velocity \text{ explicit}, InitialDist, Time, Velocity \rightarrow 5.2 \cdot m + 3 \cdot sec \cdot 5 \cdot \frac{m}{sec} \rightarrow 20.2 \cdot m$

$Distance_2 = 20.20\ m$

Mathcad allows you to hide the keyword and modifiers to simplify the variable display. To do this, right click on the keyword and select "Hide keywords." This will only show the definition as seen below. If you click on the region, the full definition will appear just as above.

$Distance_2 := InitialDist + Time \cdot Velocity \rightarrow 5.2 \cdot m + 3 \cdot sec \cdot 5 \cdot \frac{m}{sec} \rightarrow 20.2 \cdot m$

You can also change the symbolic equal sign to display as a normal equal sign. To do this, right click in the expression, hover your mouse over "View Evaluation As," and select "Equal sign."

$Distance_2 := InitialDist + Time \cdot Velocity = 5.2 \cdot m + 3 \cdot sec \cdot 5 \cdot \frac{m}{sec} = 20.2 \cdot m$

$Distance_2 = 20.20\ m$

Figure 14.17 Using the keyword "explicit"

You can also include previous results in the list following "explicit." The results vary depending on what variables are included in the modify list. See the examples below.

$E_1 := 3.0m$ $E_2 := 4.2m$ $E_3 := 5.5m$ $E_4 := 6.2m$

$F_1 := \frac{E_1}{E_2}$ $F_1 = 0.71$ $F_2 := \frac{E_3}{E_4}$ $F_2 = 0.89$

$G_1 := \frac{F_1}{F_2}$ explicit, $F_1, F_2 \rightarrow \frac{\frac{E_1}{E_2}}{\frac{E_3}{E_4}}$ $G_1 = 0.8052$

By listing F_1 and F_2, Mathcad substitutes the definitions of F_1 and F_2.

$G_2 := \frac{F_1}{F_2}$ explicit, $E_1, E_2, E_3, E_4 \rightarrow \frac{F_1}{F_2}$ $G_2 = 0.8052$

Listing only E_1, E_2, E_3, and E_4, does not have any effect, because they do not appear in the current definition.

By listing F_1 and F_2 in addition to E_1, E_2, E_3, and E_4, Mathcad substitutes all values used.

$G_3 := \frac{F_1}{F_2}$ explicit, $E_1, E_2, E_3, E_4, F_1, F_2 \rightarrow \frac{\frac{3.0 \cdot m}{4.2 \cdot m}}{\frac{5.5 \cdot m}{6.2 \cdot m}} \rightarrow .80519480519480519480$

$G_3 = 0.8052$

Repeat the above example with the keyword hidden and symbolic equal sign changed to numeric equal sign.

$G_4 := \frac{F_1}{F_2} = \frac{\frac{3.0 \cdot m}{4.2 \cdot m}}{\frac{5.5 \cdot m}{6.2 \cdot m}} = .80519480519480519480$

$G_4 = 0.8052$

Figure 14.18 Using keyword "explicit"

This Figure is very similar to Figure 14.18. In this example, the value of F_3 and F_4 are evaluated in addition to being defined.

$$F_3 := \frac{E_1}{E_2} \textbf{ explicit}, E_1, E_2 \rightarrow \frac{3.0 \cdot m}{4.2 \cdot m} \rightarrow .71428571428571428571$$

Define and evaluate F_3.

$$F_4 := \frac{E_3}{E_4} \textbf{ explicit}, E_3, E_4 \rightarrow \frac{5.5 \cdot m}{6.2 \cdot m} \rightarrow .88709677419354838710$$

Define and evaluate F_4.

Define and evaluate G_5. Because F_3 and F_4 have already been evaluated, the numeric results of F_3 and F_4 are displayed.

$$G_5 := \frac{F_3}{F_4} \textbf{ explicit}, F_3, F_4 \rightarrow \frac{.71428571428571428571}{.88709677419354838710} \rightarrow .80519480519480519480$$

Repeat the above example with the keyword hidden and symbolic equal sign changed to numeric equal sign.

$$G_6 := \frac{F_3}{F_4} = \frac{.71428571428571428571}{.88709677419354838710} = .80519480519480519480$$

Figure 14.19 Using keyword "explicit"

Using more than one keyword

Mathcad allows you to string a series of symbolic calculations together. To do this, type the symbolic expression as you normally would do. Click outside the region to display the expression. Then click on the result and insert another symbolic equal sign (**CTRL+SHIFT+PERIOD**). Enter the keyword and any modifiers. See Figures 14.20 and 14.21.

You can use several keywords in succession.

Input the first symbolic expression.

$(4x + 5y - 3x - 3y)^3$ **simplify** $\rightarrow (x + 2 \cdot y)^3$

Next, click on the result and insert another symbolic equal sign.

$(4x + 5y - 3x - 3y)^3$ **simplify** $\rightarrow (x + 2 \cdot y)^3$ **expand** $\rightarrow x^3 + 6 \cdot x^2 \cdot y + 12 \cdot x \cdot y^2 + 8 \cdot y^3$

You can also use several keywords in succession, but only display the final result. To do this, select the first keyword with the vertical selection bar on the right side, then press CTRL+SHIFT+PERIOD. This inserts a vertical bar with a placeholder. Type the next keyword in the placeholder. Mathcad only displays the result from the last keyword.

$a := 1 \quad b := 10 \quad c := 12$

$a \cdot x^2 + b \cdot x + c$ **solve, x** $\rightarrow \begin{bmatrix} (-5) + 13^{\frac{1}{2}} \\ (-5) - 13^{\frac{1}{2}} \end{bmatrix}$ **float** $\rightarrow \begin{pmatrix} -1.3944487245360107069 \\ -8.6055512754639892931 \end{pmatrix}$

$a \cdot x^2 + b \cdot x + c \left| \begin{matrix} \textbf{solve, x} \\ \textbf{float} \end{matrix} \right. \rightarrow \begin{pmatrix} -1.3944487245360107069 \\ -8.6055512754639892931 \end{pmatrix}$

Figure 14.20 Using keywords in succession

Find a numeric solution to the equation used in Figure 14.6. Use the solve keyword, and then the float keyword.

If you try to display both results, the solution extends well beyond the right margin. For this example, we will only display the final result.

$\textbf{Solution} := x^3 + x + 1 \left| \begin{matrix} \textbf{solve, x} \\ \textbf{float} \end{matrix} \right. \rightarrow \begin{pmatrix} -.68232780382801932744 \\ .34116390191400966368 - 1.1615413999972519362 \cdot i \\ .34116390191400966368 + 1.1615413999972519362 \cdot i \end{pmatrix}$

$\textbf{Solution}_1 = -0.68232780382801930$

$\textbf{Solution}_2 = 0.34116390191400964 - 1.16154139999725200i$

$\textbf{Solution}_3 = 0.34116390191400964 + 1.16154139999725200i$

Figure 14.21 Using keywords in succession

Units with symbolic calculations

The symbolic process does not recognize units. It passes units through as undefined variables. Thus, you can use units with your expressions; however, they will not be truly understood. For instance, 10 ft/s * 5 sec is returned 50 ft, but you cannot convert the 50 ft to meters. Another example is adding 10 ft + 5 sec. The symbolic processor returns 10 ft + 5 sec, not recognizing the error of adding units from different unit dimensions. In order to process the units, you need to do a numeric evaluation of the result. Do this by assigning a variable to the symbolic calculation and then numerically evaluating the variable. See Figure 14.22 for examples.

You may attach units to symbolic calculations.

$\mathbf{SampleUnits_1} := 1\mathbf{ft} + 5\mathbf{in} + 2\mathbf{m}$

The symbolic processor just passes units through. It does not evaluate them.

$\mathbf{SampleUnits_1} \rightarrow \mathbf{ft} + 5 \cdot \mathbf{in} + 2 \cdot \mathbf{m}$

$\mathbf{SampleUnits_1} = 2.43\ \mathbf{m}$

You must use the numeric processor to evaluate the units.

$10\mathbf{ft} + 5\mathbf{sec} \rightarrow 10 \cdot \mathbf{ft} + 5 \cdot \mathbf{sec}$

$\mathbf{SampleUnits_2} := 10\mathbf{ft} + 5\mathbf{sec} \rightarrow 10 \cdot \mathbf{ft} + 5 \cdot \mathbf{sec}$

The symbolic process did not recognize the error of adding units in different unit dimensions. Once a variable is assigned to the expression, the numeric processor recognizes the error.

$\mathbf{SampleUnits_2} := 10 \cdot \mathbf{ft} + 5 \cdot \mathbf{sec} \rightarrow 10 \cdot \mathbf{ft} + 5 \cdot \mathbf{sec}$

This value has units: Time,
but must have units: Length.

$10\frac{\mathbf{ft}}{\mathbf{s}} \cdot 20\mathbf{s} + \frac{1}{2}50\frac{\mathbf{m}}{\mathbf{s}^2} \cdot 20^2\mathbf{s}^2 \rightarrow 200 \cdot \mathbf{ft} + 10000 \cdot \mathbf{m}$

The symbolic processor cancels common units ((ft/s)*(s)=ft), but it does not convert ft to meters.

$\mathbf{Distance}(\mathbf{v_o}, \mathbf{a}, \mathbf{t}) := \mathbf{v_o} \cdot \mathbf{t} + \frac{1}{2}\mathbf{a} \cdot \mathbf{t}^2$

Assign a user-defined function

$\mathbf{Distance}\left(10\frac{\mathbf{ft}}{\mathbf{s}}, 50\frac{\mathbf{m}}{\mathbf{s}^2}, 20\mathbf{s}\right) \rightarrow 200 \cdot \mathbf{ft} + 10000 \cdot \mathbf{m}$

The symbolic processor does not convert units.

$\mathbf{Distance}\left(10\frac{\mathbf{ft}}{\mathbf{s}}, 50\frac{\mathbf{m}}{\mathbf{s}^2}, 20\mathbf{s}\right) = 10060.96\ \mathbf{m}$

The numeric processor does convert units.

$\mathbf{Distance}\left(10\frac{\mathbf{m}}{\mathbf{s}}, 50\frac{\mathbf{m}}{\mathbf{s}^2}, 20\mathbf{s}\right) \rightarrow 10200 \cdot \mathbf{m}$

Because all the units were consistent, the symbolic processor returned meters.

$\mathbf{Distance}\left(10\frac{\mathbf{m}}{\mathbf{s}}, 50\frac{\mathbf{m}}{\mathbf{s}^2}, 20\mathbf{s}\right) = 33464.57\ \mathbf{ft}$

The symbolic process does not have a units placeholder to convert results to other units.

Figure 14.22 Using units with symbolic calculations

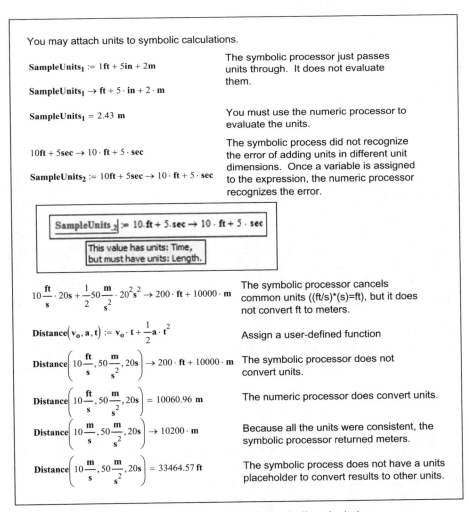

Additional topics to study

The topic of symbolic calculation could fill an entire book. In this chapter, we have only introduced you to a few symbolic calculation concepts. Even though the discussion of these concepts has been brief, these concepts will be a valuable addition to your Mathcad toolbox.

Some of the topics that we did not discuss in this chapter, that may be of interest to you for further study include:

- Partial fractions
- Calculus (see Chapter 17 for an introduction)
- Symbolic transformations
- Series expansion
- Complex input
- Assumptions
- Special functions
- Symbolic optimization

Summary

Mathcad provides two methods to symbolically solve engineering equations. The first method is to use the commands from the **Symbolics** menu. The second method is to use the symbolic equal sign. The results of the first method are static and do not change. The results of the second method are dynamic. Thus, if you change the input equation, the results are automatically updated.

The symbolic equal sign needs to use keywords. These keywords are necessary to tell Mathcad what operation is desired. In addition to the keyword, it is sometimes necessary to tell Mathcad what variable to use with the operation.

In Chapter 14 we:

- Explained how to use the commands from the Symbolics toolbar.
- Showed that the results from using the Symbolics toolbar are static.
- Introduced the symbolic equal sign.
- Illustrated how to use keywords with the symbolic equal sign.
- Emphasized that results are dynamic when using live symbolics.
- Learned how to get numeric results when using the symbolic equal sign.

- Discussed using the "explicit" keyword to display values of variables used in your calculations.
- Recommended studying the Mathcad help and Tutorial to learn more advanced features of symbolic calculations.

Practice

1. Use the following commands from the Symbolics menu on each of the equations: Evaluate (Symbolically), Solve (for each of the variables), simplify, expand, and factor.

   [a] $y = x^2 - 3$
   [b] $y = (2x + 4) * (x - 3)$
   [c] $x^2 + y^2 = 1$
   [d] $y = x^3 - 2x^2 - 5x + 6$
   [e] $y = a * x^3 + b * x^2 + c * x + d$
   [f] $y = (x - 1)(x - 2)(2x - 6)$
   [g] $2x^2 - 3x + 1/(2x - 1)$

2. Write symbolic equations to solve for x with the following functions. Assign each to a user-defined function. (See Figures 14.7 and 14.22 for examples of user-defined functions.)

   [a] $x^3 + 6x^2 - x - 30$
   [b] $y = x^2(x - 1)^2$
   [c] $3x^2 + 4y^2 = 12$

3. Write symbolic equations to evaluate the following functions. Assign each to a user-defined function.

   [a] $\cos(\pi/3) + \tan(2\pi/3)$
   [b] $[(21/11) + (5/3)]/3$
   [c] $\pi + \pi/4 + \ln(2)$

4. Write symbolic equations to expand the following functions. Assign each to a user-defined function.

   [a] $(x + 2)(x - 3)$
   [b] $(x - 1)(x + 2)(x - 5)$
   [c] $(x + 3 * i)(x - 3 * i)$ (Note: use j if that is your default for imaginary value)

5. Write symbolic equations to simplify the following functions. Assign each to a user-defined function.

   [a] $\sin(\pi/2 - x)$
   [b] $\sec(x)$
   [c] $(x^4 - 8x^3 - 7x^2 + 122x - 168)/(x - 2)$

6. Write symbolic equations to factor the following functions. Assign each to a user-defined function.

   [a] $x^3 + 5x^2 + 6x$
   [b] $x * y + y * x^2$
   [c] $2x^2 + 11x - 21$

7. Use the float keyword to evaluate the following functions. Use various levels of precision.

   [a] $\cos(\pi/6)$
   [b] $\tan(3\pi/4) + \cos(\pi/7) - \sin(\pi/5)$
   [c] $(\pi + e^3)(\tan(3\pi/8))$

8. Use the equations from exercise 7 and show intermediate results before using the float keyword.
9. From your field of study, create three expressions using assigned variables. Use the explicit keyword to show intermediate results.
10. Open the Mathcad Help and review additional topics on live symbolics from the Contents tab.

15
Solving engineering equations

Once you begin using the solving features of Mathcad, you will wonder how you ever got along without them. Solving engineering equations is one of Mathcad's most useful power tools. It ranks up with the use of units as one of Mathcad's best features. In Chapter 14, we introduced the keyword "solve" with live symbolics. In this chapter, we will add more solving tools to your Mathcad toolbox. The intent of this chapter is to illustrate how engineering problems can be solved using the Mathcad functions ***root***, ***polyroots***, and ***Find***. Because these are some of the most useful functions for engineers, this chapter will use many examples to illustrate their use. When trying to solve for an equation, it is very useful to plot the equations. This helps to visualize the solution before using Mathcad to solve the equations. Some of the functions discussed require initial guess values, and a plot is useful to help select the initial guess.

Chapter 15 will:

- Introduce the ***root*** function.
- Discuss the two different forms that it takes: unbracketed and bracketed.
- Give examples of each form.
- Show how a plot is useful to determine the initial guess.
- Discuss the ***polyroots*** function.
- Give examples of the ***polyroots*** function.
- Show how to use a "Solve Block" using ***Given*** and ***Find*** to solve multiple equations with multiple unknowns.
- Note when to use ***Minerr*** instead of ***Find***.

- Illustrate how to use units in solving equations.
- Provide several engineering examples, which illustrate the use of each method of solving equations and illustrate how each would be used.

Root function

The ***root*** function is used to find a single solution to a single function with a single unknown. In later sections, we will discuss finding all the solutions to a polynomial function. We will also discuss solving multiple equations with multiple unknowns. For now, we will focus on using the ***root*** function.

If a function has several solutions, then the solution that Mathcad finds is based on the initial guess you give Mathcad. Because of this, it is helpful to plot the function prior to giving Mathcad the initial guess.

The ***root*** function takes the form ***root***(f(var), var, [a, b]). It returns the value of var to make the function f equal to zero. The real numbers a and b are optional. If they are specified (bracketed), ***root*** finds var on this interval. The values of a and b must meet these requirements: $a < b$ and f(a) and f(b) must be of opposite signs. This is because the function must cross the x-axis in this interval. If you do not specify the numbers a and b (unbracketed), then var must be defined with an initial guess prior to using the root function.

> **Tip!** When plotting the function, use a different variable name on the x-axis than the variable you define for your initial guess. If you do not use a different variable name, the plot will not work because Mathcad will only plot the value var on the x-axis. The variable used on the x-axis needs to be a previously undefined variable.

Let's look at some simple examples of using ***root***. See Figures 15.1 and 15.2.

$\mathbf{SampleFunction(x)} := \mathbf{x}^2 + 5\mathbf{x} - 6$

Define the function.

Find the root closest to x=0

The first form of *root (bracketed)* needs two initial guesses such that a<b, and SampleFunction(a)<0 and SampleFunction(b)>0.

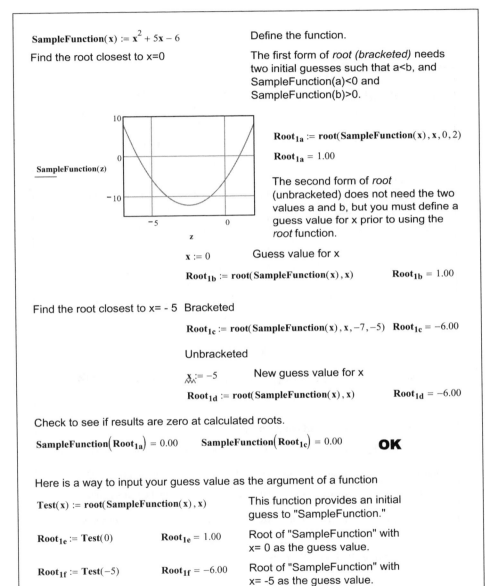

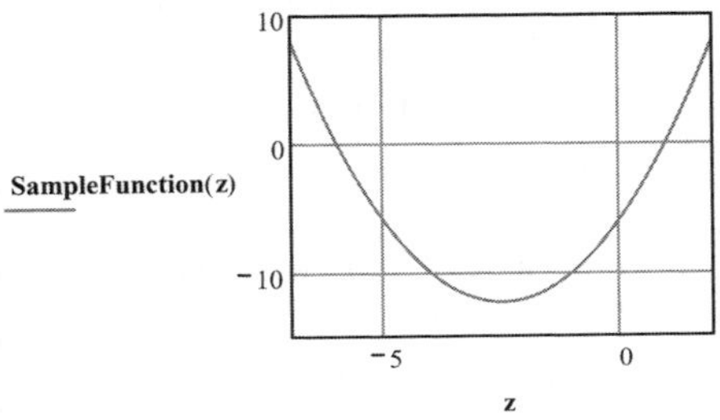

$\mathbf{Root_{1a}} := \mathbf{root(SampleFunction(x), x, 0, 2)}$

$\mathbf{Root_{1a}} = 1.00$

The second form of *root* (unbracketed) does not need the two values a and b, but you must define a guess value for x prior to using the *root* function.

$\mathbf{x} := 0$ Guess value for x

$\mathbf{Root_{1b}} := \mathbf{root(SampleFunction(x), x)}$ $\mathbf{Root_{1b}} = 1.00$

Find the root closest to x= - 5 Bracketed

$\mathbf{Root_{1c}} := \mathbf{root(SampleFunction(x), x, -7, -5)}$ $\mathbf{Root_{1c}} = -6.00$

Unbracketed

$\mathbf{x} := -5$ New guess value for x

$\mathbf{Root_{1d}} := \mathbf{root(SampleFunction(x), x)}$ $\mathbf{Root_{1d}} = -6.00$

Check to see if results are zero at calculated roots.

$\mathbf{SampleFunction(Root_{1a})} = 0.00$ $\mathbf{SampleFunction(Root_{1c})} = 0.00$ **OK**

Here is a way to input your guess value as the argument of a function

$\mathbf{Test(x)} := \mathbf{root(SampleFunction(x), x)}$ This function provides an initial guess to "SampleFunction."

$\mathbf{Root_{1e}} := \mathbf{Test(0)}$ $\mathbf{Root_{1e}} = 1.00$ Root of "SampleFunction" with x= 0 as the guess value.

$\mathbf{Root_{1f}} := \mathbf{Test(-5)}$ $\mathbf{Root_{1f}} = -6.00$ Root of "SampleFunction" with x= -5 as the guess value.

Figure 15.1 Using the function ***root***

$F_2(x) := 2^x - 9x^2 + 120$

Define the function

From the plot, use the values 7 and 10 for a and b.

$Root_{2a} := root(F_2(x), x, 7, 10)$

$Root_{2a} = 9.40$

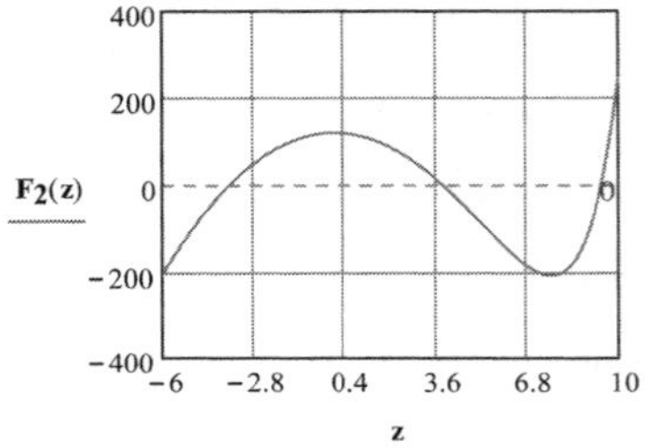

From the plot, use the values 3 and 4 for a and b.

$Root_{2b} := root(F_2(x), x, 3, 4)$

$Root_{2b} = 3.87$

From the plot, use the values -6 and -2 for a and b.

$Root_{2c} := root(F_2(x), x, -6, -2)$

$Root_{2c} = -3.65$

Check to see if results are zero at calculated roots.

$F_2(Root_{2a}) = 0.00$ $F_2(Root_{2b}) = 0.00$ $F_2(Root_{2c}) = -2.842 \times 10^{-14}$ **OK**

Input the guess value as part of a function

$TT(x) := root(F_2(x), x)$ This function provides an initial guess to "F_2".

$Root_{2d} := TT(-6)$ $Root_{2d} = -3.65$ Root of "F_2" with x=-6 as the guess value.

$Root_{2e} := TT(3)$ $Root_{2e} = 3.87$ Root of F_2" with x=3 as the guess value.

$Root_{2f} := TT(10)$ $Root_{2f} = 9.40$ Root of "F_2" with x=10 as the guess value.

Figure 15.2 Another example of using the function ***root***

If you do not use a plot to determine your initial guesses, Mathcad may not arrive at the solution you want. If your initial guesses are close to a maximum or minimum point or have multiple solutions between the bracketed initial guesses, then Mathcad may not arrive at a solution, or may arrive at a different solution than you wanted. If Mathcad is not arriving at a solution, you can refer to Mathcad Help for ideas on how to help resolve the issue.

Complex numbers may be used as an initial guess to arrive at a complex solution.

Polyroots function

The ***polyroots*** function is used to solve for all solutions to a polynomial equation at the same time. The solution is returned in a vector containing the roots of the polynomial.

In order to use this function, you need to create a vector of coefficients of the polynomial. Include all coefficients in the vector even if they are zero. The coefficients of the polynomial $f(x) = 4x^3 - 2x^2 + 3x - 5$, are (4, −2, 3, −5). For the Mathcad vector however, you begin with the constant term. The Mathcad vector looks like this:

$$\textbf{Vector} := \begin{pmatrix} -5 \\ 3 \\ -2 \\ 4 \end{pmatrix}$$

The ***polyroots*** function takes the following form: ***polyroots***(v), where v is the vector of polynomial coefficients. The result is a vector containing the roots of the polynomial. Figure 15.3 shows the solution of the same simple quadratic equation as was used in Figure 15.1.

To use the *polyroots* function, create a vector of each of the coefficients of the polynomial. Begin with the constant term as the first element. Use zero for all zero coefficients.

Use *polyroots* to find the roots of the same equation as in Figure 15.1

$\mathbf{SampleFunction_{3a}(x)} := \mathbf{x}^2 + 5\mathbf{x} - 6$

$\mathbf{V_{3a}} := \begin{pmatrix} -6 \\ 5 \\ 1 \end{pmatrix}$ Create vector of coefficients beginning with the constants term.

$\mathbf{Solution\_3a} := \mathbf{polyroots}(\mathbf{V_{3a}})$ $\mathbf{Solution\_3a} = \begin{pmatrix} -6.00 \\ 1.00 \end{pmatrix}$

Check to see if results are zero at calculated roots.

$\mathbf{SampleFunction_{3a}}(\mathbf{Solution\_3a_1}) = 0.00$ $\mathbf{SampleFunction_{3a}}(\mathbf{Solution\_3a_2}) = 0.00$

OK

Figure 15.3 Using the function ***polyroots***

Figure 15.4 illustrates a larger polynomial. Notice that this polynomial has two imaginary roots. The solution to this polynomial is displayed to 12 decimal places to illustrate two different internal solution methods that Mathcad uses to solve for roots of a polynomial. These two internal solution methods are the LaGuerre Method and the Companion Matrix Method. This book will not discuss the differences between these solution methods. It just points out that if one method does not produce satisfactory results, then there is another option available that may produce results that are slightly more accurate. The check results are displayed twice. The first result uses the General result format. The second result uses the Decimal result format with the number of decimal places set to 17. Notice that the results compare to at least the seventh decimal place.

$\mathbf{SampleFunction_{4a}(x)} := -x^5 - 2x^4 + x^3 - x^2 + 5$

Create a vector of coefficients.
Remember that the constant goes first.

$$\mathbf{V_{4a}} := \begin{pmatrix} 5 \\ 0 \\ -1 \\ 1 \\ -2 \\ -1 \end{pmatrix}$$

$\mathbf{Solution\_4a} := \mathbf{polyroots}(\mathbf{V_{4a}})$

$$\mathbf{Solution\_4a} = \begin{pmatrix} -2.436964498961 \\ -1.216885522117 \\ 0.260206634327 + 1.191574629143i \\ 0.260206643463 - 1.191574625107i \\ 1.133436743287 \end{pmatrix}$$

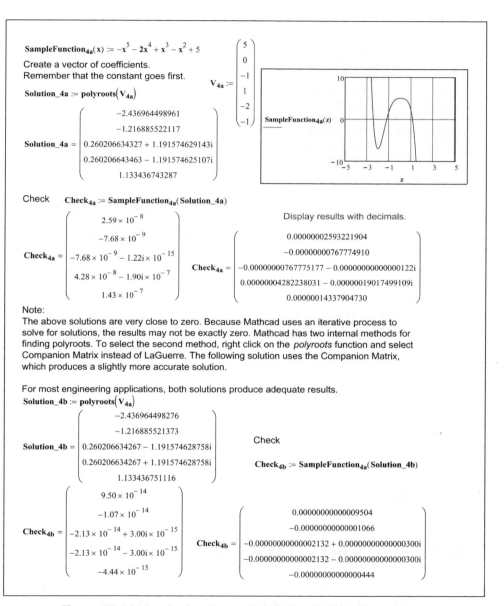

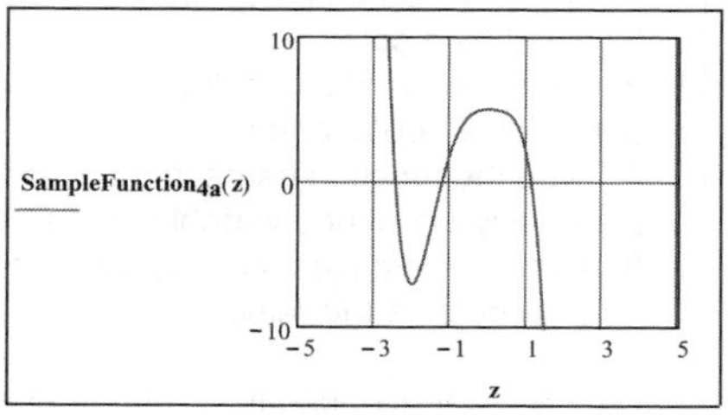

Check $\mathbf{Check_{4a}} := \mathbf{SampleFunction_{4a}(Solution\_4a)}$

$$\mathbf{Check_{4a}} = \begin{pmatrix} 2.59 \times 10^{-8} \\ -7.68 \times 10^{-9} \\ -7.68 \times 10^{-9} - 1.22i \times 10^{-15} \\ 4.28 \times 10^{-8} - 1.90i \times 10^{-7} \\ 1.43 \times 10^{-7} \end{pmatrix}$$

Display results with decimals.

$$\mathbf{Check_{4a}} = \begin{pmatrix} 0.00000002593221904 \\ -0.00000000767774910 \\ -0.00000000767775177 - 0.00000000000000122i \\ 0.00000004282238031 - 0.00000019017499109i \\ 0.00000014337904730 \end{pmatrix}$$

Note:
The above solutions are very close to zero. Because Mathcad uses an iterative process to solve for solutions, the results may not be exactly zero. Mathcad has two internal methods for finding polyroots. To select the second method, right click on the *polyroots* function and select Companion Matrix instead of LaGuerre. The following solution uses the Companion Matrix, which produces a slightly more accurate solution.

For most engineering applications, both solutions produce adequate results.

$\mathbf{Solution\_4b} := \mathbf{polyroots}(\mathbf{V_{4a}})$

$$\mathbf{Solution\_4b} = \begin{pmatrix} -2.436964498276 \\ -1.216885521373 \\ 0.260206634267 - 1.191574628758i \\ 0.260206634267 + 1.191574628758i \\ 1.133436751116 \end{pmatrix}$$

Check

$\mathbf{Check_{4b}} := \mathbf{SampleFunction_{4a}(Solution\_4b)}$

$$\mathbf{Check_{4b}} = \begin{pmatrix} 9.50 \times 10^{-14} \\ -1.07 \times 10^{-14} \\ -2.13 \times 10^{-14} + 3.00i \times 10^{-15} \\ -2.13 \times 10^{-14} - 3.00i \times 10^{-15} \\ -4.44 \times 10^{-15} \end{pmatrix}$$

$$\mathbf{Check_{4b}} = \begin{pmatrix} 0.00000000000009504 \\ -0.00000000000001066 \\ -0.00000000000002132 + 0.00000000000000300i \\ -0.00000000000002132 - 0.00000000000000300i \\ -0.00000000000000444 \end{pmatrix}$$

Figure 15.4 Using the function ***polyroots*** for a larger polynomial

Figure 15.5 uses live symbolics to have Mathcad create the polynomial vector automatically.

Mathcad can automate the generation of the coefficient vector.
Use the following steps:
1. Type the name of the variable to be used for the vector, followed by a colon.
2. Open the symbolic toolbar.
3. Click the coeffs button.
4. Type the function name in the first placeholder using a previously undefined variable name as the argument.
5. In the second placeholder, type the name of the argument.
6. Click outside the region.

For this example, use the same equation as in Figure 15.4.

$$\mathbf{SampleFunction_{4a}}(x) := -x^5 - 2x^4 + x^3 - x^2 + 5$$

Rather than redefine SampleFunction$_{4a}$ we are just displaying it. (Remember - Properties, Calculation Tab, Disable Evaluation)

$$\mathbf{V_{5a}} := \mathbf{SampleFunction_{4a}}(z)\ \mathbf{coeffs}, z \rightarrow \begin{pmatrix} 5 \\ 0 \\ -1 \\ 1 \\ -2 \\ -1 \end{pmatrix} \qquad \mathbf{V_{5a}} = \begin{pmatrix} 5.00 \\ 0.00 \\ -1.00 \\ 1.00 \\ -2.00 \\ -1.00 \end{pmatrix}$$

Mathcad generates the coefficient vector. If SampleFunction$_{4a}$ changes, the vector will update automatically.

$$\mathbf{Solution\_5a} := \mathbf{polyroots}(\mathbf{V_{5a}})$$

$$\mathbf{Solution\_5a} = \begin{pmatrix} -2.436964499 \\ -1.216885522 \\ 0.260206634 + 1.191574629i \\ 0.260206643 - 1.191574625i \\ 1.133436743 \end{pmatrix}$$

Figure 15.5 Using live symbolics to have Mathcad create the polynomial vector

Solve Blocks using *Given* and *Find*

Until now, we have been solving single equations and single variables. Mathcad is able to solve for multiple equations using multiple unknowns.

Several methods can be used to solve for multiple equations. The first method that we will use is called a Solve Block using the keyword ***Given*** and the ***Find*** function. This is illustrated in Figure 15.6.

Find the intersection points of a parabola and a line.

Use a Solve Block with *Given* and *Find*

Give initial guesses for a and b.

$\mathbf{a} := 3 \quad \mathbf{b} := 12$

Given Begin the Solve Block

List constraints.
Be sure to use the Boolean equal sign (CTRL+EQUAL).

$\mathbf{b} = \mathbf{a}^2 + \mathbf{a} + 4$ Parabola equation

$\mathbf{b} = \mathbf{a} + 8$ Line equation

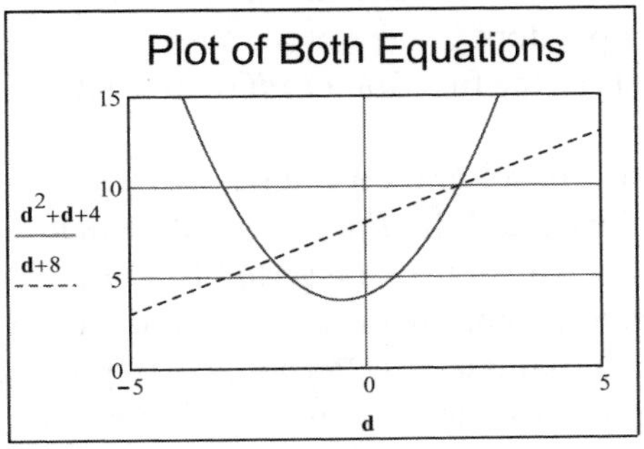

Solution

$\begin{pmatrix} \mathbf{Solution\_6a} \\ \mathbf{Solution\_6b} \end{pmatrix} := \mathbf{Find(a, b)}$

The *Find* Function ends the solve block. Use both "a" and "b" in the *Find* function, because the constraint equations use both "a" and "b."

$\mathbf{Solution\_6a} = 2.00 \qquad \mathbf{Solution\_6b} = 10.00$

Check Result

$\mathbf{Solution\_6a}^2 + \mathbf{Solution\_6a} + 4 = 10.00$ **OK**

$\mathbf{Solution\_6a} + 8 = 10.00$ **OK**

Have Mathcad solve for the second solution. Change the initial guess and try again

$\mathbf{a} := -3 \quad \mathbf{b} := 6$

Given Begins Solve Block

Gives Constraints.

$\mathbf{b} = \mathbf{a}^2 + \mathbf{a} + 4$

$\mathbf{b} = \mathbf{a} + 8$

$\begin{pmatrix} \mathbf{Solution\_6c} \\ \mathbf{Solution\_6d} \end{pmatrix} := \mathbf{Find(a, b)}$ End Solve Block

$\mathbf{Solution\_6c} = -2.00 \qquad \mathbf{Solution\_6d} = 6.00$ Mathcad selects the other solution.

Check Result

$\mathbf{Solution\_6c}^2 + \mathbf{Solution\_6c} + 4 = 6.00$ **OK**

$\mathbf{Solution\_6c} + 8 = 6.00$ **OK**

Figure 15.6 Using a Solve Block to solve two equations

This method has the following steps:

1. Give an initial guess for each variable you are solving for.
2. Type the keyword **Given**.
3. Below the word ***Given***, type a list of constraint equations using Boolean operators.
4. Type the function ***Find***() and list the desired solution variables as arguments.

If you are solving for n unknowns, you must have at least n equations between the ***Given*** and the ***Find***. Equality equations must be defined using the Boolean equals. This must be inserted from the Boolean toolbar or use **CTRL+EQUAL**. If there is more than one solution, the solution Mathcad calculates is based on your initial guesses. If you modify your guesses, then Mathcad may arrive at a different solution. If one of the solutions is negative and the other is positive, you can add an additional constraint telling Mathcad that you want the solution to be greater than zero. See Figure 15.7.

Find the intersection point of a parabola and a line.

Use a Solve Block with *Given* and *Find*

$a := -4$ Give initial guess for a.

Given Begins the Solve Block

List constraints.

$a + 8 = a^2 + a + 4$ In this case we set both equations equal to each other (b=b)

$a > 0$ We want to find the positive solution, so set "a" greater than zero.

Solution

$Result\_7 := Find(a)$ $Result\_7 = 2.00$

$a := Result\_7$ $a = 2.00$

Check Result

$a^2 + a + 4 = 10.00$ **OK**

$a + 8 = 10.00$ **OK**

The *Find* function ends the solve block. Because "a" is the only variable in the Solve Block you may use only "a" as the argument in the *Find* function. Mathcad finds the positive solution even though our initial guess was negative.

Figure 15.7 Forcing Mathcad to solve for a positive root

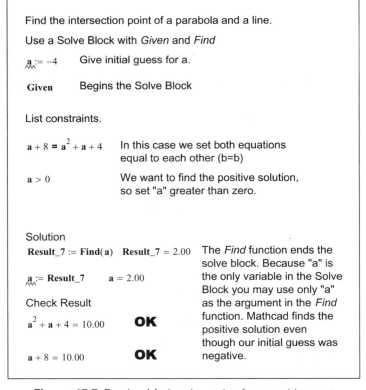

You can also assign the equations as functions and use the functions as the constraints. See Figure 15.8.

$f := -100$ — Initial guess of solution

$SampleFunction_{8a}(d) := d^2 + d + 4$ — Define two functions

$SampleFunction_{8b}(d) := d + 8$ — Remember that the argument "d" is only used to define the function. We can use "d" or any other argument when solving the functions. In this case we use the variable "f".

Given — Begin Solve Block

$SampleFunction_{8a}(f) = SampleFunction_{8b}(f)$ — Set constraints

$Result\_8a := Find(f)$ $\quad Result\_8a = -2.00$

Check

$SampleFunction_{8a}(Result\_8a) - SampleFunction_{8b}(Result\_8a) = -0.00$ **OK**

Now add an additional constraint so that "f" is greater than zero

Given

$SampleFunction_{8a}(f) = SampleFunction_{8b}(f)$

$f > 0$

$Result\_8b := Find(f)$ $\quad Result\_8b = 2.00$ — Mathcad finds the positive solution

Check

$SampleFunction_{8a}(Result\_8b) - SampleFunction_{8b}(Result\_8b) = -0.00$ **OK**

Figure 15.8 Solve Block using functions instead of equations

Solve Blocks should not contain the following:

- Range variables.
- Constraints involving the "not equal to" operator.
- Other Solve blocks. Each Solve Block can have only one ***Given*** and one ***Find***.
- Assignment statements using the assignment operator (:=).

lsolve function

The ***lsolve*** function is used similar to a Solve Block because you can solve for x equations and x unknowns. It is also similar to the ***polyroots*** function because the input arguments are vectors representing the coefficients of the equations. The use of this function is illustrated in Figure 15.9.

We can use a Solve Block to solve for four equations and four unknowns

Given $w := 1 \quad x := 1 \quad y := 1 \quad z := 1$ Initial guess values

$$2.4w + 6.1x + 6.6y + 2.1z = 2.5$$

$$-7.5w - 1.5x - 2.3y + 4.5z = 14$$

$$4.3w + 4.2x + 1.8y + 2.1z = 10.2$$

$$5.8w + 3.4x + 7.8y + 1.5z = 0.1$$

$$\mathbf{Solution_A} := \mathbf{Find(w, x, y, z)}$$

$$\mathbf{Solution_A} = \begin{pmatrix} 0.7686 \\ 0.3739 \\ -1.4480 \\ 3.7766 \end{pmatrix}$$

There is another way to solve for this condition. It is a function called *lsolve*.

This function is similar to the function polyroots, in that you use an array as input. To use this function, create a Matrix (M) of the coefficients of the four unknowns. Then create a vector of the constraints (V). The function takes the form *lsolve*(M,V)

$$\mathbf{Matrix} := \begin{pmatrix} 2.4 & 6.1 & 6.6 & 2.1 \\ -7.5 & -1.5 & -2.3 & 4.5 \\ 4.3 & 4.2 & 1.8 & 2.1 \\ 5.8 & 3.4 & 7.8 & 1.5 \end{pmatrix} \qquad \mathbf{Vector} := \begin{pmatrix} 2.5 \\ 14 \\ 10.2 \\ 0.1 \end{pmatrix}$$

With a Solve Block, you must give initial guess values. You do not need to give initial guesses with the *lsolve* function.

$$\mathbf{Solution_B} := \mathbf{lsolve(Matrix, Vector)} \qquad \mathbf{Solution_B} = \begin{pmatrix} 0.7686 \\ 0.3739 \\ -1.4480 \\ 3.7766 \end{pmatrix}$$

Check

$$\mathbf{Matrix} \cdot \mathbf{Solution_B} = \begin{pmatrix} 2.50 \\ 14.00 \\ 10.20 \\ 0.10 \end{pmatrix} \qquad \mathbf{Vector} = \begin{pmatrix} 2.50 \\ 14.00 \\ 10.20 \\ 0.10 \end{pmatrix}$$

OK

Figure 15.9 Using function ***lsolve***

TOL, CTOL, and *Minerr*

TOL and CTOL are built-in Mathcad variable names. They can be set in the Worksheet Options dialog box (**Worksheet** options from the **Format** menu). The values can also be redefined within the worksheet.

When using Solve Blocks and the ***root*** function, Mathcad iterates a solution. When the difference between the two most recent iterations is less than TOL (convergence tolerance), Mathcad arrives at a solution. The default value of TOL is 0.001. If you are not satisfied with the solutions arrived at, you can redefine TOL to be a smaller number. This will increase the precision of the result, but it will also increase the calculation time, or may make it impossible for Mathcad to arrive at a solution.

The CTOL (constraint tolerance) built-in variable, tells Mathcad how closely a constraint in a Solve Block must be met for a solution to be acceptable. The default value of CTOL is 0.001. Thus if a Solve Block Constraint is $x > 3$, this constraint is satisfied if $x > 2.9990$.

The ***Minerr*** function is used in place of the ***Find*** function in a Solve Block. You may want to use ***Minerr*** if Mathcad cannot iterate a solution. The ***Find*** function iterates a solution until the difference in the two most recent iterations is less than TOL. The ***Minerr*** function uses the last iteration even if it falls outside of TOL. This function is useful to use where ***Find*** fails to find a solution. If you use ***Minerr***, it is important to check the solution to see if the solution falls within your acceptable limits.

For more information on these topics, refer to the Mathcad Help. It has a good discussion of these topics in much greater depth.

Using units

You may use units with ***root***, ***polyroots***, and Solve Blocks. These will be illustrated in the following Engineering Examples.

Engineering Examples

Engineering Example 15.1 summarizes six different ways of solving for the time it takes to travel a specific distance.

Engineering Example 15.1 Object in motion

The general equation for the distance traveled by an object is given by the function $\mathbf{Dist} := v_0 \cdot t + \frac{1}{2} a \cdot t^2$.

Let's look at the different way of solving for this equation for the time needed to travel a specific distance.
Given: v_0=10m/s, a=6m/sec^2, and Dist=5000m

$v_0 := 10\frac{\mathbf{m}}{\mathbf{sec}}$ $\quad a := 6\frac{\mathbf{m}}{\mathbf{sec}^2}$ $\quad \mathbf{Dist} := 5000\mathbf{m}$

First, use symbolics menu.

To do this, write the equation (collect terms on the same side of the equation and set it equal to zero) using the boolean equal sign (CTRL+PERIOD).
Select the variable t.
From the Symbolics toolbar, select Variable then Solve

$$v_0 \cdot t + \frac{1}{2} \cdot a \cdot t^2 - \mathbf{Dist} = 0$$

$$\begin{bmatrix} \frac{1}{2 \cdot a} \cdot \left[(-2) \cdot v_0 + 2 \cdot \left(v_0^2 + 2 \cdot a \cdot \mathbf{Dist}\right)^{\frac{1}{2}}\right] \\ \frac{1}{2 \cdot a} \cdot \left[(-2) \cdot v_0 - 2 \cdot \left(v_0^2 + 2 \cdot a \cdot \mathbf{Dist}\right)^{\frac{1}{2}}\right] \end{bmatrix}$$

$$\mathrm{Time\_1}_1 := \frac{1}{2 \cdot a} \cdot \left[(-2) \cdot v_0 + 2 \cdot \left(v_0^2 + 2 \cdot a \cdot \mathbf{Dist}\right)^{\frac{1}{2}}\right]$$

$$\mathrm{Time\_1}_2 := \frac{1}{2 \cdot a} \cdot \left[(-2) \cdot v_0 - 2 \cdot \left(v_0^2 + 2 \cdot a \cdot \mathbf{Dist}\right)^{\frac{1}{2}}\right]$$

Hint: Copy and paste the equations rather than retype them.

$\mathrm{Time\_1}_1 = 39.19\ \mathbf{s}$

$\mathrm{Time\_1}_2 = -42.53\ \mathbf{s}$ $\quad \mathrm{Time\_1} = \begin{pmatrix} 39.19 \\ -42.53 \end{pmatrix} \mathbf{s}$

Second, use live symbolics

$$\mathrm{Time\_2a} := v_0 \cdot t + \frac{1}{2} \cdot a \cdot t^2 - \mathbf{Dist}\ \mathrm{solve}, t \rightarrow \begin{bmatrix} 10 \cdot \left(\frac{-1}{6} + \frac{1}{6} \cdot 601^{\frac{1}{2}}\right) \cdot \mathbf{sec} \\ 10 \cdot \left(\frac{-1}{6} - \frac{1}{6} \cdot 601^{\frac{1}{2}}\right) \cdot \mathbf{sec} \end{bmatrix}\ \mathrm{float} \rightarrow \begin{bmatrix} 39.192168907104209514 \cdot \mathbf{sec} \\ (-42.525502240437542848) \cdot \mathbf{sec} \end{bmatrix}$$

Hide the first solution

$$\mathrm{Time\_2b} := v_0 \cdot t + \frac{1}{2} \cdot a \cdot t^2 - \mathbf{Dist}\ \left|\begin{matrix}\mathrm{solve}, t \\ \mathrm{float}\end{matrix}\right. \rightarrow \begin{bmatrix} 39.192168907104209514 \cdot \mathbf{sec} \\ (-42.525502240437542848) \cdot \mathbf{sec} \end{bmatrix}$$

$\mathrm{Time\_2b}_1 = 39.19\ \mathbf{s}$ $\quad \mathrm{Time\_2b}_2 = -42.53\ \mathbf{s}$ $\quad \mathrm{Time\_2b} = \begin{pmatrix} 39.19 \\ -42.53 \end{pmatrix} \mathbf{s}$

Third, create a function using live symbolics

Define the function

$$\mathrm{TimeFunction\_3}(v_0, a, \mathbf{Dist}) := v_0 \cdot t + \frac{1}{2} a \cdot t^2 - \mathbf{Dist}\ \mathrm{solve}, t \rightarrow \begin{bmatrix} \frac{1}{2 \cdot a} \cdot \left[(-2) \cdot v_0 + 2 \cdot \left(v_0^2 + 2 \cdot a \cdot \mathbf{Dist}\right)^{\frac{1}{2}}\right] \\ \frac{1}{2 \cdot a} \cdot \left[(-2) \cdot v_0 - 2 \cdot \left(v_0^2 + 2 \cdot a \cdot \mathbf{Dist}\right)^{\frac{1}{2}}\right] \end{bmatrix}$$

$\mathrm{Time\_3a} := \mathrm{TimeFunction\_3}(v_0, a, \mathbf{Dist})$ Apply the arguments to the function

$$\mathrm{Time\_3a} = \begin{pmatrix} 39.19 \\ -42.53 \end{pmatrix} \mathbf{s}$$

Engineering Example 15.1 continued

This method allows you to use other input variables very easily.

$$\mathbf{Time\_3b} := \mathbf{TimeFunction\_3}\left(20\frac{m}{s}, 20\frac{m}{s^2}, 100000m\right)$$

$$\mathbf{Time\_3b} = \begin{pmatrix} 99.00 \\ -101.00 \end{pmatrix} s$$

Fourth, use the function *root*

Solving for the positive root.

$$\mathbf{Time\_4_1} := \mathbf{root}\left(v_0 \cdot t + \frac{1}{2} a \cdot t^2 - Dist, t, 0sec, 100sec\right) \qquad \mathbf{Time\_4_1} = 39.19\ s$$

$$\mathbf{Time\_4_2} := \mathbf{root}\left(v_0 \cdot t + \frac{1}{2} a \cdot t^2 - Dist, t, -100sec, 0sec\right) \qquad \mathbf{Time\_4_2} = -42.53\ s \qquad \mathbf{Time\_4} = \begin{pmatrix} 39.19 \\ -42.53 \end{pmatrix} s$$

Fifth, use the function *polyroots*

$$v_0 = 10.00\ \frac{m}{s} \qquad Dist = 5000.00\ m \qquad a = 6.00\ \frac{m}{s^2}$$

$$\mathbf{Vector} := \begin{pmatrix} \dfrac{-Dist}{m} \\ \dfrac{v_0}{\frac{m}{sec}} \\ \dfrac{\frac{1}{2} \cdot a}{\frac{m}{sec^2}} \end{pmatrix}$$

Remember that all units in a vector must be of the same unit dimension. Divide each variable by its units to create unitless numbers.

$$\mathbf{Time\_5} := \mathbf{polyroots(Vector)} \cdot sec$$

$$\mathbf{Time\_5} = \begin{pmatrix} -42.53 \\ 39.19 \end{pmatrix} s \qquad \mathbf{Time\_5_1} = -42.53\ s$$

$$\mathbf{Time\_5_2} = 39.19\ s \qquad \mathbf{Time\_5} = \begin{pmatrix} -42.53 \\ 39.19 \end{pmatrix} s$$

Sixth, use a Solve Block

$t := 100sec$ Initial Guess for time to find positive solution

Given

$$v_0 \cdot t + \frac{1}{2} \cdot a \cdot t^2 - Dist = 0$$

$\mathbf{Time\_6_1} := \mathbf{Find(t)} \quad \mathbf{Time\_6_1} = 39.19\ s$

$t := -30sec$ Initial Guess for time to find negative solution

Given

$$v_0 \cdot t + \frac{1}{2} \cdot a \cdot t^2 - Dist = 0$$

$\mathbf{Time\_6_2} := \mathbf{Find(t)} \quad \mathbf{Time\_6_2} = -42.53\ s$

Given $t := -15sec$

An initial negative guess for time would normally find a negative solution, but adding an additional constraint finds the positive solution.

$$v_0 \cdot t + \frac{1}{2} \cdot a \cdot t^2 - Dist = 0$$

$t > 0sec$ Set constraint to find positive solution

$\mathbf{Time\_6_3} := \mathbf{Find(t)} \quad \mathbf{Time\_6_3} = 39.19\ s$

Engineering Example 15.2 calculates the current flow in a multi-loop circuit.

Engineering Example 15.2 Electrical network

Using Kirchoff's current law, calculate the current flowing in this network. Kirchoff's current law states that the algebraic sum of voltage drops around any closed path within a circuit equals the sum of the voltage sources.

ΣV=Σ(I*R).

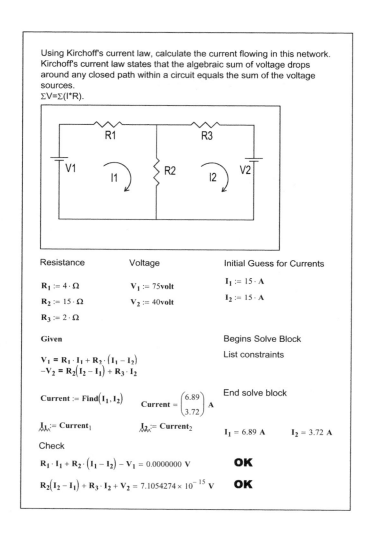

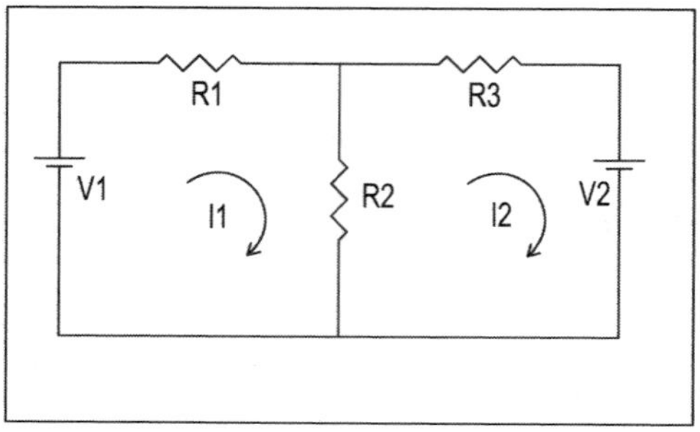

Resistance	Voltage	Initial Guess for Currents
$R_1 := 4 \cdot \Omega$	$V_1 := 75\text{volt}$	$I_1 := 15 \cdot A$
$R_2 := 15 \cdot \Omega$	$V_2 := 40\text{volt}$	$I_2 := 15 \cdot A$
$R_3 := 2 \cdot \Omega$		

Given — Begins Solve Block

List constraints

$$V_1 = R_1 \cdot I_1 + R_2 \cdot (I_1 - I_2)$$
$$-V_2 = R_2(I_2 - I_1) + R_3 \cdot I_2$$

$\text{Current} := \text{Find}(I_1, I_2)$ $\quad \text{Current} = \begin{pmatrix} 6.89 \\ 3.72 \end{pmatrix} A$ — End solve block

$I_1 := \text{Current}_1$ $\quad I_2 := \text{Current}_2$ $\quad I_1 = 6.89\ A$ $\quad I_2 = 3.72\ A$

Check

$R_1 \cdot I_1 + R_2 \cdot (I_1 - I_2) - V_1 = 0.0000000\ V$ **OK**

$R_2(I_2 - I_1) + R_3 \cdot I_2 + V_2 = 7.1054274 \times 10^{-15}\ V$ **OK**

Engineering Example 15.3 calculates the flow of water in a pipe network.

Engineering Example 15.3 Pipe network

In a pipe network, two conditions must be satisfied: 1) The algebraic sum of the pressure drops around any closed loop must be zero (The pressure at any point is the same no matter how you get there), and 2) The flow entering a junction must equal the flow leaving it.

Use the Darcy-Weisbach equation for head loss $h_L := f \cdot \frac{L}{D} \cdot \frac{V^2}{2g}$. The factor f is a function of the relative roughness of the pipe. For this example assume f=0.02.

$cfs := \frac{ft^3}{sec}$ $\quad f := 0.02$ $\quad TOL = 0.00100000000000000$ For this example, TOL=0.001 is adequate.

Calculate the flow in each pipe given that:

Flow into A is 1.2 cfs
Flow out of C is 0.8 cfs
Flow out of E is 0.4 cfs

Pipe lengths and diameters		Initial Guess
$L_{AB} := 3000ft$	$D_{AB} := 4in$	$Q_{AB} := .6cfs$
$L_{BC} := 4000ft$	$D_{BC} := 5in$	$Q_{BC} := .6cfs$
$L_{AE} := 2000ft$	$D_{AE} := 3in$	$Q_{AE} := .6cfs$
$L_{ED} := 3000ft$	$D_{ED} := 8in$	$Q_{ED} := .6cfs$
$L_{DC} := 5000ft$	$D_{DC} := 6in$	$Q_{DC} := .6cfs$
$L_{BD} := 3000ft$	$D_{BD} := 7in$	$Q_{BD} := .6cfs$

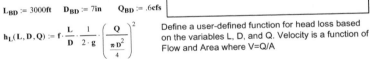

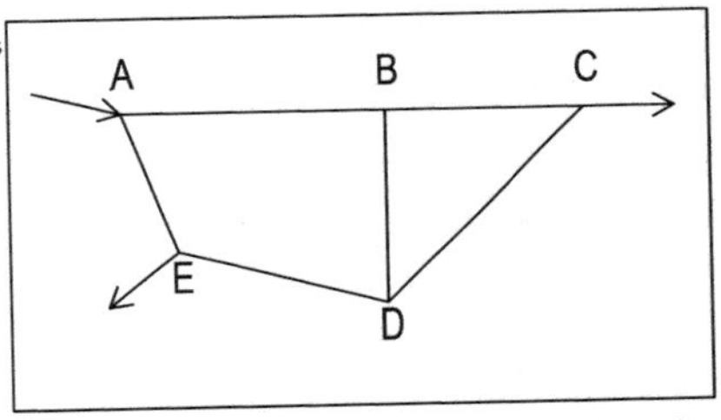

$$h_L(L, D, Q) := f \cdot \frac{L}{D} \cdot \frac{1}{2 \cdot g} \cdot \left(\frac{Q}{\frac{\pi \cdot D^2}{4}} \right)^2$$

Define a user-defined function for head loss based on the variables L, D, and Q. Velocity is a function of Flow and Area where V=Q/A

Given Begin Solve Block

Write equation for head loss at Point C.

$$\left(h_L(L_{AB}, D_{AB}, Q_{AB}) + h_L(L_{BC}, D_{BC}, Q_{BC})\right) - \left(h_L(L_{AE}, D_{AE}, Q_{AE}) + h_L(L_{ED}, D_{ED}, Q_{ED}) + h_L(L_{DC}, D_{DC}, Q_{DC})\right) = 0$$

Write equation for head loss at Point D.

$$\left(h_L(L_{AB}, D_{AB}, Q_{AB}) + h_L(L_{BD}, D_{BD}, Q_{BD})\right) - \left(h_L(L_{AE}, D_{AE}, Q_{AE}) + h_L(L_{ED}, D_{ED}, Q_{ED})\right) = 0$$

Write equation for head loss at Point B.

$$\left(h_L(L_{BC}, D_{BC}, Q_{BC})\right) - \left(h_L(L_{BD}, D_{BD}, Q_{BD}) + h_L(L_{DC}, D_{DC}, Q_{DC})\right) = 0$$

Write relationship between various flows into each of the points.

$1.2cfs = Q_{AB} + Q_{AE}$ $\quad Q_{AB} = Q_{BD} + Q_{BC}$

$Q_{AE} = 0.4cfs + Q_{ED}$ $\quad Q_{ED} + Q_{BD} = Q_{DC}$

$Q_{BC} + Q_{DC} = 0.8cfs$

Continued

Engineering Example 15.3 continued

$$\begin{pmatrix} Q_{AB} \\ Q_{BC} \\ Q_{AE} \\ Q_{ED} \\ Q_{DC} \\ Q_{BD} \end{pmatrix} := \mathbf{Find}(Q_{AB}, Q_{BC}, Q_{AE}, Q_{ED}, Q_{DC}, Q_{BD})$$

End Solve Block

$Q_{AB} = 0.749210\,\mathbf{cfs}$

$Q_{BC} = 0.351147\,\mathbf{cfs}$

$Q_{AE} = 0.450790\,\mathbf{cfs}$

$Q_{ED} = 0.050790\,\mathbf{cfs}$

$Q_{DC} = 0.448853\,\mathbf{cfs}$

$Q_{BD} = 0.398063\,\mathbf{cfs}$

Check results

$$(h_L(L_{AB}, D_{AB}, Q_{AB}) + h_L(L_{BC}, D_{BC}, Q_{BC})) - (h_L(L_{AE}, D_{AE}, Q_{AE}) + h_L(L_{ED}, D_{ED}, Q_{ED}) + h_L(L_{DC}, D_{DC}, Q_{DC})) = 5.93 \times 10^{-6}\,\mathbf{ft}$$

$$(h_L(L_{AB}, D_{AB}, Q_{AB}) + h_L(L_{BD}, D_{BD}, Q_{BD})) - (h_L(L_{AE}, D_{AE}, Q_{AE}) + h_L(L_{ED}, D_{ED}, Q_{ED})) = 4.30 \times 10^{-7}\,\mathbf{ft}$$

$$(h_L(L_{BC}, D_{BC}, Q_{BC})) - (h_L(L_{BD}, D_{BD}, Q_{BD}) + h_L(L_{DC}, D_{DC}, Q_{DC})) = 5.50 \times 10^{-6}\,\mathbf{ft}$$

$Q_{AB} + Q_{AE} = 1.200000\,\mathbf{cfs}$ **Equals 1.2 cfs OK**

$Q_{AE} - Q_{ED} = 0.400000\,\mathbf{cfs}$ **Equals 0.4 cfs OK**

$Q_{BC} + Q_{DC} = 0.800000\,\mathbf{cfs}$ **Equals 0.8 cfs OK**

Summary

The Mathcad solve features are essential power tools to have in your toolbox. They should be kept sharp and used often.

In Chapter 15 we:

- Introduced the ***root*** function and showed how it can be used to solve for a single variable with a single function.
- Showed how the ***root*** function can be used with an initial guess prior to using ***root***, or how two guesses can be included as arguments in the ***root*** function.
- Encouraged the use of a plot to help define the initial guess values for the ***root*** function.
- Introduced the ***polyroots*** function and showed how it can be used to solve for all the roots of a polynomial equation.

- Briefly introduced the two solution methods used by the ***polyroots*** function (LaGuerre and Companion Matrix Methods).
- Introduced Solve Blocks and showed how they can be used to find solutions to multiple unknowns using multiple equations.
- Emphasized the rules of using ***Given*** and ***Find***.
- Discussed the built-in variables TOL and CTOL.
- Told how the ***Minerr*** function can be used in place of the ***Find*** function if Mathcad is not finding a solution.
- Provided several engineering examples to illustrate the use of the solve features.

Practice

1. Create ten polynomial equations. Use at least 2nd- and 3rd-order equations. Plot each equation. Use both the bracketed ***root*** function and unbracketed ***root*** function to solve for each equation.
2. Use the ***polyroots*** function to solve for each equation used in Exercise 1.
3. Create three 5th-order equations, and use the ***polyroots*** function to solve for all roots. Use the live symbolics keyword "coeffs" to create the polynomial vector.
4. From your field of study, create four problems where you can use a Solve Block to solve for at least two unknowns with at least two constraints.

16
Advanced programming

Mathcad programs can be much more powerful than the logic programs discussed in Chapter 12. This chapter expands the topic of programming.

Chapter 16 will:

- Introduce the local definition symbol, and gives examples of its use.
- Discuss the use of the other programming commands such as: ***for***, ***while***, ***break***, ***continue***, ***return***, and ***on error***. Several examples will be given for the use of these features.
- Illustrate the use of programs within programs and the use of subroutines.
- Give examples which combine the features of programming, user-defined functions, and solving.

Local definition

Mathcad allows you to define new variables within a program. These variables will be local to the program, meaning they will be undefined outside of the program. These are called local variables, and they are assigned with the local assignment operator. The local assignment operator is represented by an arrow pointing left $\leftarrow$. It can be found on the Programming toolbar, or inserted by pressing [{]. Let's look at a simple example of a program that uses local variables. See Figure 16.1.

Create a user-defined function that will find the two roots of a quadratic equation. Use local variables to simplify the program.

Use the left arrow button on the Programming toolbar to insert the local variable operator.

The local variables are only defined within the program. They are not defined outside the program.

$$\mathbf{QuadRoots(a,b,c)} := \left| \begin{array}{l} \mathbf{Part} \leftarrow \sqrt{\mathbf{b}^2 - 4 \cdot \mathbf{a} \cdot \mathbf{c}} \\ \mathbf{x_1} \leftarrow \dfrac{-\mathbf{b} + \mathbf{Part}}{2\mathbf{a}} \\ \mathbf{x_2} \leftarrow \dfrac{-\mathbf{b} - \mathbf{Part}}{2\mathbf{a}} \\ \begin{pmatrix} \mathbf{x_1} \\ \mathbf{x_2} \end{pmatrix} \end{array} \right.$$

Define a local variable "Part."

Define a local variable "x_1."

Define a local variable "x_2."

Create an output vector comprising local variables x_1 and x_2. The values of x_1 and x_2 will be the output results of the function, but the local variables will not be recognized outside of the function.

$$\mathbf{QuadRoots}(1,5,6) = \begin{pmatrix} -2.00 \\ -3.00 \end{pmatrix}$$

Results of Function

$\mathbf{x_1} := \blacksquare$ $\qquad$ $\mathbf{Part} := \blacksquare$

The local variables are not recognized outside of the program.

$$\mathbf{Roots} := \mathbf{QuadRoots}(1,5,6)$$

$$\mathbf{Roots} = \begin{pmatrix} -2.00 \\ -3.00 \end{pmatrix} \qquad \mathbf{Roots_1} = -2.00 \qquad \mathbf{Roots_2} = -3.00$$

In order to make the results of the function available later on, the variable "Roots" was defined equal to the function QuadRoots.

Figure 16.1 Local variables

As you can see from Figure 16.1, local variables can help simplify a program by breaking the program down into smaller pieces. In Figure 16.2, the function is very long, but each numerator is the same. By assigning a local variable, the function definition can be simplified.

Local variables can help simplify complex equations

The numerators in this function are all the same.

$$G_1(x,y,z) := \frac{\left(\frac{x}{y}\right)+\left(\frac{y}{2x}\right)^2+\left(\frac{x}{3z}\right)^3+\left(\frac{z}{4x}\right)^4}{x} + \frac{\left(\frac{x}{y}\right)+\left(\frac{y}{2x}\right)^2+\left(\frac{x}{3z}\right)^3+\left(\frac{z}{4x}\right)^4}{y} \ldots$$
$$+ \frac{\left(\frac{x}{y}\right)+\left(\frac{y}{2x}\right)^2+\left(\frac{x}{3z}\right)^3+\left(\frac{z}{4x}\right)^4}{z}$$

$G_1(2,6,3) = 2.61$

Note: Use CTRL+ENTER to break the expression at an addition operator.

If you define a local variable, the function is much simpler to follow.

$$G_2(x,y,z) := \left| \begin{array}{l} a \leftarrow \left(\frac{x}{y}\right)+\left(\frac{y}{2x}\right)^2+\left(\frac{x}{3z}\right)^3+\left(\frac{z}{4x}\right)^4 \\ \frac{a}{x}+\frac{a}{y}+\frac{a}{z} \end{array} \right.$$

$G_2(2,6,3) = 2.61$

Figure 16.2 Local variables

You can also use local variable as counters to count how many times certain things are done in a program, or to count how many true statements are in a program. Figure 16.3 illustrates this concept.

Local variables can be used as counters.

```
Counter(x) := | m ← 0
              | m ← m + 1  if  x > 4
              | m ← m + 1  if  x > 3
              | m ← m + 1  if  x > 2
              | m ← m + 1  if  x > 1
              | m ← m + 1  if  x > 0
              | m
```

Define a local variable "m" equal to zero.

This program adds 1 to the local variable "m" for every true statement.

$\mathbf{Counter}(-1) = 0.00$

$\mathbf{Counter}(1) = 1.00$

$\mathbf{Counter}(3) = 3.00$

$\mathbf{Counter}(5) = 5.00$

Check the program for different arguments.

In order to check a program for several different arguments, you can use a range variable or a vector. The example below use both a range variable and a vector to provide different arguments to the function "Counter."

Using a Range Variable

$\mathbf{i} := -1, 0 .. 7$

i =	Counter(i) =
-1.00	0.00
0.00	0.00
1.00	1.00
2.00	2.00
3.00	3.00
4.00	4.00
5.00	5.00
6.00	5.00
7.00	5.00

Using a Vector

The vectorize operator (CTRL+MINUS SIGN) must be used in order to use the vector with this program.

$$\mathbf{b} := \begin{pmatrix} -1 \\ 0 \\ 1 \\ 2 \\ 3 \\ 4 \\ 5 \\ 6 \\ 7 \end{pmatrix} \qquad \overrightarrow{\mathbf{Counter}(\mathbf{b})} = \begin{pmatrix} 0.00 \\ 0.00 \\ 1.00 \\ 2.00 \\ 3.00 \\ 4.00 \\ 5.00 \\ 5.00 \\ 5.00 \end{pmatrix}$$

Figure 16.3 Local variables as counters

Looping

Looping allows program steps to be repeated a certain number of times, or until a certain criteria is met. This book will introduce the concept of looping. An in-depth study of looping is beyond the scope of this book. The Mathcad Help and Tutorial provide excellent coverage on this subject.

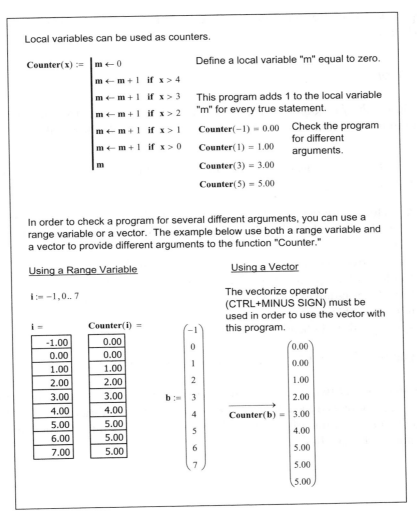

There are two types of loop structures. The first allows a program to execute a specific number of times. This is called a ***for*** loop. The second type of loop structure will execute until a specific condition is met. This is called a ***while*** loop.

For loops

You insert the ***for*** loop into a program by clicking on the **for** button on the Programming toolbar. Do not type the word "for." A ***for*** loop takes the form

for $x \in y$
  z

. Mathcad evaluates z for each value of x over the range y. The placeholders have these meanings:

- x is a local variable within the program defined by the ***for*** symbol. It initially takes the first value of y. The value of x changes for each step in the value of y.
- y is referred to as an iteration variable. It is usually a range variable, but it can be a vector, or a series of scalars or arrays separated by commas. Each time the program loops, the next value of y is used and this value is assigned to the variable x.
- z is any valid Mathcad expression or sequence of expressions.

The number of times a ***for*** loop executes is controlled by y.

Let's first look at two examples using range variables. In the examples, think of the ∈ symbol as a local variable definition. The local variable "i" is a range variable, and the value of variable "i" increments for each loop. There are two examples in Figure 16.4.

$j := 1, 2 .. 5$

$$\text{ForLoop}_1(x) := \left| \begin{array}{l} a \leftarrow 0 \\ \text{for } i \in 1, 2 .. x \\ \quad a \leftarrow a + 1 \end{array} \right.$$

Define a local variable "a" equal to zero.

"i" is a local range variable that takes on the values 1,2,3... until x. For this example, i is not used anywhere. It only controls when the loop stops.

j =	ForLoop₁(j) =
1.00	1.00
2.00	2.00
3.00	3.00
4.00	4.00
5.00	5.00

For the first loop, "a" takes the value 0+1=1.
For the second loop "i" increments to 2, and "a" becomes 1+1=2.
For the third loop "i" increments to 3, and "a" becomes 2+1=3.
The loop continues until "i" reaches the value of x.

For this next example, increment "a" by the value of i.

$$\text{ForLoop}_2(x) := \left| \begin{array}{l} a \leftarrow 0 \\ \text{for } i \in 1, 2 .. x \\ \quad a \leftarrow a + i \end{array} \right.$$

Define "a" as a local variable.

"i" is a local range variable that takes on the values 1,2,3... until x.

j =	$\text{ForLoop}_2(j)$ =
1.00	1.00
2.00	3.00
3.00	6.00
4.00	10.00
5.00	15.00

For the first loop, "i" is one, and "a" takes the value 0+1=1.
For the second loop, "i" is 2 and "a" becomes 1+2=3
For the third loop, "i" is 3, and "a" becomes 3+3=6.
For the fourth loop, "i" is 4, and "a" becomes 6+4=10.
For the fifth loop, "i" is 5, and "a" becomes 10+5=15.

Figure 16.4 ***For*** loop using range variables

Figure 16.5 shows a *for* loop using a list as the iteration variable. Notice that the list of numbers is not consecutive, and the numbers do not increment by the same value every time. When Mathcad loops, it moves to the next value in the list and assigns the value in the list to the local variable "i." Figure 16.6 is the same as Figure 16.5 except that the iteration variable is a previously defined vector containing the same numbers as the list in Figure 16.5.

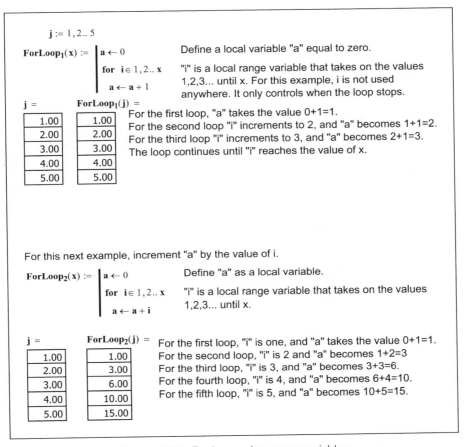

$$\mathbf{ForLoop_3(x)} := \left| \begin{array}{l} \mathbf{a} \leftarrow 0 \\ \mathbf{for}\ \ \mathbf{i} \in 1,2,5,3 \\ \quad \mathbf{a} \leftarrow \mathbf{x} \cdot (\mathbf{a}+\mathbf{i}) \\ \mathbf{a} \end{array} \right.$$

In this example the iteration variable is a list that controls the number of loops. The list also sets the assigned values of "i" .

The local variable "a" is incremented by the value of "i" and then multiplied by the argument "x".

$\mathbf{ForLoop_3}(2) = 58.00$

For x=2, the result is 2(0+1),2*(2+2),2*(8+5),2*(26+3), The final result is 2*(26+3)=58

$\mathbf{j} =$	$\mathbf{ForLoop_3(j)} =$
1.00	11.00
2.00	58.00
3.00	189.00
4.00	476.00
5.00	1015.00

Figure 16.5 ***For*** loop using a list

$$\mathbf{Vector} := \begin{pmatrix} 1 \\ 2 \\ 5 \\ 3 \end{pmatrix}$$

$$\mathbf{ForLoop_4(x)} := \left| \begin{array}{l} \mathbf{a} \leftarrow 0 \\ \mathbf{for}\ \ \mathbf{i} \in \mathbf{Vector} \\ \quad \mathbf{a} \leftarrow \mathbf{x}(\mathbf{a}+\mathbf{i}) \end{array} \right.$$

This is exactly the same as in the previous figure except that the iteration is controlled by a previously defined vector.

$\mathbf{ForLoop_4}(2) = 58.00$

$\mathbf{j} =$	$\mathbf{ForLoop_4(j)} =$
1.00	11.00
2.00	58.00
3.00	189.00
4.00	476.00
5.00	1015.00

Figure 16.6 ***For*** loop using a vector

Figure 16.7 uses two local variables. It uses the second variable "b" as a means to sum all the different values of "a."

This example sums each value of "a."

```
ForLoop5(x) :=  | a ← 0
                | b ← 0
                | for i ∈ 1, 2.. x
                |     | a ← a + i
                |     | b ← b + a
                | b
```

j =	$ForLoop_5(j)$ =
1.00	1.00
2.00	4.00
3.00	10.00
4.00	20.00
5.00	35.00

Define "a" and "b" as a local variables.

"i" is a local variable that takes on the values 1,2,3 until "x."

For the first loop "i" is one, and "a" takes the value 0+1=1, Thus b=0+1=1.
For the next loop "i" is 2 and "a" becomes 1+2=3 and b=1+3=4.
For the third loop "i" is 3, and "a" becomes 3+3=6 and b=4+6=10.
For the fourth loop "i" is 4, and "a" becomes 6+4=10 and b=10+10=20.
For the fifth loop "i" is 5, and "a" becomes 10+5=15 and b=20+15=35.

Figure 16.7 ***For*** loop using two local variables

Chapter 13 showed two methods of converting range variables to vectors. The third method was not discussed because ***for*** loops had not yet been introduced. We now provide the third method of converting a range variable to a vector. This is illustrated in Figure 16.8.

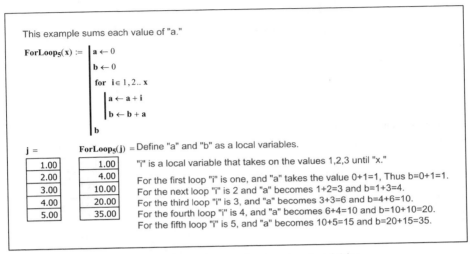

$$\mathbf{Range2Vec}(s_1, s_2, e) := \left| \begin{array}{l} \mathbf{Count} \leftarrow \mathbf{ORIGIN} \\ \mathbf{for}\ \ i \in s_1, s_2 .. e \\ \quad \left| \begin{array}{l} \mathbf{Vec}_{\mathbf{Count}} \leftarrow i \\ \mathbf{Count} \leftarrow \mathbf{Count} + 1 \end{array} \right. \\ \mathbf{Vec} \end{array} \right.$$

This function converts a range variable into a vector. Once it is a vector, each element can be accessed. This is based on a function in the February 2002 Mathcad Advisor Newsletter. Refer to Chapter 13 for two other methods of converting range variables to vectors. The ***for*** loop creates a range variable for iterating from the three input arguments. The vector "Vec" is a local variable.

Convert the following range variable to a vector beginning at ORIGIN.

RangeVariable_1 := 0, 0.1 .. 1.5

NewVector_1 := **Range2Vec**(0, 0.1, 1.5)

Refer to the discussion of range variables and vectors in Chapters 9 and 13.
These two variables appear to be identical, but they behave very differently.

$\mathbf{RangeVariable\_1}_3 = \blacksquare$

$\mathbf{RangeVariable\_1}_3 = \blacksquare\blacksquare$

This value must be a vector.

$\mathbf{NewVector\_1}_3 = 0.20$

RangeVariable_1 =

	1
1	0.00
2	0.10
3	0.20
4	0.30
5	0.40
6	0.50
7	0.60
8	0.70
9	0.80
10	0.90
11	1.00
12	1.10
13	1.20
14	1.30
15	1.40
16	1.50

NewVector_1 =

	1
1	0.00
2	0.10
3	0.20
4	0.30
5	0.40
6	0.50
7	0.60
8	0.70
9	0.80
10	0.90
11	1.00
12	1.10
13	1.20
14	1.30
15	1.40
16	1.50

Figure 16.8 ***For*** loop to convert a range variable to a vector

While loops

The ***while*** loop is different than the ***for*** loop. A ***for*** loop executes a fixed number of times, based on the size of the iteration variable. A ***while*** loop continues to execute while a specified condition is met. Once the condition is not true, the execution stops. If the condition is always true, Mathcad will go into an infinite loop. For this reason, you must be careful in using a ***while*** loop. You insert the ***while*** loop into a program by clicking on the **while** button on the Programming toolbar. Do not type the word "while."

Figure 16.9 shows a simple ***while*** loop. The program continues to loop until "a" is equal to or greater than the argument "x." When this happens, the loop stops and Mathcad returns the value of "a."

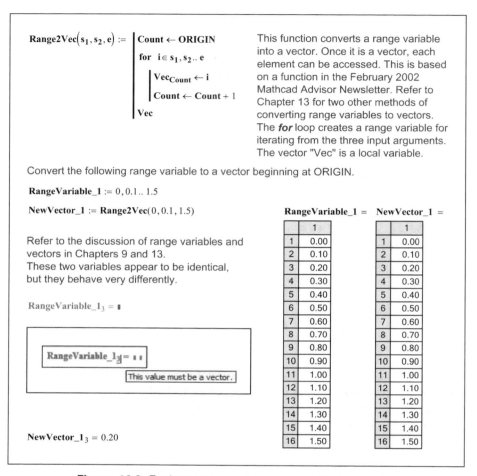

```
ForLoop1(x) := | a ← 0                Defines local variable.
               | while  a < x         Checks value of local variable.
               |   a ← a + 1          Increments local variable by one.
               | a                    Displays the value of "a" after the program stops.
```

j =	ForLoop$_1$(j) =
1.00	1.00
2.00	2.00
3.00	3.00
4.00	4.00
5.00	5.00

For the first loop "a" is zero, so the "while" statement is true and the program continues to the next line where "a" is incremented to the value 1.
On the second loop "a" is now 1 and the "while" statement is checked again. If it is true, then the program continues to the next line, and "a" is incremented to two. If the statement is false, then the program execution stops and the value of "a" is returned (1).
The loop continues until a >= x.

Figure 16.9 ***While*** loop

Figure 16.10 uses local variable "a" to control the number of loops. Local variable "b" is multiplied by 2 every time the program loops. Once "a" becomes equal to or greater than the argument "x," the loops stops and the program returns the value of "b."

```
WhileLoop1(x) := | a ← 0              Defines local variables.
                 | b ← 1
                 | while  a < x       Checks value of local variable "a".
                 |   | a ← a + 1      Increments "a" by one.
                 |   | b ← b · 2      Every time the program loops it multiplies "b" by 2
                 | b                  Displays the value of "b" after the program stops.
```

j =	WhileLoop$_1$(j) =
1.00	2.00
2.00	4.00
3.00	8.00
4.00	16.00
5.00	32.00

For the first loop "a" is zero, so the "while" statement is true and the program continues to the next lines where "a" is incremented to the value 1 and b is multiplied by 2.
On the second loop "a" is now 1 and the "while" statement is checked again. If a is less than x, then the program continues to the next line, and "a" is incremented to two and "b" becomes 2*2=4. If a is not less than x, then the program execution stops and the value of "b" is returned (4).

Figure 16.10 ***While*** loop

Figure 16.11 shows three ways to create a user-defined function for factorial. Mathcad already has the factorial operator on the Math toolbar, but it is interesting to use factorial to demonstrate the ***while*** loop.

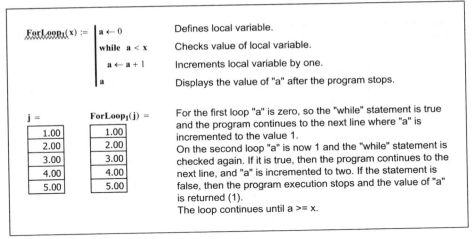

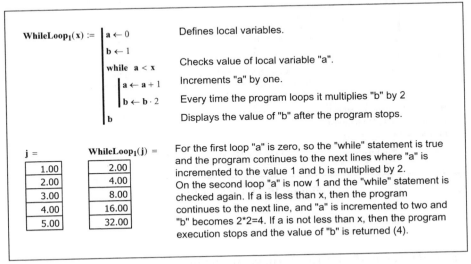

Mathcad already has a built in operator to calculate factorial, but let's look at a way to create a user-defined function for factorial.

```
FactorialNew(x) := | f ← x − 1          Assigns local variable "f" as one less than the argument "x"
                   | while f > 0        Checks if "f" is greater than zero
                   |   | x ← x·f        If "f" is greater than zero, reassigns "x" the value x*f
                   |   | f ← f − 1      Reduces value of "f" by one
                   | x                  Returns x if "f" is not greater than zero
```

j =	FactorialNew(j) =
1.00	1.00
2.00	2.00
3.00	6.00
4.00	24.00
5.00	120.00

Assuming x=4:
On the first loop, "f" is assigned f=4-1=3.
Since "f" is greater than zero, x=4*3=12, and "f" is reduced by one to f=3-1=2, and the program returns to the while statement.
On the next loop, f=2 is greater than zero, so x=12*2=24, and f=2-1=1.
On the next loop, f=1 is greater than zero, so x=24*1=24, and f=1-1=0.
On the next loop, "f" is not greater than zero (the statement is false), so the execution stops and the value of x=24 is returned.

The above function is not perfect. If a negative number is entered, then the value of the negative number is returned. Some **if** statements are need to resolve this issue.

Note:
There are additional ways to create the factorial function.
The following are listed in Mathcad Help.

```
Fac1(n) := | f ← 1
           | while n ← n − 1
           |   f ← f·(n + 1)
           | f

Fac1(4) = 24.00
```

This while statement is not a conditional statement, but the loop stops when the value of n=0 (a false condition).

Assuming n=4:
On the first loop the argument "n" is reduced by 1 and becomes 3. The value of f becomes f=1*(3+1)=4.
On the second loop, n=3-1=2, and f=4*(2+1)=12.
On the third loop, n=2-1=1, and f=12*(2+1)=24.
On the fourth loop, n=1-1=0. This means that this is a false condition for the while loop and the loop stops and returns the value of f=24.

The following is a recursive function, which means that the function refers to itself. Refer to "Recursion" in Mathcad Help.

```
Fac2(n) := | n · Fac2(n − 1)  if n > 1
           | 1  otherwise                 Fac2(5) = 120.00     5! = 120.00
```

Figure 16.11 Factorial function

The ***while*** loop in Figure 16.12 will cause an infinite loop because the condition will always be true. If this condition happens, you can interrupt the loop by pressing the escape [esc] key. Beginning with Mathcad 13, you can also hover your mouse over **Debug** on the **Tools** window and click **Interrupt**.

The following causes an infinite loop because "a" can never be anything but zero.

$$\mathbf{WhileLoop_2(x)} := \left| \begin{array}{l} \mathbf{a} \leftarrow 0 \\ \mathbf{while}\ \ \mathbf{a} \le \mathbf{x} \\ \quad \mathbf{a} \leftarrow \mathbf{a} \cdot 2 \end{array} \right.$$

If this condition happens, press escape (esc) to stop the execution of the loop. Beginning with Mathcad 13, you can also click **Interrupt** by hovering your mouse over **Debug** on the **Tools** menu.

Figure 16.12 Avoid infinite loops

Break and ***continue*** operators

The ***break*** and ***continue*** operators give more control to programs. They tell Mathcad what to do if specific conditions are met and usually precede an ***if*** statement. These operators are used to check for specific conditions during the time that Mathcad is looping with either the ***for*** loop or the ***while*** loop.

The ***break*** operator stops the loop if the condition is true, while the ***continue*** operator allows Mathcad to skip the current loop iteration if a true statement is encountered. The continue operator is useful if you want to continue looping even if a condition is true. For example, you may want to extract only even numbers. By using the ***continue*** operator, you can skip over odd numbers. Without the ***continue*** operator, Mathcad would stop if it encountered an odd number.

See Figure 16.13 for an example of the ***break*** operator. See Figure 16.14 for an example of the ***continue*** operator. These examples are relatively simple, and are intended to introduce the concept of using these two operators. More advanced examples are given in Mathcad Help.

This example is similar to Figures 16.9 and 16.10, but it allows you to give starting and ending numbers. In order to prevent an infinite loop, if the starting number is less than or equal to zero, the program is stopped by the break statement.

```
WhileLoop3(w, x) := | i ← 1                    Define local variable.
                    | while  w ≤ x             Continue looping as long as w <= x
                    |   | break  if  w ≤ 0     Note: To add the break operator, add
                    |   | a_i ← w              the if statement first, then click on
                    |   | w ← w · 2            the first placeholder and add the
                    |   | i ← i + 1            break operator. The break operator
                    |                          stops the loop if x<=0.
                    | a
```

This creates a vector using the value of "i" as the vector subscript and assigns it the value of "w".

Changes the values of "w" and "i".

Returns the vector "a".

$$\mathbf{WhileLoop_3}(1,32) = \begin{pmatrix} 1.00 \\ 2.00 \\ 4.00 \\ 8.00 \\ 16.00 \\ 32.00 \end{pmatrix}$$

$$\mathbf{WhileLoop_3}(0,32) = 0.00$$

$$\mathbf{WhileLoop_3}(-3,32) = 0.00$$

$$\mathbf{WhileLoop_3}(5,100) = \begin{pmatrix} 5.00 \\ 10.00 \\ 20.00 \\ 40.00 \\ 80.00 \end{pmatrix}$$

Figure 16.13 Using the ***break*** operator to prevent infinite loops

This example illustrates how the continue operator can skip over certain conditions within a loop. In this example, the program skips over elements that are not integers. The function uses argument x to divide into integers from 1 to argument y, and selects only the integers.

```
WhileLop4(x, y) := | i ← 0
                   | while i ≤ y
                   |   | j ← (i + 1)/x
                   |   | i ← i + 1
                   |   | continue  if floor(j) ≠ j
                   |   | a_i ← j
                   | a
```

Local variable "i" begins with 0, in order to increment before the continue statement.

Local variable "j" begins with j=(0+1)/x

Increment "i" by 1 to i=0+1=1

When you see the continue operator it means, "Continue with the loop if this statement is false. If the statement is true, then stop this loop and go back to the top of the loop." This statement is checking to see if "j" is an integer.

The local variable "a" is a vector with all the values of j.

Assume x=2, y=5:
On the first loop, i=0 and is less than 5 so the while loop is executed. The local variable j=(0+1)/2=0.5. Increment "i" by 1 to i=0+1=1. Check the continue statement. Floor(j)=0 is not equal to 0.5 so the statement is true. Since the statement is true, the continue operator tells Mathcad to stop the current loop and begin at the top of the loop.
On the second loop, i=1 and is less than 5 so the while loop is executed again. j=(1+1)/2=1.0. i=1+1=2. (Note: We had to increment i above the continue statement, because if the continue statement is true, the rest of the loop is skipped and "i" would not have been incremented. This would have created an infinite loop.) Floor(1)=1 is equal to 1.0 so the statement is false, and the loop continues. The vector "a" has element a_2 assigned the value 1.0. Because we did not assign element a_1, it is left at 0.0.

On the third loop, i=2 and is less than 5 so the while loop is executed again. j=(2+1)/2=1.5. i=2+1=3. Floor(1.5)=1 is not equal to 1.5. so the continue statement is true and the loop begins again at the top.
On the fourth loop, i=3 and is less than 5 so the while loop is executed again. j=(3+1)/2=2.0. i=3+1=4. Floor(2.0)=2 is equal to 2.0, so the continue statement is false, and the loop continues. The vector "a" has element a_4 assigned the value 2.0. Element a_3 is left at zero.

On the fifth loop, i=4 and is less than 5 so the while loop is executed again. j=(4+1)/2=2.5. i=4+1=5. Floor(2.5)=2 is not equal to 2.5. so the continue statement is true and the loop begins again at the top.
On the sixth loop, i=5 is less than or equal to 5 so the while loop is executed again. j=(5+1)/2=3.0. i=5+1=6. Floor (3.0)=3 is equal to 3.0, so the continue statement is false, and the loop continues. The vector "a" has element a_6 assigned the value 3.0. Element a_5 is left at zero.

On the next loop i=6 is not less than or equal to 5, so the while loop is stopped. The vector "a" is returned.

WhileLoop4(2, 5) =

	1
1	0.00
2	1.00
3	0.00
4	2.00
5	0.00
6	3.00

WhileLoop4(1, 9) =

	1
1	1.00
2	2.00
3	3.00
4	4.00
5	5.00
6	6.00
7	7.00
8	8.00
9	9.00
10	10.00

WhileLoop4(3, 9) =

	1
1	0.00
2	0.00
3	1.00
4	0.00
5	0.00
6	2.00
7	0.00
8	0.00
9	3.00

Figure 16.14 Using the ***continue*** operator

Return operator

We introduced the ***return*** operator in Chapter 12. This operator is used in conjunction with an **if** statement. It stops program execution when the **if** statement is true, and returns the value listed. See Figure 12.5 for an example.

On error operator

This operator is very useful to tell Mathcad what to do if it encounters an error. A very common error occurs when Mathcad tries to divide by zero. There will be many times when an input value to a function causes this to occur. The ***on error*** operator gives you the ability to tell Mathcad what to do when this occurs. See Figure 16.15 for several examples of using the ***on error*** operator.

The *on error* operator has the form x *on error* y. If y causes an error then x is evaluated.

▮ on error ▮

This figure give three methods of dealing with the function 24/x.

Result := 0 $\qquad$ j := −1, 0 .. 3

Example 1

$\mathbf{Example_1(x)} := \frac{24}{x}$

$\mathbf{Example_1(Result)}$ = ▮

Example 2

$\mathbf{Example_2(x)}$:= "Division by zero" $\mathbf{on\ error}\ \frac{24}{x}$

$\mathbf{Example_2(Result)}$ = "Division by zero"

Example 3

$\mathbf{Example_3(x)} := \infty\ \ \mathbf{on\ error}\ \frac{24}{x}$

$\mathbf{Example_3(Result)} = 1.00 \times 10^{307}$

Note: To insert the built-in constant infinity type CRTL+SHIFT+z.

$\mathbf{Example_1(j)}$ = ▮

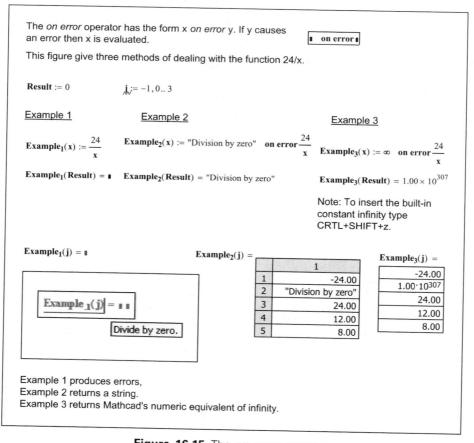

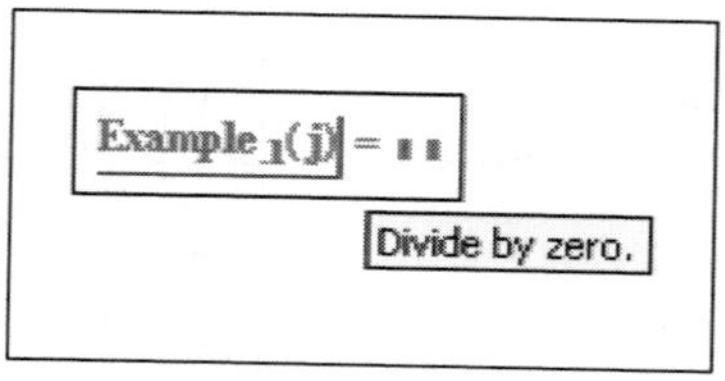

$\mathbf{Example_2(j)}$ =

	1
1	-24.00
2	"Division by zero"
3	24.00
4	12.00
5	8.00

$\mathbf{Example_3(j)}$ =

-24.00
$1.00 \cdot 10^{307}$
24.00
12.00
8.00

Example 1 produces errors,
Example 2 returns a string.
Example 3 returns Mathcad's numeric equivalent of infinity.

Figure 16.15 The ***on error*** operator

Summary

The two main concepts covered in this chapter are local variables and looping. These, added with the logical programming discussed in Chapter 12, can make some very powerful programs.

Local variables are useful to help simplify complex expressions. A local variable is a variable that is only defined within a program. It is not available outside the program. These local variables are essential when using loops, because they help control the loops.

There are two types of loops: ***for*** loops and ***while*** loops. The ***for*** loop loops a specific number of times and is controlled by an iteration variable. The ***while*** loop will loop until a specific condition is met. If the condition is never met, the loop will go indefinitely.

The ***break*** and ***continue*** operators are used to control when and how a program loops. The ***break*** operator stops the execution when a specific condition is met. The ***continue*** operator allows Mathcad to skip over a portion of a loop and move back to the beginning of the loop.

The ***return*** operator is used in conjunction with an **if** statement. It stops the execution of the program and returns the value following the return ***operator***.

The ***on error*** operator is very useful when a particular operation could cause an error, because you can tell Mathcad what to do when the error occurs.

Practice

1. From your field of study, create 10 programs that use the features discussed in this chapter.
2. Create five functions or expressions that use the ***on error*** operator.

17 Useful information—Part III

This chapter is a collection of topics and information related to the chapters in Part III. There will not be a summary or practice exercises.

Calculus

Mathcad will easily perform calculus operations using either the numeric or symbolic processors. The following is a brief discussion if you are interested in using Mathcad to perform calculus.

Differentiation

Let's start by using the symbolic processor. You can use the symbolics menu if you are only performing the operation once. If you are performing the operation more than once, or if you want to use the result for further calculations, then use the live symbolics operator.

To differentiate an expression using the symbolics menu, write the expression, select the variable and then click **Variables>Differentiate** from the **Symbolics** menu. The result is added below the expression. Remember that the result is static. It does not change if you revise the expression. In order to keep the result dynamic, use the live symbolics operator. To do this, you need to use the derivative operator from the Calculus toolbar and the symbolics operator.

Figure 17.1 compares the Symbolics menu and live symbolics operator.

Using the Symbolics Menu

To differentiate using the Symbolics menu, write the expression, select the variable, and select Tools>Variable>Differentiate. The result is static. It does not change if you change the expression.

$$x^3 + 3x^2 + 4x - 2 \qquad 2^x \qquad \sin(4x+1) \qquad \frac{x^2+2}{x^3-1}$$

$$3 \cdot x^2 + 6 \cdot x + 4 \qquad 2^x \cdot \ln(2) \qquad 4 \cdot \cos(4 \cdot x + 1) \qquad 2 \cdot \frac{x}{x^3-1} - 3 \cdot \frac{x^2+2}{\left(x^3-1\right)^2} \cdot x^2$$

Using Live Symbolics Operator

To differentiate using the live symbolics operator, use the Calculus tool bar to get the derivative operator, write the expression, type **CTRL+PERIOD**, then click outside the expression.

$$\mathrm{Deriv_1}(x) := \frac{d}{dx}\left(x^3 + 3x^2 + 4x - 2\right) \rightarrow 3 \cdot x^2 + 6 \cdot x + 4 \qquad \mathrm{Deriv_1}(2) = 28.00$$

$$\mathrm{Deriv_2}(x) := \frac{d}{dx}2^x \rightarrow 2^x \cdot \ln(2) \qquad \mathrm{Deriv_2}(2) = 2.77$$

$$\mathrm{Deriv_3}(x) := \frac{d}{dx}\sin(4x+1) \rightarrow 4 \cdot \cos(4 \cdot x + 1) \qquad \mathrm{Deriv_3}(2) = -3.64$$

$$\mathrm{Deriv_4}(x) := \frac{d}{dx}\frac{x^2+2}{x^3-1} \rightarrow 2 \cdot \frac{x}{x^3-1} - 3 \cdot \frac{x^2+2}{\left(x^3-1\right)^2} \cdot x^2 \qquad \mathrm{Deriv_4}(2) = -0.90$$

Figure 17.1 Symbolic differentiation

Now let's look at using the numeric processor. This processor only gives the derivative at a single point. It does not calculate the equation, as does the symbolic operator. Figure 17.2 shows how to use the numeric processor.

You can also take the *n*th derivative of an expression, and you can plot the results. See Figure 17.3.

$x := 2$ Compare to the results of Figure 17.1

Use the Calculus toolbar to get the derivative operator.

$$\mathbf{Deriv_5} := \frac{d}{dx}\left(x^3 + 3x^2 + 4x - 2\right) \qquad \mathbf{Deriv_5} = 28.00$$

$$\mathbf{Deriv_6} := \frac{d}{dx}2^x \qquad \mathbf{Deriv_6} = 2.77$$

$$\mathbf{Deriv_7} := \frac{d}{dx}\sin(4x + 1) \qquad \mathbf{Deriv_7} = -3.64$$

$$\mathbf{Deriv_8} := \frac{d}{dx}\frac{x^2 + 2}{x^3 - 1} \qquad \mathbf{Deriv_8} = -0.90$$

Figure 17.2 Numeric differentiation

$$F_1(x) := \frac{x^3 + 1}{x^3 - 8}$$

$$\mathbf{FirstDeriv}(x) := \frac{d}{dx}F_1(x) \rightarrow 3 \cdot \frac{x^2}{x^3 - 8} - 3 \cdot \frac{x^3 + 1}{\left(x^3 - 8\right)^2} \cdot x^2 \text{ simplify } \rightarrow (-27) \cdot \frac{x^2}{\left(x^3 - 8\right)^2}$$

$$\mathbf{SecondDeriv}(x) := \frac{d^2}{dx^2}F_1(x) \rightarrow 6 \cdot \frac{x}{x^3 - 8} - 18 \cdot \frac{x^4}{\left(x^3 - 8\right)^2} + 18 \cdot \frac{x^3 + 1}{\left(x^3 - 8\right)^3} \cdot x^4 - 6 \cdot \frac{x^3 + 1}{\left(x^3 - 8\right)^2} \cdot x \text{ simplify } \rightarrow 108 \cdot x \cdot \frac{x^3 + 4}{\left(x^3 - 8\right)^3}$$

$a := -4, -3.99 .. 4$ Set plot range

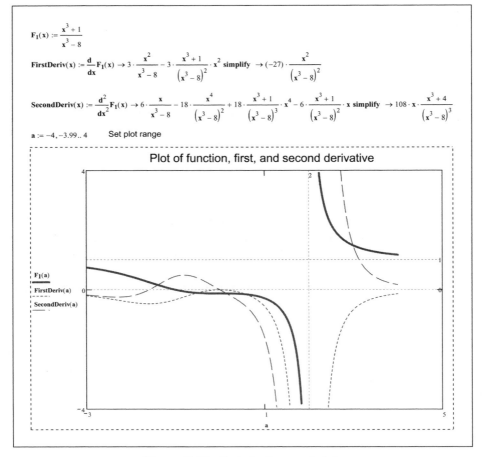

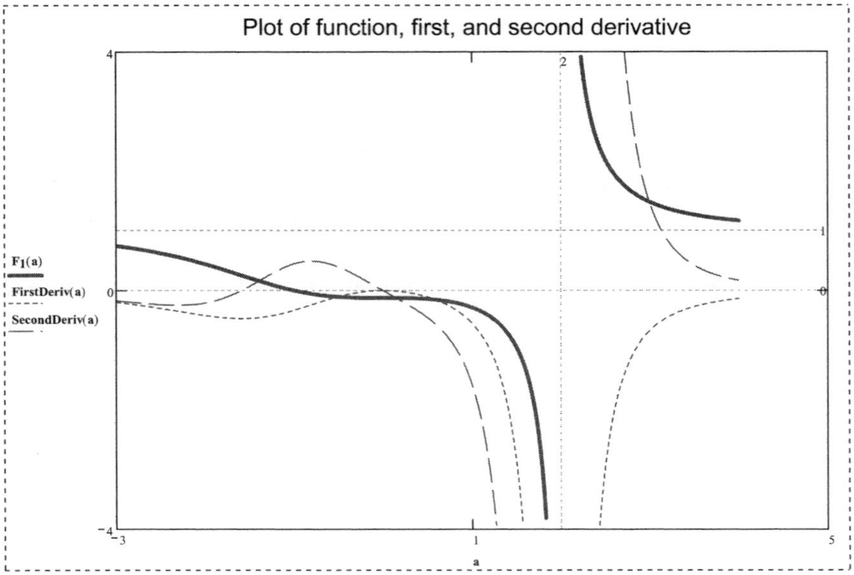

Figure 17.3 *n*th derivatives and plots

Integration

The numeric processor will perform only definite integrals. The symbolic processor will perform both definite integrals and indefinite integrals. The examples here are just an introduction. Mathcad can perform very sophisticated calculus operations. Refer to Mathcad Help for additional information.

Figure 17.4 gives examples of indefinite integrals using both the Symbolics menu and the live symbolics operator. Figure 17.5 gives examples of definite integrals using the live symbolics operator and using the numeric processor.

Using the Symbolics Menu

To integrate using the Symbolics menu, write the expression, select the variable, and select Tools>Variable>Integrate. The result is static. It does not change if you change the expression.

$x^2 \qquad x^3 + 2x^2 + x - 3 \qquad \sin(x) \qquad (30 - x^2)^{\frac{3}{2}}$

$\frac{1}{3} \cdot x^3 \qquad \frac{1}{4} \cdot x^4 + \frac{2}{3} \cdot x^3 + \frac{1}{2} \cdot x^2 - 3 \cdot x \qquad -\cos(x) \qquad \frac{1}{4} \cdot x \cdot (30 - x^2)^{\frac{3}{2}} + \frac{45}{4} \cdot x \cdot (30 - x^2)^{\frac{1}{2}} + \frac{675}{2} \cdot \mathrm{asin}\left(\frac{1}{30} \cdot 30^{\frac{1}{2}} \cdot x\right)$

Using the Live Symbolics Operator

$$\mathrm{Integrate}_1(x) := \int x^2 \, dx \rightarrow \frac{1}{3} \cdot x^3$$

$I_{1a} := \mathrm{Integrate}_1(5) \qquad I_{1a} = 41.67 \qquad I_{1b} := \mathrm{Integrate}_1(1) \qquad I_{1b} = 0.33$

$I_{1c} := I_{1a} - I_{1b} \qquad I_{1c} = 41.33$

$$\mathrm{Integrate}_2(x) := \int x^3 + 2x^2 + x - 3 \, dx \rightarrow \frac{1}{4} \cdot x^4 + \frac{2}{3} \cdot x^3 + \frac{1}{2} \cdot x^2 - 3 \cdot x$$

$I_{2a} := \mathrm{Integrate}_2(5) \qquad I_{2a} = 237.08 \qquad I_{2b} := \mathrm{Integrate}_2(1) \qquad I_{2b} = -1.58$

$I_{2c} := I_{2a} - I_{2b} \qquad I_{2c} = 238.67$

$$\mathrm{Integrate}_3(x) := \int \sin(x) \, dx \rightarrow -\cos(x)$$

$I_{3a} := \mathrm{Integrate}_3(5) \qquad I_{3a} = -0.28 \qquad I_{3b} := \mathrm{Integrate}_3(1) \qquad I_{3b} = -0.54$

$I_{3c} := I_{3a} - I_{3b} \qquad I_{3c} = 0.26$

$$\mathrm{Integrate}_4(x) := \int (30 - x^2)^{\frac{3}{2}} \, dx \rightarrow \frac{1}{4} \cdot x \cdot (30 - x^2)^{\frac{3}{2}} + \frac{45}{4} \cdot x \cdot (30 - x^2)^{\frac{1}{2}} + \frac{675}{2} \cdot \mathrm{asin}\left(\frac{1}{30} \cdot 30^{\frac{1}{2}} \cdot x\right)$$

$I_{4a} := \mathrm{Integrate}_4(5) \qquad I_{4a} = 527.97 \qquad I_{4b} := \mathrm{Integrate}_4(1) \qquad I_{4b} = 161.59$

$I_{4c} := I_{4a} - I_{4b} \qquad I_{4c} = 366.38$

Figure 17.4 Indefinite integrals

Using the Live Symbolics Operator

Compare from Figure 17.4

$$\int_1^5 x^2\,dx \rightarrow \frac{124}{3}\ \mathbf{float} \rightarrow 41.333333333333333333 \qquad I_{1c} = 41.33$$

$$\int_1^5 x^3 + 2x^2 + x - 3\,dx \rightarrow \frac{716}{3}\ \mathbf{float} \rightarrow 238.66666666666666667 \qquad I_{2c} = 238.67$$

$$\int_1^5 \sin(x)\,dx \rightarrow (-\cos(5)) + \cos(1)\ \mathbf{float} \rightarrow .25664012040491345293 \qquad I_{3c} = 0.26$$

$$\int_1^5 \left(30 - x^2\right)^{\frac{3}{2}} dx \rightarrow \frac{125}{2}\cdot 5^{\frac{1}{2}} + \frac{675}{2}\cdot \mathrm{asin}\left(\frac{1}{6}\cdot 30^{\frac{1}{2}}\right) - \frac{37}{2}\cdot 29^{\frac{1}{2}} - \frac{675}{2}\cdot \mathrm{asin}\left(\frac{1}{30}\cdot 30^{\frac{1}{2}}\right)\ \mathbf{float}, 6 \rightarrow 366.379$$

$$I_{4c} = 366.38$$

Using Numeric integration

$a := 1 \quad b := 5$

Compare from Figure 17.4

$$\mathrm{Integrate}_5(a,b) := \int_a^b x^2\,dx \qquad I_5 := \mathrm{Integrate}_5(a,b) \quad I_5 = 41.33 \qquad I_{1c} = 41.33$$

$$\mathrm{Integrate}_6(a,b) := \int_a^b x^3 + 2x^2 + x - 3\,dx \qquad I_6 := \mathrm{Integrate}_6(a,b) \quad I_6 = 238.67 \qquad I_{2c} = 238.67$$

$$\mathrm{Integrate}_7(a,b) := \int_a^b \sin(x)\,dx \qquad I_7 := \mathrm{Integrate}_7(a,b) \quad I_7 = 0.26 \qquad I_{3c} = 0.26$$

$$\mathrm{Integrate}_8(a,b) := \int_a^b \left(30 - x^2\right)^{\frac{3}{2}} dx \qquad I_8 := \mathrm{Integrate}_8(a,b) \quad I_8 = 366.38 \qquad I_{4c} = 366.38$$

Figure 17.5 Definite integrals

Functions

Maximize and Minimize in Solve Blocks

The ***Maximize*** and ***Minimize*** functions are used in Solve Blocks. They find the values of variables in a function that satisfy the constraints in a Solve Block so that the function takes its largest or smallest value. The examples for the ***Maximize*** and ***Minimize*** functions can become rather complex. There are several examples in the Mathcad Help, QuickSheets and Tutorial. The Mathcad Advisor Newsletter also has articles in the April 2004 and June 2003 issues.

Sources of additional information

Resources toolbar and My Site

The Resources toolbar is your one-stop place to access Mathcad information. See Figure 17.6. From the drop-down box, you can access Mathcad Tutorials, QuickSheets and any Mathcad E-books or Extension Packs you have installed. The My Site is a treasure chest of information. You can use the default site, or you can select a different site from the Preferences dialog box on the **Tools** menu. From My Site, you can access the Mathsoft Website, the Mathcad User Forum, the Mathcad Web Library, Mathcad Download Site and the Mathcad Knowledge Base. If you are looking for additional information on a Mathcad topic, My Site is the place to begin your search. See Figure 17.7.

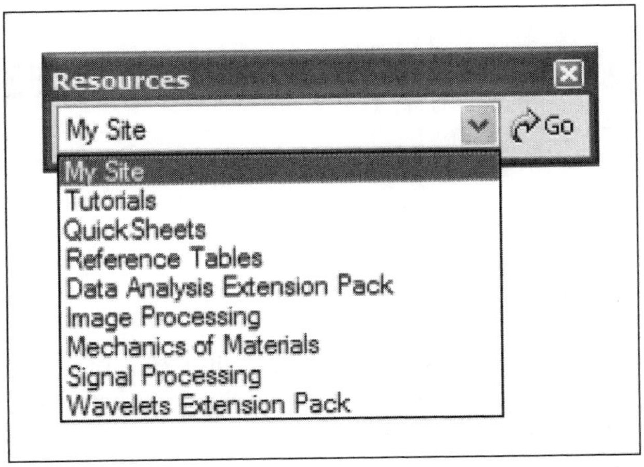

Figure 17.6 Resources toolbar

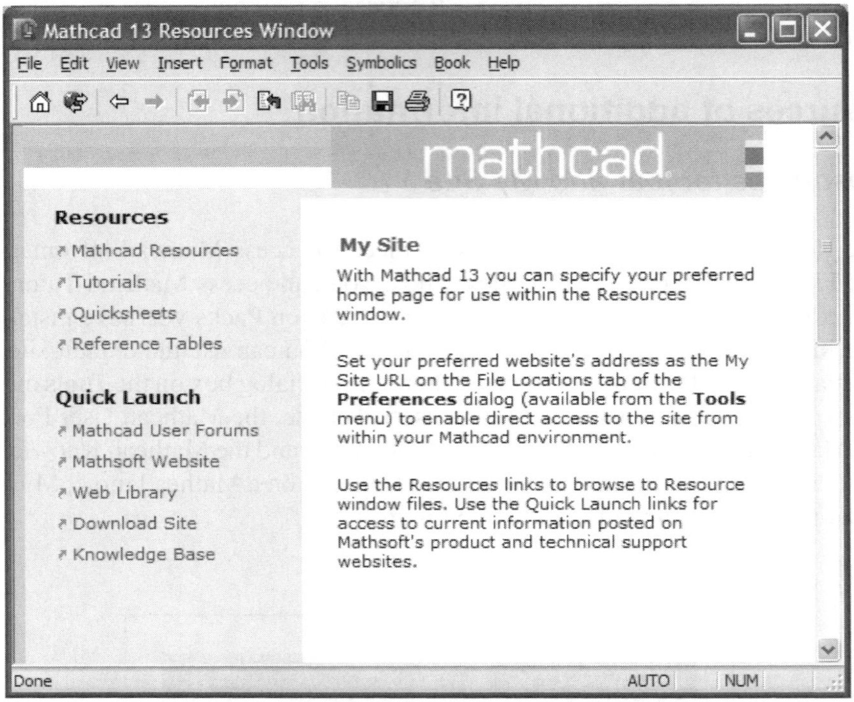

Figure 17.7 My Site

Reference Tables

The Mathcad Reference Tables contain almost 40 tables of scientific and engineering data. You will find information about physics, chemistry, mechanics of materials, mathematics, electronics, and more. You can access these tables from the Resources toolbar. See Figure 17.8.

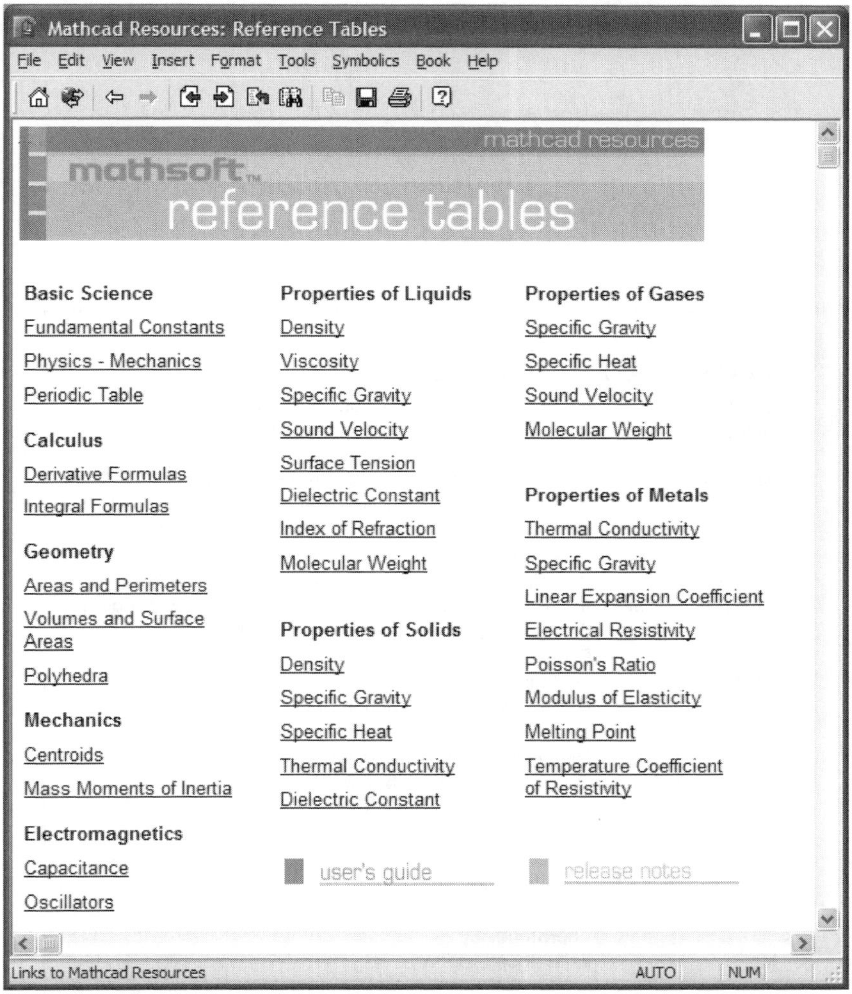

Figure 17.8 Reference tables

Keyboard shortcuts

For your convenience Figures 17.9–17.11 provide keyboard shortcuts for editing, adding Mathcad operators, and controlling worksheets.

Keys for Editing

Enter	Insert blank line. In text, begin a new paragraph.
Delete	Delete blank line. In text or math, remove character to the right of the insertion line.
Shift+Enter	In text, begin a new line within a paragraph.
Ctrl+Enter	Insert a page break. In text, set the width of the text region. In math, insert addition with linebreak operator.
Ctrl+A	In text, select all the text in the text region.
Ctrl+A	In a blank spot, select all regions in the worksheet.
Ctrl+Shift+Enter	In a region, move the cursor out of and below a region.
Ctrl+F	Find a character or string of characters.
Ctrl+H	Replace a character or string of characters.
Ctrl+Z	Undo last edit.
Ctrl+Y	Redo. Reverses the action of Undo.
Ctrl+C	Copy selection to clipboard.
Ctrl+V	Paste clipboard contents into worksheet.
Ctrl+X	Cuts selection to clipboard.
Ctrl+E	Open the Insert Function dialog box.
Ctrl+U	Open the Insert Unit dialog box.
Ctrl+Shift+J	Allows you to type characters inside brackets as for chemistry notation.
Ctrl+Shift+K	Allows you to type characters that usually insert operators.
Ins	Enter insert mode.

Figure 17.9 Keys for editing

Keys for Creating Operators

The following keys are used to insert operators. They behave the same way in blank space as they do in an equation, with one exception, the " key.

Keystroke	Operator
!	factorial
"	complex conjugate
#	range product
$	range sum
&	integration
'	Matched pair of parentheses
*	multiplication or inner (dot) product
+	addition
,	Separates arguments in a function
,	Separates expressions to be plotted on the same axis
,	Precedes 2nd number in range
;	Precedes last number in range
-	negation or subtraction
/	division
<	less than
>	greater than
?	differentiation
[	vector subscript or matrix subscript
\	square root
^	exponentiation or matrix inverse
|	magnitude or determinant
Ctrl+1	transpose
Ctrl+3	not equal
Ctrl+4	sum of elements in vector
Ctrl+9	less than or equal
Ctrl+0	greater than or equal
Ctrl+8	cross product
Ctrl+-	vectorize
Ctrl+=	equal to
Ctrl+Shift+=	mixed-number operator
Ctrl+6	superscript
Ctrl+Shift+4	summation
Ctrl+Shift+3	product
Ctrl+Shift+?	*n*th derivative
Ctrl+\	*n*th root
Ctrl+Enter	addition with linebreak
Ctrl+.	symbolic evaluation
Ctrl+Shift+.	symbolic evaluation using keywords
Ctrl+I	Indfinite integral
Ctrl+L	limit
Ctrl+A	right-hand limit
Ctrl+B	left-hand limit

Figure 17.10 Keys for creating operators

The following keys are used for manipulating windows and worksheets as a whole.

Ctrl+F4	Close worksheet
Ctrl+F6	Make next window active
Ctrl+K	Insert or edit hyperlink
Shift+Enter	Exit a text region
Ctrl+N	Create new worksheet
Ctrl+O	Open worksheet
Ctrl+P	Print worksheet
Ctrl+S	Save worksheet
Alt+F4	Quit
Ctrl+R	Redraw screen
F1	Open Help window
F9	Recalculate screen
Shift+F1	Enter or exit context sensitive Help
Esc	Exit context sensitive Help or interrupt a calculation.

Figure 17.11 Keys for worksheet and window management

Part IV

Creating and Organizing your Engineering Calculations with Mathcad

Part IV is what sets *Engineering with Mathcad* apart from all other books written about Mathcad. This part is unique because it discusses the use of Mathcad to create a complete set of engineering calculations.

You now have a toolbox full of simple tools and power tools. There are many other tools that still can be added, but it is time to discuss how to apply the tools you have. It is time to start building Mathcad calculations.

Part IV will discuss embedding other programs into Mathcad (OLE). It will also discuss Mathcad components, especially the use of Microsoft Excel, including the use of your existing Microsoft Excel spreadsheets. It will discuss how to assemble calculations from multiple Mathcad files, and how to input and output data to your worksheets. This part will also include adding hyperlinks and table of contents to your worksheets.

18 Putting it all together

Calculations from traditional engineering projects use a combination of various computer programs, spreadsheets and hand calculations. The printout from these computer programs are combined with the hand calculation sheets and placed into calculation books. The paper books then become the official calculations that are saved with the project.

Mathcad can be used to replace the book that contains the paper. All that information can reside within a Mathcad file: computer input and output files, printouts from computer programs, spreadsheets, PDF scans of hand calculations (why would you want to do hand calculations when you can use Mathcad?), and of course—your Mathcad calculations. The chapters in Part IV will discuss how to do all of this.

Creating an entire set of engineering project calculations with Mathcad is much more complex than writing a one or two page worksheet. Many issues need to be considered when creating a full set of engineering project calculations. The purpose of Chapter 18 is to briefly review some of the topics covered in earlier chapters and to introduce the topics to be discussed in Part IV.

Chapter 18 will:

- Discuss the concept of the Mathcad toolbox and the tools in it.
- Introduce the concept of using the tools to build a project.
- Review the topics discussed in all the chapters up to this point.
- Paint a picture of what will be accomplished in Part IV, by briefly discussing the topics to be covered in Chapters 19–24.

Introduction

Let's go back to the analogy used at the beginning of this book—teaching you how to build a house. We have helped you build a toolbox—a place where you can store your tools. Your Mathcad toolbox is a thorough understanding of Mathcad basics.

We have also collected many tools and put them into your toolbox. We have taken classes in hammers, screwdrivers, pliers, power saws, power drills, etc. In Part IV, we teach you how to use these tools to build the house.

Mathcad toolbox

Let's quickly review the skills and understanding you should have mastered from Part I.

Variables

By now, you are completely familiar with how to define and use variables, but do you remember how to use the special text mode (CTRL+SHIFT+k)? Do you remember the Chemistry Notation (CTRL+SHIFT+j)? If not, you may want to review these concepts from Chapter 1.

As we move into Part IV of this book, the naming guidelines discussed in Chapter 1 become most important. Let's review the variable naming guidelines discussed in Chapter 1:

1. Make your variable names more than just a single letter.
2. Make your variable names mean something.
3. Use a combination of upper case and lower case letters to help make your variable names easier to read.
4. Use underscore to separate different names in your variable names.
5. Make good use of subscripts in your variable names.
6. Use the (′) key if you need to use a "prime" (single apostrophe) in your variable name.

As you begin creating and assembling calculations, the variable names you chose to use will begin to be very important. When you create a simple one or two page worksheet, it usually does not matter what variable names you chose to use. Single letter variables are more than adequate because the entire worksheet can be taken in context and you usually do not run the risk of redefining the variable.

When you create a complete set of project calculations that can be over 100 pages long, then variable names are critical. You may have many conditions where you want to use the same variable name, but you want each variable to be unique, and you do not want to redefine a previously used variable. The use of subscripts to make each variable name unique is very useful in this case.

Descriptive variable names are also very useful. They help you remember what the variable refers to, and they also help those reviewing the calculations know what the variable name means. For example, is it easier to understand d:=5 ft or DepthOfWater:= 5 ft? In a single page worksheet either one would be adequate. When your calculations are 50 pages long, if you defined "d" on page 5 and do not use it again until page 50, then it may be more useful to have a descriptive variable name.

Descriptive variable names also help prevent you from redefining a variable name. In the previous example, if you defined the depth of water as d=5 ft and then later defined the depth of soil as d=6 ft, you have redefined the variable d. This will cause an error in your calculations if later on you use the variable "d" thinking that it is the depth of water. It is much better to use DepthOfWater:=5 ft and DepthOfSoil:=6 ft.

> **Tip!** It will be helpful to establish some personal or company naming guidelines. There have been many times when I have been very careful to use descriptive variable names, only to be caught later on not remembering what I named a particular variable. Did I name it DepthOfWater or WaterDepth? Was it BuildingLength or LengthOfBuilding or Building_Length? It will be beneficial for you to create some guidelines for your specific needs so that the names used in your calculations have a consistent usage.

The figures for each chapter in this book are created in a single Mathcad file. This required the use of unique variable names for each figure. This book uses a variety of variable naming methods in the figures. This has been intentional. It is to expose you to several different ways to name variables.

Editing

You could not have gotten this far in the book without understanding how to edit an expression. We will assume that you have a good understanding of editing techniques.

User-defined functions

The discussion of user-defined functions in Chapter 3 was very basic. This was intentional. The intent was to establish a solid understanding of user-defined functions prior to introducing the more powerful and complex features.

Each chapter has been built on this foundation of user-defined functions. In Chapter 4, we introduced units to user-defined functions. In Part II and Part III, we added arrays, programming, and live symbolics to user-defined functions.

When you define an expression, all the variables need to be defined prior to using the expression. If you want to use the same expression later in your worksheet with different variables, you need to define new variables, and then redefine the expression. This can cause a problem with engineering calculations because you are reusing the same equation. If the equation changes, you will need to change every occurrence of the equation. If you forget to change one of the occurrences, then you will have incorrect calculations. Creating a user-defined function helps to prevent this problem because you only need to change the equation once, and all future uses of the equation will be changed automatically.

Units!

Units are powerful! They are essential to engineering calculations! Use them consistently!

You should be very comfortable with using units. If not, go back and review Chapter 4. Units are very easy to use with most engineering equations. For the few equations where units will not work (such as empirical equations), divide the variable by the desired unit to make it a unitless number of the expected value.

Remember to include units in the argument list for user-defined equations. When plotting, remember to divide the expressions by the units you want to plot. For example, if you want to plot in feet, divide the expression by feet. If you want to plot in meters, divide the expression by meters.

Mathcad settings

Now that you are more familiar with many of the Mathcad features, Chapter 5 may make more sense. Now may be a good time to refresh your understanding of the Mathcad settings. If you have not made the suggested changes, please consider making them now.

One highly recommended change is to change the value of the built-in variable ORIGIN, from the Mathcad default of 0 to 1. This change needs to be made in a customized Mathcad style and saved in the Normal.xmct file. This is discussed in the next section. All examples used in this book have ORIGIN set at 1.

Customizing Mathcad with templates

In Chapter 6, we discussed styles for text and math variables. We recommended changing some default Mathcad styles. We also recommended adding some new styles. We also discussed the use of headers and footers. In Chapter 7, we discussed saving these customizations into a template. When you first read these two chapters, you may not have had any opinions about what to include in your Mathcad templates. Now that you have used Mathcad, you may have some specific ideas about what to include. If you have created previous templates, you may want to revisit the settings and consider new changes. If you have not created a customized template, review Chapter 7 and create the EWM Metric and EWM U.S. template.

> **Tip!** If you have not changed the default file Normal.xmct to a customized template, I recommend doing so now. Follow the directions as described in Chapter 7.

Hand tools

Arrays are an essential tool for engineering calculations. We have used vectors in numerous examples in order to display results from multiple input variables. Range variables have been used in dozens of examples. If you have forgotten about the vectorize operator, review Chapter 9 for its use. These are essential tools as you begin to build engineering calculations.

Chapter 10 discussed a few selected functions. These are only the beginning of the functions you can add to your Mathcad toolbox. Review the functions and operators discussed in Chapter 10, and see which functions will be most useful to you.

Plotting is a very useful tool. If you are not comfortable with creating simple X-Y plots or simple Polar plots, review Chapter 11.

A thorough understanding of simple logic programming is essential. Engineering calculations require the use of logic programs to choose appropriate actions. If you are not yet comfortable with the use of programs in expressions or user-defined programs, review Chapter 12.

Power tools

The Chapters 14, 15, and 16 introduced some topics that may seem confusing at first. That is one reason that their discussion was held off until Part III.

The symbolic calculations, solving tools, and programming tools are extremely powerful. These tools will be very beneficial in your engineering calculations. When it comes to using Mathcad to solve for various equations, there are many ways to arrive at the same answer. We have tried to show you the various ways of solving for solutions. The engineering examples shown in Chapter 15 provide a good summary of how to use the solving functions.

Let's start building

Now that you have your toolbox full of tools, it is time to start building. One of the primary purposes of this book is to teach you how to create and organize engineering calculations. The principals used, can be used for any technical calculations, not just engineering calculations.

Mathcad is perfectly suited to create and to organize your project calculations. The reasons for this include:

- It speaks our language—math.
- Calculations and equations are visible, and not hidden in spreadsheet cell.
- The calculations are electronic and can now be printed, archived, shared with coworkers around the world, searched, and reused.
- Mathcad calculations can be used over and over again.
- If input variables change, the calculations are automatically updated.
- Results from other software programs can be incorporated into your calculations.
- Mathcad can share information with other programs such as Microsoft Excel.

What is ahead

The chapters in Part IV, will help you as you begin using Mathcad to create and organize your engineering calculations. In the next chapter, we discuss how to save and reuse worksheets. It is most likely that Mathcad will not be the only software program you use. Therefore, it is important to discuss how to bring information from other software programs into your Mathcad worksheets. We discuss this in Chapter 20. A very powerful feature of Mathcad is its ability to communicate with and transfer information between other programs. These programs are called "Components." In Chapters 21 and 22, we discuss the use of components—especially the use of Microsoft Excel. Finally, in Chapters 23 and 24 we discuss different ways of inputting and outputting information from your calculations, as well as ways to create hyperlinks to information in your current worksheet as well as other worksheets.

Summary

This chapter is a review of the topics covered in previous chapters. We emphasized the importance of choosing suitable variable names in your calculations. We also encouraged you to go back and review any chapter with which you are not comfortable.

We then looked to the future and highlighted the topics we are about to discuss.

Practice

1. Create five variable names using the special text mode discussed in Chapter 1. Use different names than you used for the practice in Chapter 1.
2. Create five variable names using Chemistry Notation discussed in Chapter 1. Use different names than you used for the practice in Chapter 1.
3. Open the custom template you created for the practice in Chapter 7. Make changes to the result format, the header and footer. Does your template set the value of ORIGIN to 1 (Worksheet Options dialog box, Built-in Variable tab)?

4. If you have not changed the Normal.xmct template from its Mathcad default, do so now (refer to Chapter 7). Use your customized template from exercise 3, or use the EWM Metric or EWM U.S. that were created in Chapter 7.
5. Go back and briefly review the topic covered in each chapter. Look at the practice exercises to see how well you remember the topics.

19
Assembling calculations from standard calculation worksheets

Throughout this book, we have seen the benefits of using Mathcad to solve engineering problems. In this chapter, we will focus on the benefits of using Mathcad to organize your engineering calculations.

Once Mathcad worksheets are developed, they can be used over and over again. It is very useful to develop standard Mathcad calculation worksheets. These standard calculation worksheets can be used as independent files, or they can be copied and pasted into new Mathcad calculations. There are times when the Mathcad calculation file becomes so large that it may be necessary to divide the calculation into several different files. These files can be linked together, so they behave as if they were in one file.

Chapter 19 will:

- Explore ways of creating worksheets that can be used repeatedly.
- Demonstrate how to copy and paste these standard worksheets into new worksheets.
- Discuss the Mathcad features available to keep track of information such as: Annotate, Comment, and Provenance.
- Show different ways of locking worksheets to protect some information from being changed while allowing other information to be edited.
- Explain the procedures for locking worksheets.

- Discuss ways of storing files and using files in web-based repositories.
- Show how to use the ***reference*** command to include information from worksheets located in different files.
- Discuss when it is appropriate to break calculations into different files.
- Discuss the potential problems that may occur when assembling calculations from different files.
- Show how variable definitions may change after pasting portions of another Mathcad worksheet.
- Provide examples showing some of the problems that occur after pasting other Mathcad worksheets into the calculations.
- Present different solutions to the problems discussed.
- Make recommendations as to how the calculations can be organized so that critical information is not redefined unexpectedly.
- Provide guidelines and examples to help avoid problems associated with variable redefinitions.
- Demonstrate the use of ***Find*** and ***Replace*** to help solve problems associated with variable names.

Copying regions from other Mathcad worksheets

The purpose of Part IV of *Engineering with Mathcad* is to teach you how to use Mathcad to create and organize your engineering calculations. In this chapter, we focus on Mathcad's ability to use previously created Mathcad calculations. You can easily copy and paste regions from previously created worksheets. You can copy a single region or an entire worksheet.

To reuse information from a previous Mathcad worksheet, open the worksheet, select the regions you want to copy, copy the regions, and then paste the regions into your current worksheet.

> Tip! If you are pasting between existing regions in your worksheet, you must add enough blank lines to make room for the pasted regions. Mathcad does not automatically add new lines. If you do not have enough space, Mathcad will paste regions on top of one another. If this occurs, undo the paste and add more blank lines.

It is possible to drag and drop regions from one worksheet to another, but this method is not recommended. The reasons will be discussed in the next section.

XML and metadata

Beginning with Mathcad 12, Mathcad worksheets are stored in XML file format. This file format is a text-based format that allows information to be tagged inside the Mathcad file. This data is called metadata, and it can be attached to the entire worksheet, a region or portions of a region.

Worksheet metadata can include such things as:

- Worksheet title.
- Worksheet comments.
- Author.
- Department.
- Dates.
- Any other custom data.

Region metadata—called annotations—can be added to math regions, and can include such things as:

- Comments.
- Notes about where the data in the region was derived.
- Notes about the formulas being used.

To attach metadata to a file, select **Properties** from the **File** menu. This opens the File Properties dialog box. The Summary tab is shown in Figure 19.1, and the Custom tab is shown in Figure 19.2. The information you add in this dialog box becomes permanent metadata stored with the file. Any text in the Description box is automatically tagged to regions that are copied and pasted into another worksheet. To add information in the Custom tab, click the down arrow under Name and select one of the options, or type your own description. The next thing you need to do is tell Mathcad what type of data to expect: Text, Date, Number, or Yes or No. You can then input your custom data in the Value box. Click the **Add** button to add the information to the Custom Properties box.

To attach metadata to a math region, select the entire region, right-click on the region and select **Annotate Selection** from the pop-up window. This opens the Selection Annotation dialog box. You can now add a comment about the selected region. It is also possible to annotate a portion of a region such as a variable name, a constant or a constant with units. To do this, place the portion of the region you want to annotate between the blue editing lines and right-click on the region. Select **Annotate Selection** from the pop-up window. Each region may have several annotations.

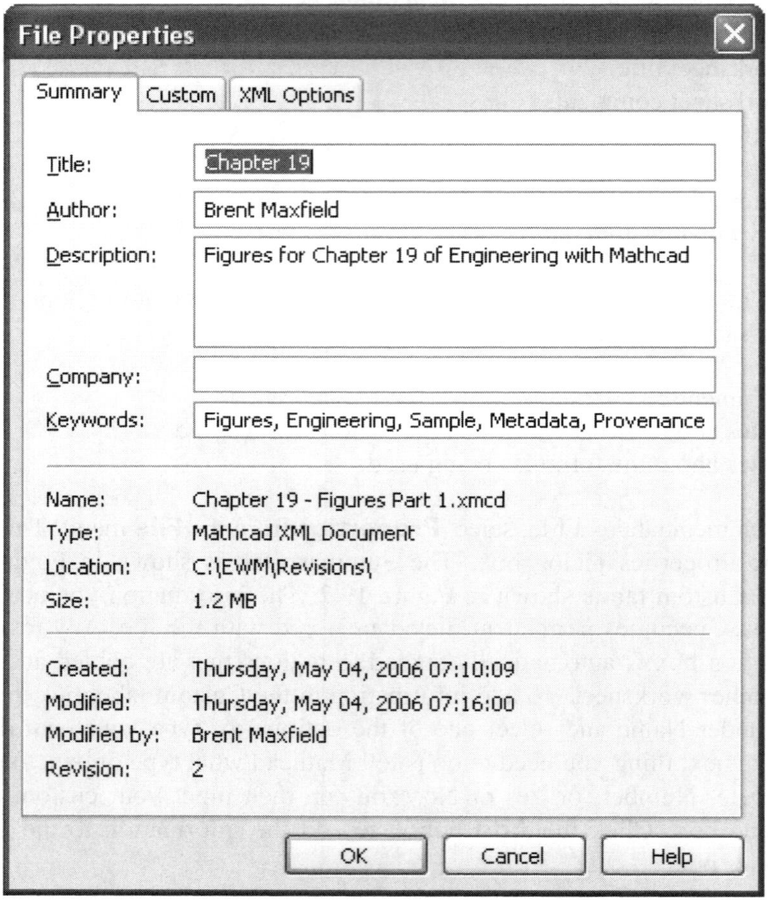

Figure 19.1 File Properties dialog box—Summary tab

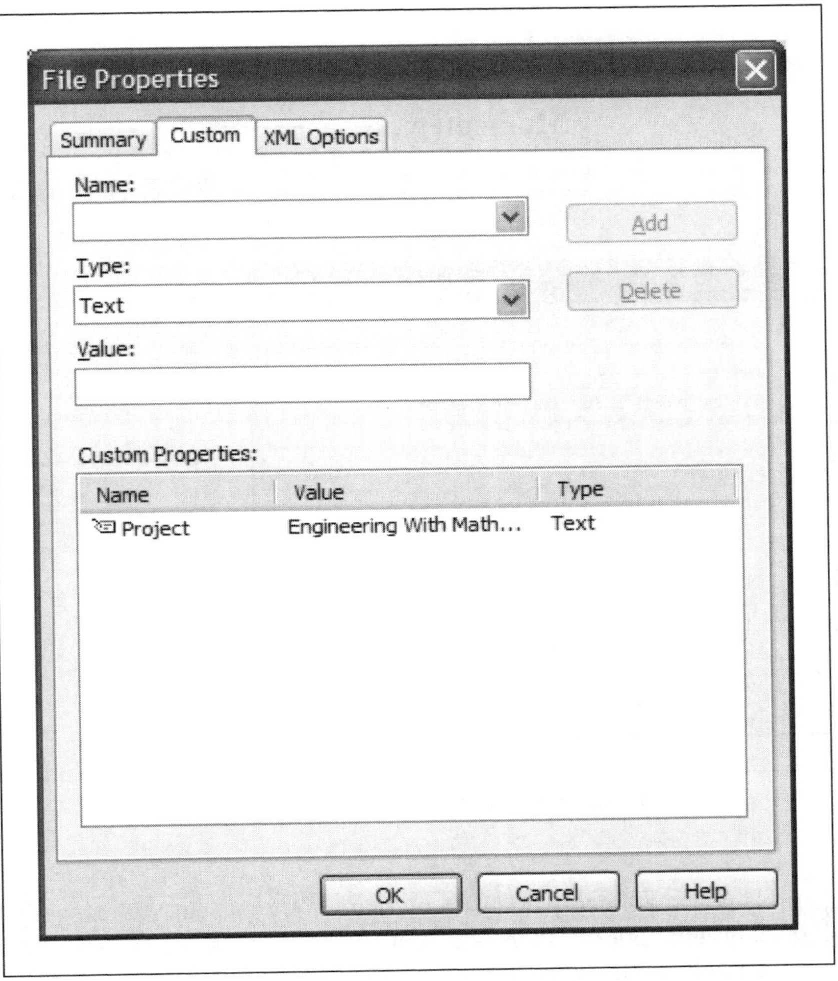

Figure 19.2 File Properties dialog box—Custom tab

When you click on a region that has annotations added, green parentheses appear around the region or portion of a region. These parentheses indicate the portion of the region that has the annotation. In order to view the annotation, enclose the parentheses between the blue editing lines, and right-click on the region. If you have properly enclosed the parentheses, you should see a **View/Edit Annotation** menu option from the pop-up window. Clicking this will open the Selection Annotation dialog box. You can now view and/or edit the annotation. See Figure 19.3 for an example of creating an annotation, and see Figure 19.4 for an example of viewing an annotation.

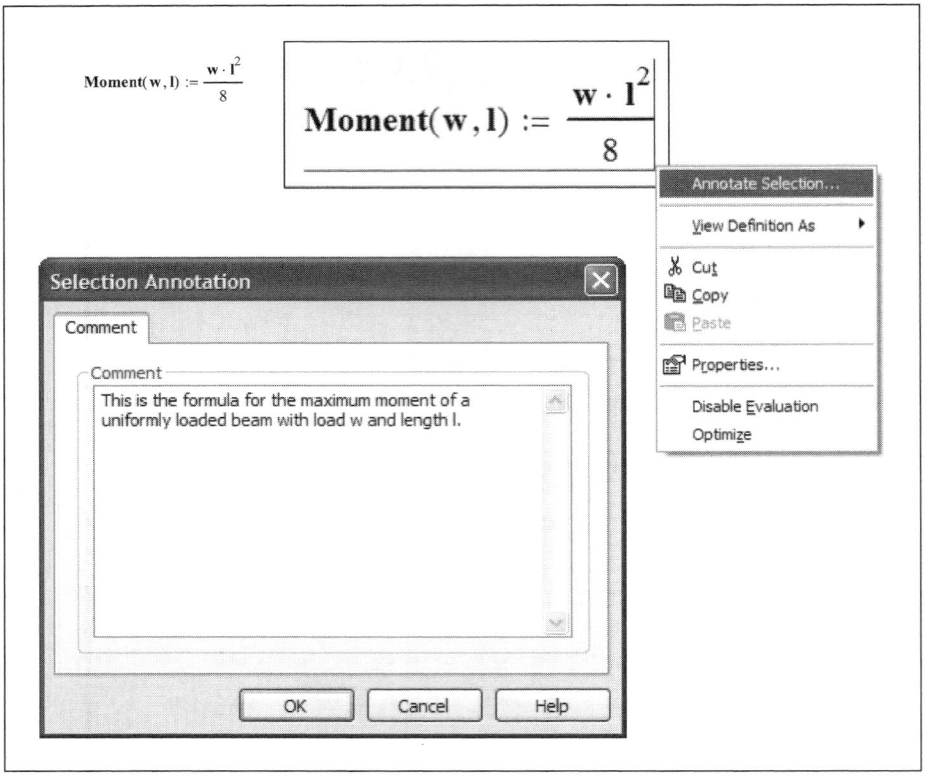

Figure 19.3 Adding comments to a region

Tip! If you see green parentheses around a region, but you cannot find any annotations, then you probably do not have the parentheses completely between the editing lines.

To view all annotation marks in a worksheet, select Annotations from the View menu. When this is turned on, all regions that have tagged annotations will have green parentheses showing. You can also change the color of the parentheses by selecting **Color** from the **Format** menu and then clicking **Annotation**.

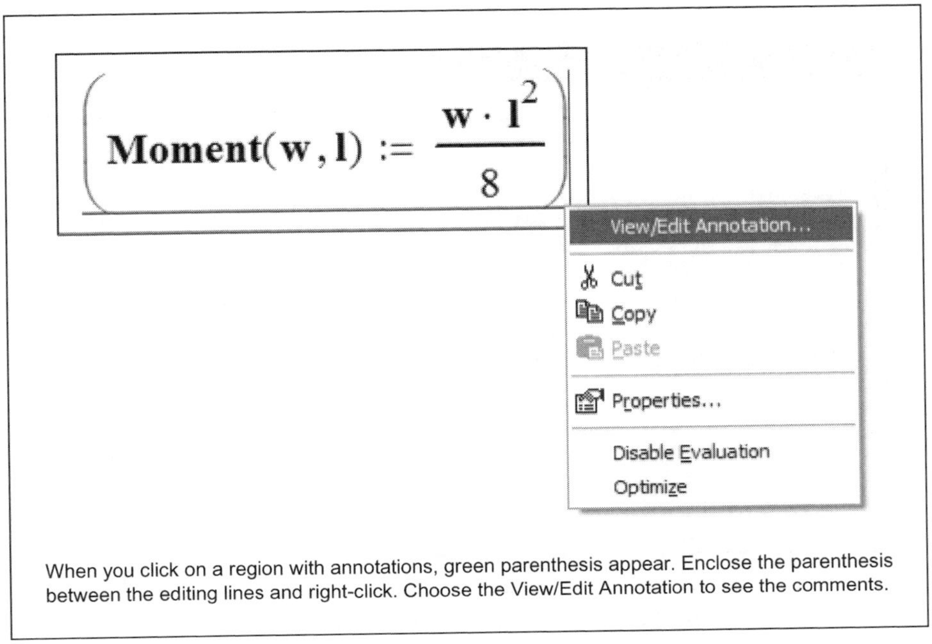

Figure 19.4 Viewing annotations

It is also possible to attach custom metadata to a math region. To do this, right-click on the region and click **Properties**. Select the Custom tab from the Properties dialog box. The information here is similar to the worksheet Custom tab, except the data is only added to the selected region. See Figure 19.5. This custom data seems to be redundant. The comment annotations should usually be adequate.

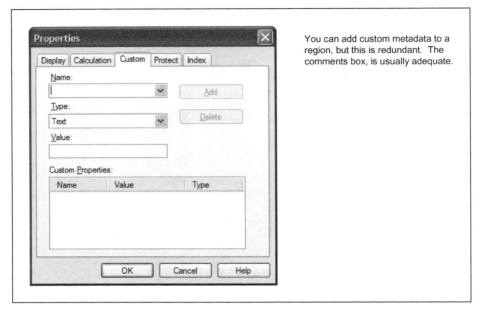

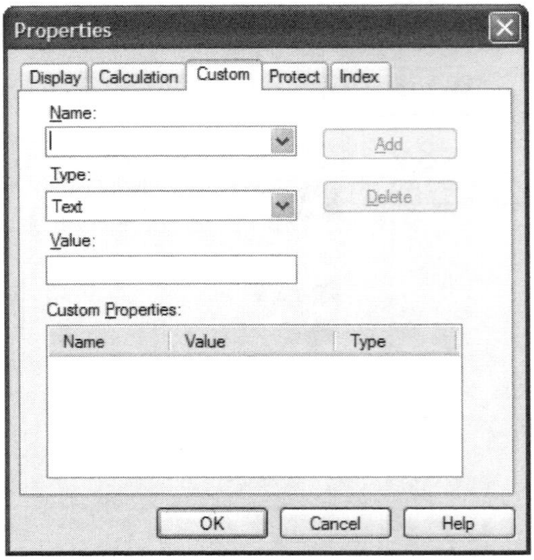

You can add custom metadata to a region, but this is redundant. The comments box, is usually adequate.

Figure 19.5 Adding comments to a specific region

Provenance

When you copy and paste regions from one worksheet to another worksheet, Mathcad automatically attaches specific metadata information. The attached metadata includes the path and filename of the file from which the region was copied. It also includes annotations attached to the region. If there are comments attached to the worksheet, these worksheet comments are also attached. If the copied region was originally from a previous file, then the original worksheet path and filename are also included in the metadata. The custom metadata attached to specific regions is not copied. See Figure 19.6.

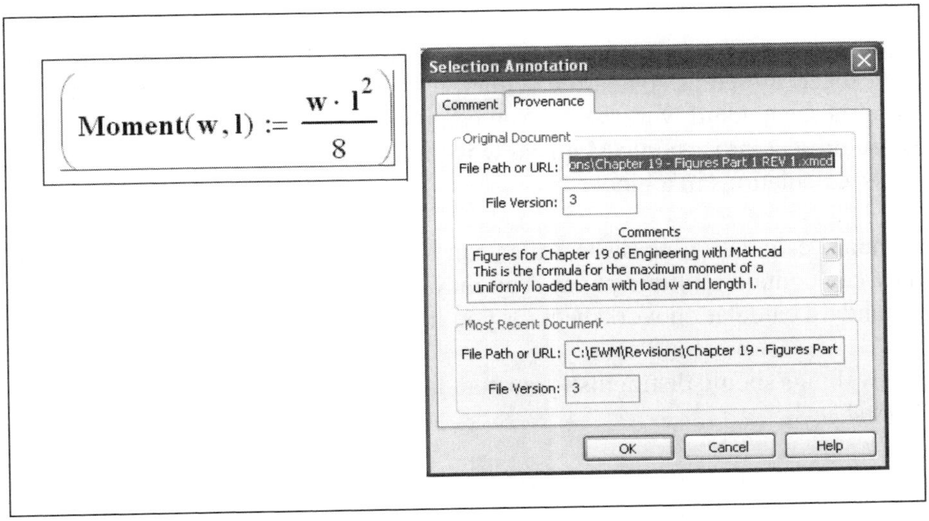

Figure 19.6 Provenance

Mathcad calls this transfer of metadata—Provenance. It is a valuable way of tracing information in your calculations because it provides a history of the source data for copied regions. The Provenance metadata always includes the path and filename of the original source, as well as the path and filename of the most recent source. In some cases, these are the same.

Tip! If you drag and drop regions, this Provenance data is not created. For this reason, it is much better to copy and paste regions when copying them from one worksheet to another.

If you modify a region after pasting it from another worksheet, the region is no longer exactly as it was in the previous worksheet, and Mathcad automatically removes the Provenance information, including any comments attached to the region.

Creating standard calculation worksheets

You may have already created calculation worksheets that you use repeatedly. The results from these worksheets are undoubtedly printed and placed in your calculation binder. The power of Mathcad is that you can assemble all of these calculation worksheets into a single calculation file.

Mathcad makes it easy to create and organize the project calculations because you can reuse existing calculations and insert them anywhere into your current worksheet. Standard calculation worksheets can make it much easier to create a complete set of engineering project calculations. If you have commonly used worksheets stored as standard calculation worksheets, you can copy and paste from these standard worksheets. You can then assemble the calculations for an entire project into a single Mathcad worksheet, just as you used to assemble the paper calculations in a binder.

Standard calculation worksheet files can be as small as only a few regions, or they can contain multiple pages. You may only need to use a few regions from a standard calculation worksheet, or you may use the entire worksheet.

A few things should distinguish your standard calculation worksheets from other worksheets:

- They are designated as standard calculation worksheets.
- They have been thoroughly checked.
- They include a worksheet comment, listing who checked the calculation and date that the calculation was checked.
- Additional metadata can be added such as Author, Department, Dates, etc.
- The file is locked to allow read access only, so that changes are not accidentally made to the checked worksheet file.
- They are stored in a common folder location, accessible to everyone in your organization.
- They are organized with thought about how they will be copied into other calculations.
- Portions of the worksheet may be protected or locked. (See discussion in next section.)

What type of worksheets should be saved as standard calculation worksheets?

- Repeatedly used equations and functions.
- Repeatedly used calculation worksheets.
- Cover sheets for project calculations.
- Project design criteria.
- Code equations.
- Reference data.
- Almost anything that will be used more than once.

Protecting information

After working hard to create and verify standard calculations, you need to be able to prevent unwanted changes from being made to your standard worksheets.

The first thing you need to do is write-protect the standard calculation files. This way you do not accidentally overwrite the standard calculation file. The easiest way to write-protect a file is to find the file in My Computer, then right-click on the file and click **Properties**. On the General tab of the Properties dialog box, place a check mark in the "Read-only" box under Attributes. This will allow the file to be opened, but not saved.

Most standard calculation worksheets will have some variables that need to be changed and many equations and regions that you want to remain unchanged. There are two ways of protecting regions from being changed. First, you can lock complete areas of your calculations. Second, you can protect specific regions (similar to protecting cells in a spreadsheet). It is also possible to hide specific areas of the worksheet so that they are not visible. Each of these protection means can be protected by a password.

Locking areas

Let's first discuss Mathcad areas.

A Mathcad area is a portion of your worksheet defined between two horizontal lines. To insert an area into your worksheet, click **Area** from the **Insert** menu. This places two horizontal lines in your worksheet.

These lines behave as regions and can be moved up or down in your worksheet. If you move the top line up or the bottom line down, the area expands and includes additional regions. The regions between these lines constitute an area.

Once the area has been defined, you can then do several things with the area. You can lock the area so that no changes can be made to regions within the area. You can also collapse the area so that all the regions in the area are not visible. You can even hide the lines that define the collapsed area. (You can select the hidden lines by dragging your mouse over the region where the hidden lines are located.) There are also several formatting things that can be done to the area definition icons. These will be discussed shortly.

To lock an area, right-click on one of the area boundaries and choose **Lock**. This opens the Lock Area dialog box. See Figure 19.7. From this dialog box you can enter a password, tell Mathcad to collapse the area when it is locked, tell Mathcad whether or not to allow the area to be collapsed or expanded while it is locked, and tell Mathcad to show a lock timestamp. The lock timestamp displays the date and time that the worksheet was locked. To unlock an area, right-click on an area boundary and choose unlock.

> **Tip!** If you enter a password and then forget the password, the area will be permanently locked. I usually do not enter a password. I only lock areas to prevent the accidental change of regions. If you do use a password, be sure to keep careful track of the password.

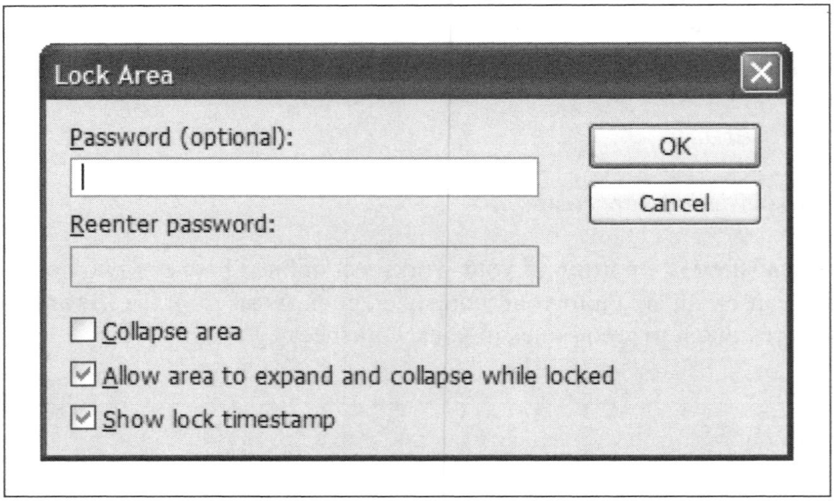

Figure 19.7 Lock Area dialog box

If you right-click on an area boundary and choose **Collapse**, then the area between the boundaries disappears. A single horizontal line is shown, and all regions below the area are moved upward. All the regions within the collapsed area continue to function as before. You can use the collapse feature either to save space or to hide regions you do not want others to see.

Let's look at the formatting options associated with areas. If you right-click an area boundary and choose **Properties**, it opens a Properties dialog box. The Display tab is shown in Figure 19.8. If you place a check in the "Highlight Region" box, then a colored band will appear on each of the area boundaries. The Mathcad default color is yellow, but you can change the color by selecting the **Choose Color** button. The "Show Border" check box will place a box around each of the area boundaries.

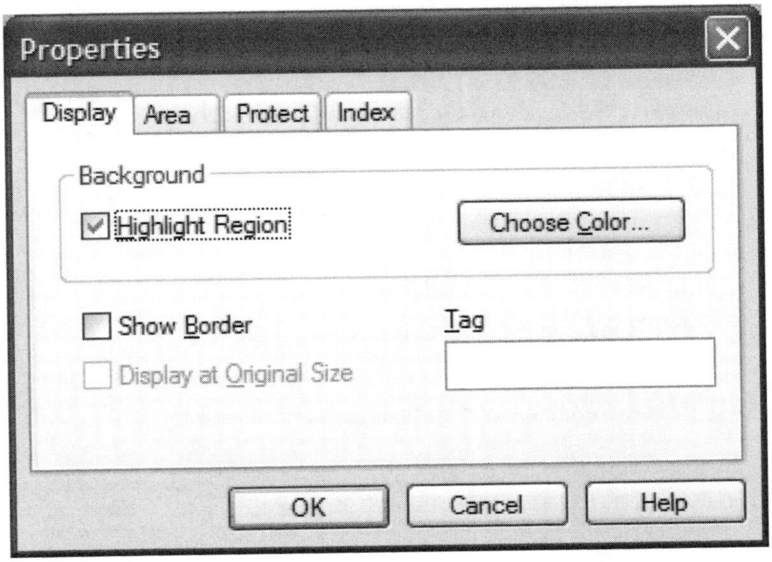

Figure 19.8 Display tab from area Properties dialog box

The Area tab is shown in Figure 19.9. This tab allows you to name the area. Using a name provides the opportunity to provide a description of what is being done within the area. This is especially helpful if the area is collapsed. You are not required to use a name. This tab also allows you to show four items. If you uncheck the "Line" box, then the horizontal area boundary lines are not displayed. If you uncheck the "Icon" box, then the small boxes with arrows are not displayed. If you uncheck the "Name" box, then the name will not be used, even if you have used an area name. When the "Timestamp" box is checked, then the date and time that the area is locked will be displayed. If you have a collapsed region, and you uncheck all four boxes (assuming you have not used a highlight color and do not have a border), then the collapsed region will not be visible. The only way you would see that there is a collapsed area is if you have "View Region" turned on (**Regions** from the **View** menu), or if you drag your mouse over the collapsed area. The last check box is similar to the check box in the Lock Area dialog box. It tells Mathcad whether or not to allow regions to be collapsed or expanded while they are locked.

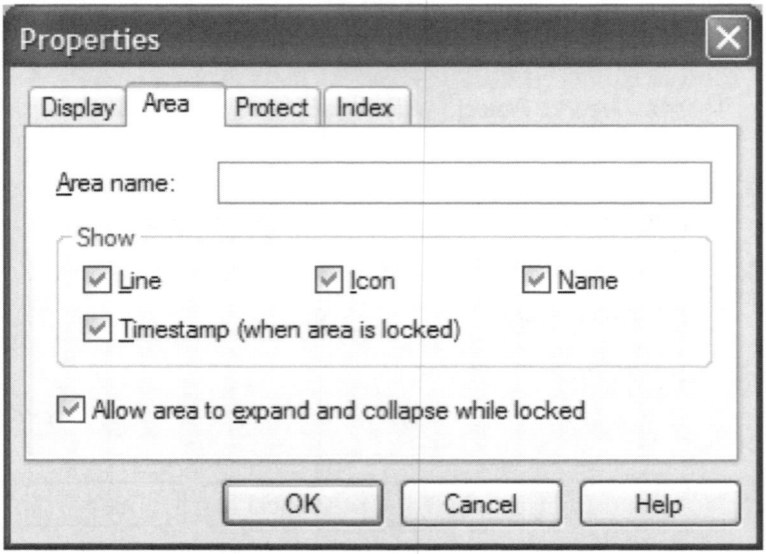

Figure 19.9 Area tab from the area Properties dialog box

Tip! In the context of this book, we are discussing areas as a way to protect standard calculations from being changed. I do not recommend using the collapse feature for standard calculations because it hides the regions that will need to be checked and reviewed before the project is finished.

You can delete unlocked areas by selecting area boundary and deleting it. You can copy a locked region from one worksheet to another worksheet. The area will still be locked after it is pasted into the new worksheet.

Protecting regions

Protecting regions in Mathcad is similar to protecting cells in a spreadsheet. Each region in Mathcad by default has an attribute that tells Mathcad to protect the region from editing when the worksheet protection is turned on. To prevent a region from being locked, you must tell Mathcad not to protect the region. You do this in the Properties dialog box. Select one or more regions, then right-click and choose **Properties**. Select the Protect tab and uncheck the "Protect region from editing" box.

In order to protect your worksheet it must be saved as a compressed XML file. To save a worksheet in this format choose **Mathcad Compressed XML (*.xmcdz)** from the "Save as type" drop-down box of the Save As dialog box. Once you have saved your worksheet in the xmcdz format, you can protect your worksheet. There is a reason for this. Mathcad XML format can be opened and edited in programs other than Mathcad. The xmcdz format cannot be opened (and thus edited) in these other programs.

To protect your worksheet, choose **Protect Worksheet** from the **Tools** menu. This opens the Protect Worksheet dialog box. See Figure 19.10. You can choose a password, but the same rule applies as with Mathcad areas—if you forget the password, your worksheet will be permanently locked. There are three levels of protection that you can choose from.

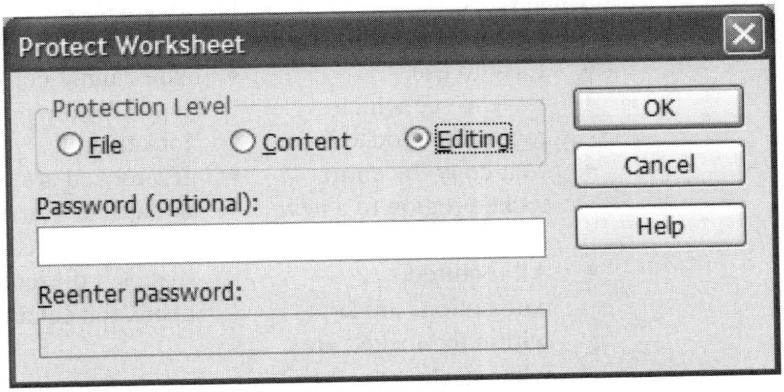

Figure 19.10 Protect Worksheet dialog box

The most restrictive level of protection is "Editing." When this option is selected from the Protect Worksheet dialog box, the worksheet cannot be edited in any way. You cannot move regions, add regions, or select any region, except those regions where the properties were changed to allow editing. This level of protection works well on worksheets that will be used by themselves. Because you cannot select regions, this type of protection does not allow you to copy the locked regions to another worksheet. It does not work well for a standard calculation worksheet.

The next level of protection is "Content." This level of protection allows locked regions to be selected so that they can be copied, but you cannot change or move the selected regions. You can also add additional regions to the worksheet. This is the best level of protection to use for standard calculation worksheets.

Because you can select regions, you can also copy them and paste them into other worksheets. When you paste the regions into another worksheet, the regions retain their same properties. This means that if the worksheet where you paste the regions has protection turned on, then the pasted regions will also be locked. If the worksheet does not have protection turned on, then the regions will be unlocked until the protection is turned on.

The lowest level of protection is "File." This means that the worksheet can only be saved in "Mathcad Compressed XML (*.xmcdz)" format, or similar formats. It does not allow the worksheet to be saved in XML format.

Advantages and disadvantages of locking areas versus protecting worksheets

Locked Area	Advantages	Disadvantages
	• Easy to use. • Easy to see which regions are locked. • Can copy the entire locked region to a new region. • All required calculations are kept within the locked area when pasting to a new worksheet.	• You cannot copy individual regions in a locked area. • If a locked area is collapsed, you cannot see the regions. This makes it difficult to check the calculations.
Protected Worksheet("File" level of protection)	**Advantages**	**Disadvantages**
	• More similar to a spreadsheet. • When you copy to a new worksheet, the regions are not automatically locked. • You do not need to copy all the regions, as you would need to do in a locked area.	• Cannot visually see which regions are unlocked. • When you copy to a new worksheet, the regions are not automatically locked.

Potential problems with inserting standard calculation worksheets and recommended solutions

There are many advantages to using standard calculation worksheets. However, there is also one very large disadvantage. When you paste a standard calculation worksheet that contains the same variable name(s) as your current worksheet, you redefine the variable definition(s). This can have disastrous consequences in your calculations.

When creating project calculations, it is wise to define the project criteria only once. This allows you to be able to change the criteria and have the entire project result updated automatically. For example, assume that you need to use the yield strength of steel in your calculations. You define this to be Fy:= 36 ksi on the first page of your calculations. Each time you use Fy in your calculations you do not need to redefine Fy:= 36 ksi. You simply use the variable Fy, and Mathcad knows that Fy = 36 ksi. Now, if you paste a standard calculation worksheet into your project calculations that has Fy:= 36 ksi you have a new variable definition in your worksheet. If the value is Fy:= 36 ksi, then it appears that there is no problem—all your results are correct. But what happens if at some later time, you change the value of steel from Fy:= 36 ksi to Fy:= 50 ksi? You change this at the beginning of your calculation, but do not change the new definition added when you pasted the standard calculation worksheet. Now, all the results following the new definition are not updated to Fy = 50 ksi. They are still using Fy = 36 ksi.

Another scenario is, if the standard calculation worksheet uses the value Fy:= 50 ksi. For our discussion, let's assume that Fy = 50 ksi is appropriate for the specific standard calculation worksheet so you do not change the value of Fy from 50 ksi to 36 ksi after pasting into your project calculation. If you happen to insert this standard calculation worksheet into the middle of your project calculation, what happens to all the expressions that use Fy for the remainder of the worksheet? They all use the value of Fy = 50 ksi. Now all the project calculation results for the remainder of the worksheet are incorrect.

We have been discussing only a single variable. It may be easy to solve this problem when you are dealing with only a single variable, but if the project calculations and the standard calculation worksheet have many identical variable names, then it becomes very difficult to ensure that you do not have a problem.

This important issue needs to be considered whenever you create project calculations. It needs careful attention. The advantages of using Mathcad outweigh this disadvantage. Let's now look at some ways to avoid the problems discussed.

Guidelines

The following guidelines are suggested as a means to prevent unwanted changes to your project calculations.

1. Use Mathcad's redefinition warnings. To turn these on, click **Preferences** from the **Tools** menu, and then click on the Warnings tab. See Figure 19.11. Place a checkmark in all the boxes.
2. After pasting a standard calculation worksheet into your project calculations, scan the pasted regions and look for redefinition warnings. If you find a redefinition warning, then you need to decide if the variable that is being redefined has the exact same meaning as a previously defined variable.

   a. If the redefined variable has exactly the same definition and value as a previously defined variable, then delete the definition operator := and replace it with the equal sign. This way the value of the earlier defined variable can be seen, but not overwritten.

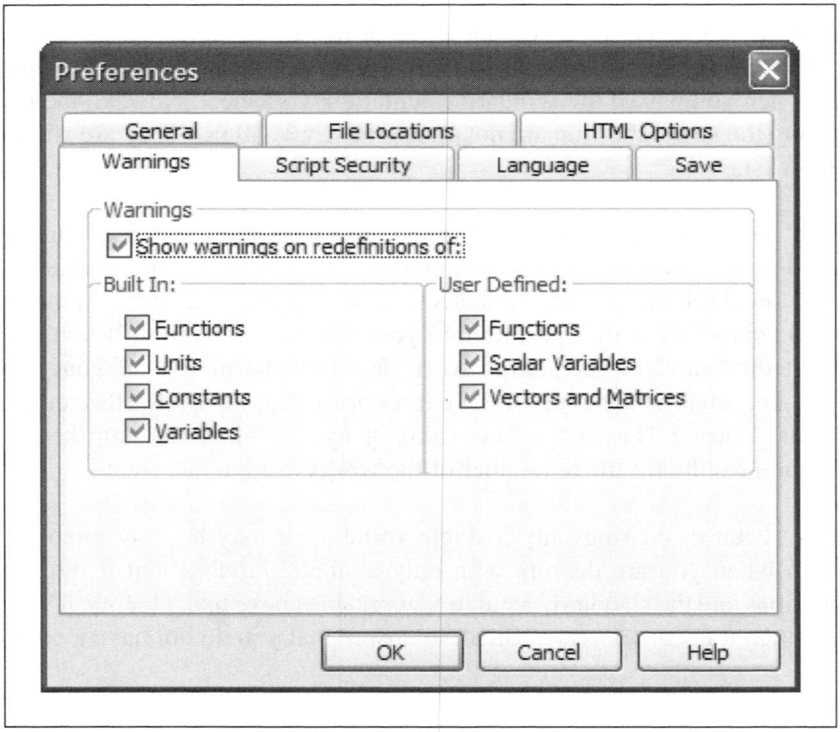

Figure 19.11 Turn on all redefinition warnings

b. If the redefined variable has a different value or a different meaning than a previously defined variable, then you need to revise either the project calculations or the standard calculation worksheet so that each variable has a unique name. See the section below on using **Find** and **Replace**.
c. There may be a situation where redefined variables do not cause a problem. See the discussion in the following section.

3. If possible, place all variable definitions at the top of your standard calculation worksheets. This makes it easier to check for redefinitions after pasting into your project calculations. The most critical variables to place at the top are the ones that may overwrite your project design criteria.
4. Whenever you use previously defined variables in an expression, always display the value of the variables near the expression. For example, if you have an expression Area:= b*d, then always type b= and d= near the expression. (This is assuming that b and d were defined on a previous page.) This makes it easier to check the calculations, and it helps ensure that the proper values are being used in the expression. You may also use the "explicit" keyword with the symbolic equal sign. See Chapter 14 for a discussion of how to use this method of displaying previously defined variables.
5. Always display the resulting value of expressions or functions. For example, if you have an expression Area:= b*d, then always type Area= following the definition. This makes it easier to check the calculations, and it helps to ensure that the results are what you expect.
6. Once you define a variable, do not redefine it (usually). Use subscripts or other means to make each variable unique. See the discussion in the next section on reusing variables.
7. Once a variable has been defined, use the variable name. In future expressions, do not use the variable value. This helps ensure that the calculations are correctly updated if the variable changes.

How to use redefined variables in project calculations

There are times when you need to use redefined variables in your project calculations. This section will give you some ideas on how to do this without causing problems in your project calculations.

Pasting the same standard calculation worksheet into your project calculations multiple times will cause redefinitions to occur. This could cause problems as discussed above. It is not practical to change the variables used in a standard calculation worksheet each time it is pasted into your project calculations. How do you prevent critical project calculation variables from being redefined?

Let's look at an example. Figure 19.12 shows a standard calculation worksheet. Figure 19.13 shows a project calculation with a standard calculation pasted into it two times. Notice that every variable in System 2 is redefining a previously used variable. None of the variables from System 1 can be reused later in the calculations. Figure 19.14 shows the same condition, but demonstrates how you can keep the variables unique in each System.

This standard calculation calculates the total head (Energy/Density of fluid) in a water system using the Bernoulli equation.

Input Variables:

Pressure	$\mathbf{P} := 80\mathbf{psi}$
Density	$\rho := 62.4\mathbf{pcf}$
Velocity of fluid	$\mathbf{v} := 5\frac{\mathbf{ft}}{\mathbf{sec}}$
Height above datum	$\mathbf{z} := 25\mathbf{ft}$

Calculate total head in system

$$\mathbf{H} := \frac{\mathbf{P}}{\rho} + \frac{\mathbf{v}^2}{2\mathbf{g}} + \mathbf{z} \qquad \mathbf{H} = 210.00\,\mathbf{ft}$$

Figure 19.12 Example of a standard calculation worksheet

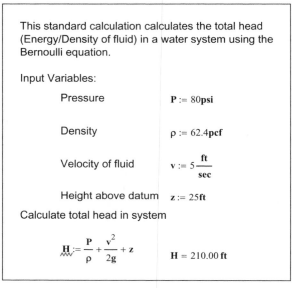

Calculate the head for System 1

Input Variables:

Pressure $\mathbf{P} := 20\mathbf{psi}$

Density $\rho := 62.4\mathbf{pcf}$

Velocity of fluid $\mathbf{v} := 10\frac{\mathbf{ft}}{\mathbf{sec}}$

Height above datum $\mathbf{z} := 40\mathbf{ft}$

Calculate total head in system

$$\mathbf{H} := \frac{\mathbf{P}}{\rho} + \frac{\mathbf{v}^2}{2\mathbf{g}} + \mathbf{z} \qquad \mathbf{H} = 87.71\ \mathbf{ft}$$

Calculate head for System 2

Input Variables:

Pressure $\mathbf{P} := 30\mathbf{psi}$

Density $\rho := 62.4\mathbf{pcf}$

Velocity of fluid $\mathbf{v} := 20\frac{\mathbf{ft}}{\mathbf{sec}}$

Height above datum $\mathbf{z} := 10\mathbf{ft}$

Calculate total head in system

$$\mathbf{H} := \frac{\mathbf{P}}{\rho} + \frac{\mathbf{v}^2}{2\mathbf{g}} + \mathbf{z} \qquad \mathbf{H} = 85.45\ \mathbf{ft}$$

In this example the variables for System 2 redefined every variable for System 1. This may not be a problem. It only becomes a problem if one of the variables in System 1 needs to be used later on in the project calculations. Figure 19.14 shows one method of using standard calculation worksheets and still keeping unique variable names.

Figure 19.13 Inserting the same standard calculation worksheet multiple times

This example is similar to Figure 19.13 except we attempt to prevent the redefinition of critical variables. Let's assume that the variables P, ρ, v, z, and H are not critical variables. They are only useful within the context of the standard calculation. If any of these variable names is critical to your project calculations, then you will need to change the variable names in either the project calculations or in the standard calculation worksheet.

For this example, let's assume that you either calculated or were given the input variables to be used in calculating the total head for System 1 and System 2. Let's also assume that you need to reuse these variables later on in the project so you do not want to redefine them.

Calculate the head for System 1

$P_{System1} := 20psi$ $WaterDensity := 62.4\frac{lbf}{ft^3}$ $V_{System1} := 10\frac{ft}{sec}$ $Height_{System1} := 40ft$

We can now rewrite the standard calculation equation and use the above variables, but that would make the equation much more complex looking. A better way is to assign the above variables to standard calculation variables.

Input Variables:

Pressure	$P := P_{System1}$	$P = 20.00\,psi$
Density	$\rho := WaterDensity$	$\rho = 62.40\,pcf$
Velocity of fluid	$v := V_{System1}$	$v = 10.00\frac{ft}{s}$
Height above datum	$z := Height_{System1}$	$z = 40.00\,ft$

Calculate total head in system

$H := \frac{P}{\rho} + \frac{v^2}{2g} + z$ $H = 87.71\,ft$ $Head_{System1} := H$ $Head_{System1} = 87.71\,ft$

Calculate the head for System 2

$P_{System2} := 30psi$ $WaterDensity = 62.40\,pcf$ $V_{System2} := 20\frac{ft}{sec}$ $Height_{System2} := 10ft$

Note: We do not want to redefine the variable WaterDensity so we only display it.

Input Variables:

Pressure	$P := P_{System2}$	$P = 30.00\,psi$
Density	$\rho := WaterDensity$	$\rho = 62.40\,pcf$
Velocity of fluid	$v := V_{System2}$	$v = 20.00\frac{ft}{s}$
Height above datum	$z := Height_{System2}$	$z = 10.00\,ft$

Calculate total head in system

$H := \frac{P}{\rho} + \frac{v^2}{2g} + z$ $H = 85.45\,ft$ $Head_{System2} := H$ $Head_{System2} = 85.45\,ft$

Using this method, we are able use reuse the standard calculations, and still maintain the unique variable names from System 1 and System 2.

Figure 19.14 Inserting the same standard calculation worksheet multiple times and still keeping unique results

Resetting variables

There is another problem that may occur in your calculations when you reuse standard calculation worksheets. If for some reason, a redefined variable is accidentally deleted, then Mathcad goes up and gets the previously defined value. This may occur without you knowing it. There is no error in Mathcad, but your result will be incorrect. Here is another case that can cause errors. If your displayed result is just slightly higher in the worksheet than the definition for the result, Mathcad will go up and get the result from the previous variable instead of the adjacent variable. See Figure 19.15. There is a way to prevent this from happening. At the top of your standard calculation worksheet, you can define each variable to be a text string. Then, if a variable is deleted, Mathcad will use the text string instead of an unwanted number above. This will cause an error or display the text string. This way you will be notified that there is a problem, and it prevents Mathcad from using unwanted previously defined variables. See Figure 19.16.

Look for the inaccurate displayed value of H for System 2

Calculate the head for System 1

Input Variables:

Pressure $P := 20\mathbf{psi}$

Density $\rho := 62.4\mathbf{pcf}$

Velocity of fluid $v := 10\dfrac{\mathbf{ft}}{\mathbf{sec}}$

Height above datum $z := 40\mathbf{ft}$

Calculate total head in system

$H := \dfrac{P}{\rho} + \dfrac{v^2}{2g} + z$ $\quad H = 85.45\,\mathbf{ft}$

Calculate head for System 2

Input Variables:

Pressure $P := 30\mathbf{psi}$

Density $\rho := 62.4\mathbf{pcf}$

Velocity of fluid $v := 20\dfrac{\mathbf{ft}}{\mathbf{sec}}$

Height above datum $z := 10\mathbf{ft}$

Calculate total head in system

If H is just slightly above the redefined definition of H, then Mathcad selects the previous definition of H. This can cause errors to occur in your displayed calculations.

$H = 87.71\,\mathbf{ft}$

$H := \dfrac{P}{\rho} + \dfrac{v^2}{2g} + z$

The correct result should be $H = 85.45\,\mathbf{ft}$. Figure 19.16 shows one way to prevent this from occurring.

Figure 19.15 Potential problem to avoid when using standard calculations multiple times

In Figure 19.15 we showed how the displayed value of H for system 2 was incorrect. In order to catch this error, you would need to do a very thorough check of the calculations. To prevent this error from occurring, reset the variables at the beginning of each System.

Calculate the head for System 1

Input Variables: $P := "*"$ $\quad \rho := "*"$ $\quad v := "*"$ $\quad z := "*"$ $\quad H := "*"$

Pressure $P := 20\mathbf{psi}$

Density $\rho := 62.4\mathbf{pcf}$

Velocity of fluid $v := 10\frac{\mathbf{ft}}{\mathbf{sec}}$

Height above datum $z := 40\mathbf{ft}$

Calculate total head in system

$$H := \frac{P}{\rho} + \frac{v^2}{2g} + z \qquad H = 87.71\,\mathbf{ft}$$

Calculate head for System 2

Input Variables: $P := "*"$ $\quad \rho := "*"$ $\quad v := "*"$ $\quad z := "*"$ $\quad H := "*"$

Pressure $P := 30\mathbf{psi}$

Density $\rho := 62.4\mathbf{pcf}$

Velocity of fluid $v := 20\frac{\mathbf{ft}}{\mathbf{sec}}$

Height above datum $z := 10\mathbf{ft}$

Calculate total head in system

Now, if H is just slightly above the redefined definition of H, then Mathcad gives a text message instead of the previous definition of H. This way Mathcad will not use any previous definitions, and you can detect an error. If any of the variables in System 2 get deleted, you will get an error instead of using the value from System 1.

$$H := \frac{P}{\rho} + \frac{v^2}{2g} + z \qquad H = "*"\ \mathbf{ft}$$

The correct result should be $H = 85.45\,\mathbf{ft}$.

Figure 19.16 Resolving a potential problem to avoid when using standard calculations multiple times

Using user-defined functions in standard calculation worksheets

When you define variables and equations in your standard worksheets, then these variables and equations are pasted into your projects calculations. One way to avoid having all these additional variables added to your project calculations is to use user-defined functions.

By doing this, the standard worksheets become more compact, and the project calculations will not have so many additional variables. Let's relook at Figure 19.12 and see how much cleaner it is to use a user-defined function. See Figures 19.17 and 19.18.

This standard calculation calculates the total head (H) (Energy/Density of fluid) in a water system using the Bernoulli equation.

Input Variables: Pressure (P), Density (D), Velocity of fluid (v), and Height above datum (z)

Calculate total head in system

$$H(P,\rho,v,z) := \frac{P}{\rho} + \frac{v^2}{2g} + z$$

$$H\left(80\text{psi}, 62.4\text{pcf}, 5\frac{\text{ft}}{\text{sec}}, 25\text{ft}\right) = 210.00\,\text{ft}$$

Figure 19.17 Example of a revised Figure 19.12

Calculate the head for System 1 (Values from Figure 19.14)

$P_{System1} = 20.00\ \mathbf{psi}$ **WaterDensity** $= 62.40\ \mathbf{pcf}$ $V_{System1} = 10.00\frac{\mathbf{ft}}{\mathbf{s}}$ $\mathbf{Height}_{System1} = 40.00\ \mathbf{ft}$

Input Variables: Pressure (P), Density (D), Velocity of fluid (v), and Height above datum (z)

Calculate total head in system

$$H(P,\rho,v,z) := \frac{P}{\rho} + \frac{v^2}{2g} + z$$

This formula is copied from the standard calculation worksheet in Figure 19.17.

$\mathbf{Head}_{system1} := H(P_{System1}, \mathbf{WaterDensity}, V_{System1}, \mathbf{Height}_{System1})$

$\mathbf{Head}_{system1} = 87.71\ \mathbf{ft}$

Calculate the head for System 2 (Values from Figure 19.14)

$P_{System2} = 30.00\ \mathbf{psi}$ **WaterDensity** $= 62.40\ \mathbf{pcf}$ $V_{System2} = 20.00\frac{\mathbf{ft}}{\mathbf{s}}$ $\mathbf{Height}_{System2} = 10.00\ \mathbf{ft}$

$\mathbf{Head}_{system2} := H(P_{System2}, \mathbf{WaterDensity}, V_{System2}, \mathbf{Height}_{System2})$

$\mathbf{Head}_{system2} = 85.45\ \mathbf{ft}$

Notice how much cleaner this worksheet is than Figure 19.14. The standard calculation only contained the user-defined function. There were not extra variables to copy.

Using this method, we are able use reuse the standard calculations, and still maintain the unique variable names from System 1 and System 2.

Figure 19.18 Inserting standard calculation containing user-defined equations

Using the *reference* function

As you assemble your project calculations, you do not need to have all of the calculations included in a single Mathcad file. The ***reference*** function allows Mathcad to get variable definitions from another Mathcad file and use them as if they were included in the current worksheet. For example, you could have a project design criteria worksheet where all the key variables for a project are located. Then each project worksheet will reference this one worksheet for all project related criteria. You can also have a corporate worksheet that contains standard definitions and functions. This corporate worksheet can be referenced from your corporate template. Figures 19.19 and 19.20 give an example of using the ***reference*** function.

The values from this worksheet will be used in Figure 19.20.

Yield Strength of Steel $\mathbf{F_y} := 50\,\mathbf{ksi}$

Compressive Strength of Concrete $\mathbf{f'_c} := 4000\,\mathbf{psi}$

Figure 19.19 Using the *reference* function

To access the Insert Reference dialog box, click **Reference** on the **Insert** menu.

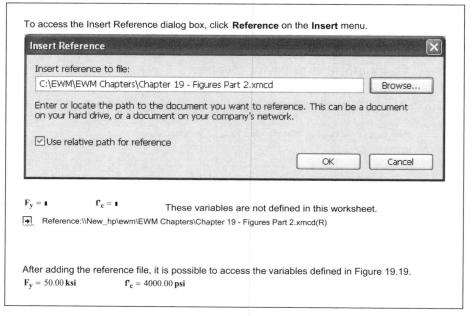

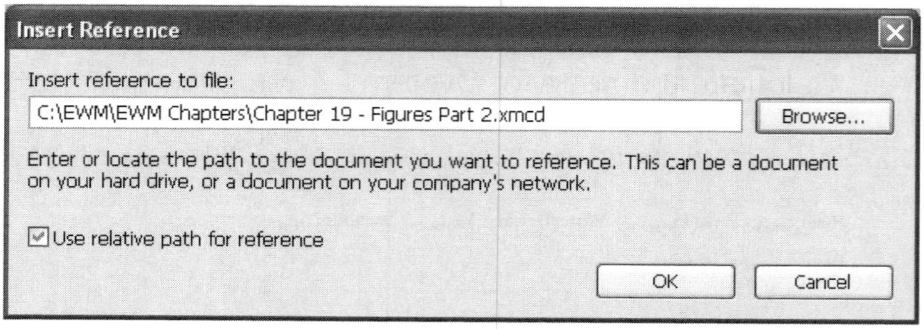
Insert Reference
Insert reference to file:
C:\EWM\EWM Chapters\Chapter 19 - Figures Part 2.xmcd
Browse...
Enter or locate the path to the document you want to reference. This can be a document on your hard drive, or a document on your company's network.
Use relative path for reference
OK
Cancel

$\mathbf{F_y} = \blacksquare$ $\mathbf{f'_c} = \blacksquare$ These variables are not defined in this worksheet.

Reference:\\New_hp\ewm\EWM Chapters\Chapter 19 - Figures Part 2.xmcd(R)

After adding the reference file, it is possible to access the variables defined in Figure 19.19.

$\mathbf{F_y} = 50.00\,\mathbf{ksi}$ $\mathbf{f'_c} = 4000.00\,\mathbf{psi}$

Figure 19.20 Using the *reference* function

To insert the ***reference*** function, click **Reference** from the **Insert** menu. You can then browse to the file location, and select the file to be referenced. It is a good idea to place a check in the "Use relative path for reference" check box. When this box is checked, it allows you to move both the worksheet and the referenced file together, and they will remain linked together. In order to use this feature, worksheets must reside on the same local or network drive, and both files must be stored prior to using the ***reference*** function.

When to separate project calculation files

When is a project worksheet so large that it needs to be separated into two or more files? Try to stay with one file unless one or more of the following occurs:

- It takes too much time to recalculate the worksheet.
- Your worksheet has many graphics, which makes the files over 10 or 15 MB in size.
- You worksheet gets to be over 100 pages in length.
- It makes sense to separate the project calculations into separate worksheets with related topics.

If you separate your project calculations, here are a few suggestions:

- Create a master worksheet that contains the critical project design criteria.
- Reference the master worksheet to get the critical project design criteria. Do not define the project design criteria in each worksheet because there is a risk of one worksheet using different criteria than another worksheet.
- If you need to use results from another file, then use the ***reference*** function to add a reference to the file. Do not redefine the data in the new worksheet. If the data in the other file changes, you want to have up-to-date results in the current worksheet.
- After referencing a file, display key variables below the reference line. This will aid in checking calculations.
- Add page numbers to either the header or footer. Add prefixes to the page numbers for each file, so that paper copies of the calculations will have unique page numbers.

Using *Find* and *Replace*

If you find yourself in a situation where you have conflicting variable names that need to be changed, you can use the ***Find*** and ***Replace*** features to simplify the effort. The ***Replace*** feature finds the requested variable and replaces it with another variable. The ***Replace*** feature is on the **Edit** menu. See Figure 19.21.

There are a few things to keep in mind when using the ***Replace*** feature. The "Find in Math Regions" is turned off by default. Be sure to place a check in this box so that Mathcad will search in Math regions. If you are searching for the variable "If," then you may want to check the "Match whole word only" box, so that you are not stopped in every word that contains the letters "if." If "Match Case" is checked, then Mathcad will not stop at "if." If your variable name uses

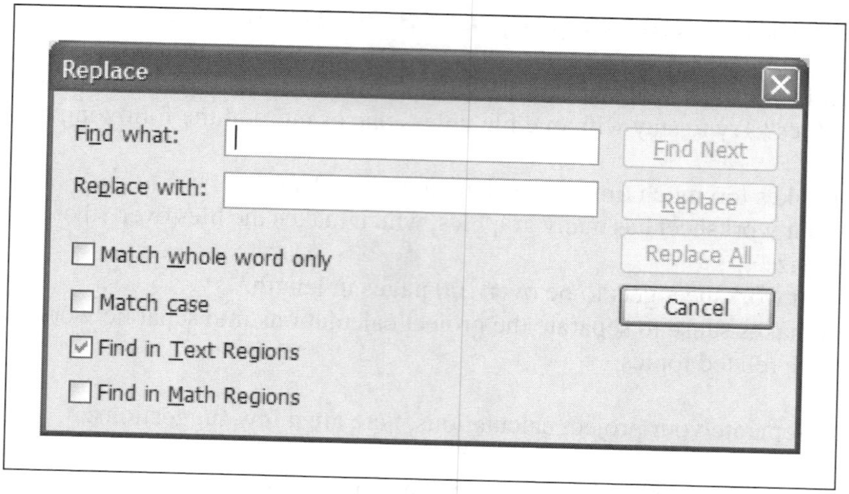

Figure 19.21 Replace dialog box

a literal subscript, then be sure to include the period. For example F_y should be typed `f.y`.

Mathcad Application Server and Designate

If you are in a large organization that has offices over many locations, or if you need to distribute your calculations to several different locations, Mathsoft has two products that may interest you.

Mathcad Application Server

The Mathcad Application Server is a web-based application that can distribute Mathcad worksheets to many people through an intranet or over the internet. Users do not need to have Mathcad installed on their computers to use this product. This could be a valuable way to distribute an entire project calculations or portions of calculations to others in your organization, or to contractors, consultants or clients.

Designate

Designate is a web-based application that provides a centralized means to access standard calculation worksheets. All your standard calculation worksheets and final project calculations can be stored in a central file library. These calculations are cataloged and you can do a full-text search or a metadata search of all the stored calculations. You can copy information from calculations in the library, and Designate knows who used it and for what project. That way if there is a change in data, you will know where that data was used.

For more information on these products, contact Mathsoft or visit their Web site at www.mathsoft.com.

Summary

This chapter focused on using Mathcad to assemble project calculations. It showed how to use information from standard calculation worksheets. It also showed the advantages and disadvantages of pasting regions from standard calculation worksheets.

In Chapter 19 we:

- Demonstrated how to copy and paste regions from one worksheet to another.
- Discussed how to add comments to worksheets and regions.
- Introduced Provenance.
- Discussed creating standard calculation worksheets.
- Showed two ways of locking and protecting worksheets.
- Warned of potential problems that may occur when you paste regions from other worksheets.
- Gave suggestions on how to avoid the potential problems.
- Provided guidelines for using standard calculation worksheets.
- Showed how to use the ***reference*** function.
- Discussed when to split project calculations into separate files.
- Introduced the Mathcad Application Server and Designate.

20
Importing files from other programs into Mathcad

The ability of Mathcad to import information from other programs is what makes Mathcad such an ideal program to not only create engineering calculations, but to also organize the calculations. Most project calculations will be comprised of information generated from many different computer programs. It is possible to import and display the information from these programs within the Mathcad calculations.

Chapter 20 will:

- Discuss the linking and embedding of information from other software applications into your Mathcad worksheet.
- Identify the differences between linking and embedding.
- Discuss how the information from various software applications can be brought into Mathcad.
- Show how Mathcad can be used to store input files for software applications.
- Illustrate how to include output from other engineering programs into Mathcad.
- Show how to use Mathcad to calculate the input for engineering programs.

Introduction

In Chapter 18, we introduced the concept of using Mathcad to replace the binder that was used to hold all of your paper calculations. In this chapter, we will show you how to accomplish that goal. Most project calculations are comprised of output from multiple software programs. Your Mathcad project calculations can be used to assemble the output from these programs.

The data we discuss in this chapter is not useable by Mathcad. In other words, Mathcad cannot use the data in its calculations. It can only display the data. If you want Mathcad to use any of the displayed data, you must manually input it into Mathcad. In the next chapter, we discuss the intelligent transfer of information between Mathcad and other software programs.

Object linking and embedding (OLE)

Object linking and embedding allows Mathcad to display information from other software programs within the Mathcad worksheet. When you double-click on the displayed information, the application that created the result actually opens, and you can work on the result in the originating software program. If you are working on a worksheet that was created on another computer, you must have the application loaded on your computer in order to edit the result in the original application.

An embedded object means that the displayed data resides within the Mathcad file that you are using. It is like having a photocopy of the original file. If you make changes to the copied data, the changes are not made to the original file. If you make a change to the original file, the changes are not reflected in the Mathcad file.

A linked object means that the source of the displayed data resides in another file. You have created a link to the original file, and the data does not reside in the Mathcad file. If the original file is changed, the displayed data in the Mathcad worksheet is also updated. If you double-click on the data to open the application, you are actually changing the original data file. If you copy the Mathcad worksheet to a location where the linked object is not available, you will lose the displayed data. You can update, change, or delete links by selecting **Links** from the **Edit** menu.

The advantage of using a linked file is that the data in Mathcad is always up-to-date. The disadvantage is that if the original file gets deleted, then Mathcad no longer has access to the data. You must also keep track of the original file if you are going to archive the Mathcad project calculations. The advantage to using an embedded file is that it always stays with the Mathcad worksheet. You do not need to worry about keeping track of it. The disadvantage is that if the original file is updated, then the Mathcad file will be out-of-date until an updated file is embedded in the project calculations.

Bring objects into Mathcad

There are two ways you can use embedded and linked objects. You can display the data from the object, or you can display an icon on the Mathcad worksheet. The icon can be either an embedded file stored within the Mathcad worksheet file, or it can be a link to an external file. When you double-click the icon, it opens the associated application and then opens the embedded file, or opens the linked file.

There are several ways of bringing objects into Mathcad. The easiest way is to copy the information from one application and then paste it into the Mathcad worksheet. You can highlight the information from one application, and then drag and drop it onto the Mathcad worksheet. You can also use the Insert Object dialog box. To access this dialog box, click **Object** from the **Insert** menu. See Figure 20.1

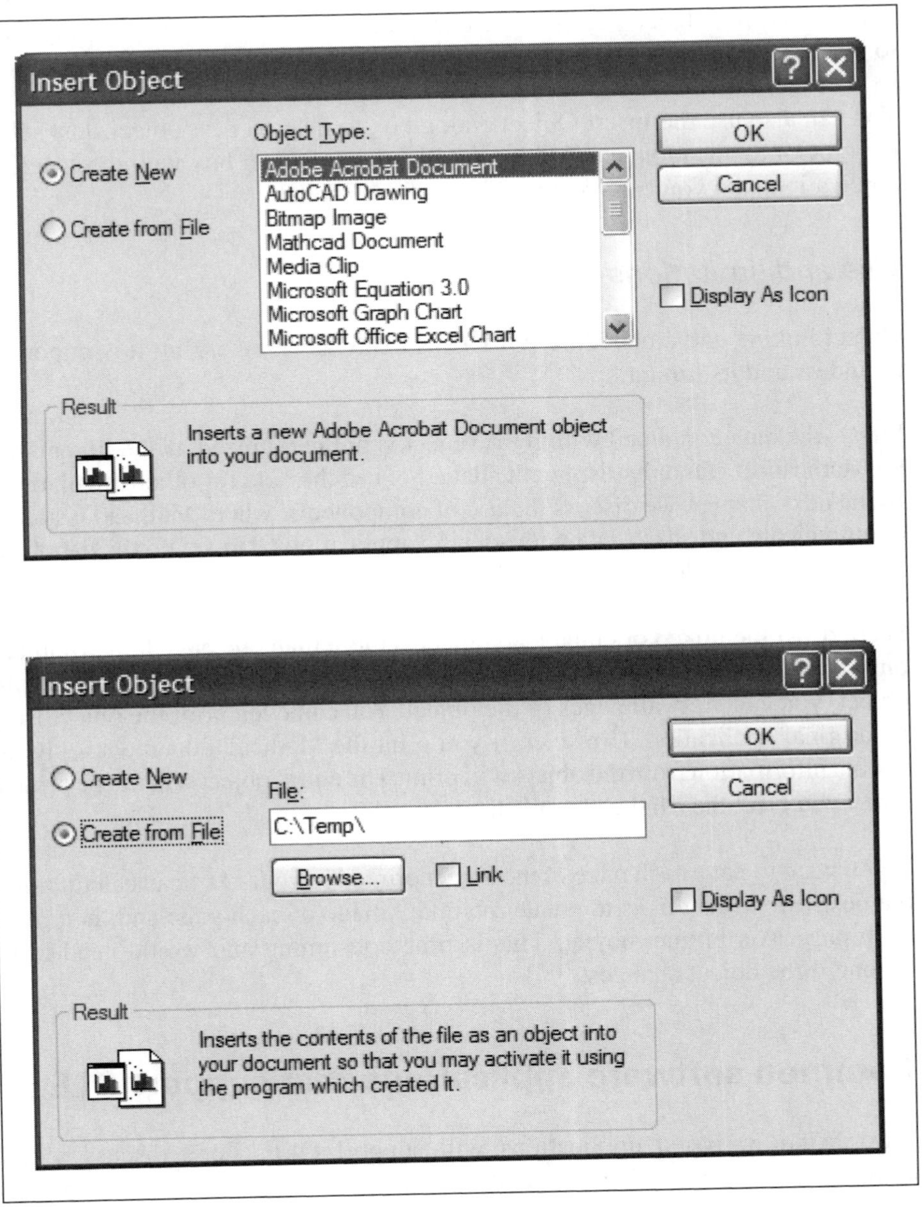

Figure 20.1 Insert Object dialog box

From this dialog box, you can insert an object from an existing file. Just click **Create from File** then browse to the file, and open the file. Mathcad will insert the object into your Mathcad worksheet. If you want to link to the object, rather than embed the object, then place a check in the "Link" box. If you only want to insert an icon, then place a check in the "Display As Icon" box.

You can also use the Insert Object dialog box to insert a new object. Just select the type of application, and Mathcad will open an object box with the specified application, and you can create an object from scratch.

Use and limitations of OLE

Object linking and embedding (OLE) has many advantages, but it is important to understand its limitations.

First—the data contained within the object is not recognized as intelligent data by Mathcad. It can only display the data, not use the data for other calculations. In the next chapter, we discuss the use of components, where Mathcad is able to communicate and share data with specific applications. However, the data from linked and embedded objects is only displayed.

Second—if the linked or embedded object is more than one page long, Mathcad only displays the first page of the document. When you double-click on the object, you can view all pages of the object. You can even print the object from its original application. However, if you print the Mathcad document, only the visible information from the object will print. The entire object will not be printed when you print the Mathcad worksheet.

If you want to have each page of the object printed in your Mathcad calculations, the best way to do this is to create a bitmap image of each page and then paste each page as a bitmap object. This is time consuming and would need to be redone if the object changes.

Common software applications that support OLE

Most Windows-based applications will support OLE. Let's discuss a few common software applications and how they can be used in your project calculations.

> **Tip!** When you copy data from these applications, it is always best to use the Paste Special command. This gives you control over what type of object Mathcad will create. The Paste Special command can be used by clicking **Paste Special** from the **Edit** menu, or you can right-click and choose **Paste Special**.

Microsoft Excel

This is a very popular application used for engineering calculations. Microsoft Excel files can be embedded or linked in your project calculations. However, the real power of Excel files is being able to have the Excel files communicate with your Mathcad worksheet. We devote an entire chapter to the discussion of Microsoft Excel spreadsheets in Chapter 22.

Microsoft Word or Corel WordPerfect

These word processing programs can be used if you have large text files that you want to include in your calculations. You can also bring text files from computer output into a word processor, and then paste the results into Mathcad. This way you can include the results from other programs in your Mathcad project calculations. The easiest way to bring data from a word processing program is to copy and paste. This will embed the information in Mathcad. If you want to link to a word processor file, you will need to insert the link using the Insert Link dialog box.

Adobe Acrobat

Many software applications do not provide output in a format that can be read into a word processor. It is difficult to include output from these programs into your Mathcad project calculations unless you have Adobe Acrobat. With Adobe Acrobat you can print your results to a PDF file. This PDF file can then be linked or embedded into Mathcad. If the PDF file is longer than one page, only the first page will print with the Mathcad worksheet, but the entire PDF file will be available to view when you double-click on the object. If you want to show all pages of the PDF file in Mathcad, then you can copy each page from the PDF file (use the Snapshot tool) and paste it into Mathcad as a bitmap object. The size of the bitmap image is a function of what zoom level Acrobat is set at when you copy the page.

If you do not have Adobe Acrobat and want to bring printed output into Mathcad, there are a couple of options. First, scan the printed output into a tif, jpg, or pdf format and embed these objects in the Mathcad worksheet. The second method is to do a print screen and paste the bitmap into Mathcad. If you first paste the object into another program, you can crop the object prior to pasting into Mathcad.

AutoCAD

AutoCAD files are very useful to bring into Mathcad calculations. There are unlimited uses to having AutoCAD drawings included in your calculations.

Here are a few tips to help bring AutoCAD images into your calculations.

1. In AutoCAD, select and copy the items you want to paste into Mathcad, and then paste the items into Mathcad.
2. The size of the AutoCAD window determines the size of the object brought into Mathcad. If the AutoCAD window is full-screen, then a full-screen object will be created in Mathcad, even if the items you copy only occupy a small portion of the AutoCAD window.
3. It is best to size the AutoCAD window to the size of the object you want to paste into Mathcad. Once the AutoCAD window is the size you would like, then zoom into the item you want to copy. You can select the items you want to copy before or after you zoom into the objects. Only the objects visible in the AutoCAD window will be copied into Mathcad, even though additional items are selected.

You can copy and paste bitmap images from many different programs. There are many times when you just want to create a simple quick bitmap sketch to include with your project calculations. You can do this by inserting a bitmap image from the Insert Object dialog box. This will open a small object box in Mathcad and give you a few tools to create a simple sketch.

Multimedia

You can even include photos, video clips, sound clips, Microsoft PowerPoint presentations, and other multimedia in your project calculations. Many of these are identified by icons. When you double-click the icon the application runs. You can copy and paste photos from any photo editing program.

Data files

You can store data files in your Mathcad worksheet. Simply drag and drop a data file from My Computer or Windows Explorer onto your Mathcad worksheet.

This creates an icon in your worksheet, and this file then becomes a part of the Mathcad project calculations. If you double-click the file, Mathcad will either open the data file in the associated application, or ask you which application you want to use to open the file. If you have included output results in your Mathcad project calculations (either as a displayed object or as an icon), it is an excellent idea to include the input file as well. Your Mathcad worksheet can contain numerous data files.

Summary

Mathcad makes it possible to include graphics, photos, and CAD drawings in your project calculations. You can also include MS Excel, Word, and PowerPoint files. This makes it possible to bring the results from other engineering software into your Mathcad project calculations. You can also include input and output data files within Mathcad.

In Chapter 20 we:

- Introduced object linking and embedding (OLE).
- Gave a brief description of different software applications that can be used in engineering calculations.
- Discussed the benefits and limitations of including output from other programs within Mathcad.

21 Communicating with other programs using components

Components are a powerful way to place data from Mathcad into other programs and to place data from other programs into Mathcad. The previous chapter discussed object linking and embedding (OLE). The information displayed by these objects is useful, but it is unintelligent. Mathcad can only display it, not use it. Mathcad does not know what the information is. However, by using components, Mathcad can give intelligent information to another program. That program can then use the input information, process it, and then return output values to Mathcad. Once Mathcad receives the output, the data becomes incorporated into the Mathcad worksheet. Components greatly expand the capability of Mathcad, by dynamically including other programs within the Mathcad calculations.

Mathcad also has an automation Application Programming Interface (API) that allows you to insert Mathcad into other software applications, and have Mathcad communicate with these programs.

Chapter 21 will:

- Introduce the concepts of components.
- Discuss the following components: Excel, ODBC, SmartSketch, MathWorks MATLAB.
- Show how to incorporate these components into Mathcad.

- Discuss the different types of components: Application components, Data components, Controls.
- Discuss how to read from data files.
- Introduce the power of writing scripted objects.
- Show how scripting can add power to communications with other programs.
- Discuss Mathcad Add-ins that allow Mathcad to be added as a component in other programs.

What is a component?

Application components are similar to the simple OLE objects discussed in Chapter 20. The biggest difference between an application component and a simple OLE object is that components allow the intelligent transfer of data between Mathcad and another application. This greatly expands the capability of using Mathcad for your project calculations. You can now have all the benefits of Mathcad in addition to the power and advantages of other software applications. In order to use an application component, the application must be installed on your computer.

Mathcad comes with several pre-built easy-to-use components. There are three types of pre-built components: Application components, Data components, and Scriptable Object components.

If you are familiar with C++ programming then you can build your own components using the Mathcad Software Development Kit (SDK). This can be downloaded from www.mathcad.com. The SDK is .NET compliant and is intended for use with Microsoft's Visual Studio .NET.

We will first discuss the Application components and then the Data components. Since these are pre-built components, they are easy to use and do not require any programming experience. We will then briefly introduce the Scriptable Object components. These components provide much added power to your project calculations, but they require some knowledge of scripting and Visual Basic programming.

Application components

The pre-built Application components are Microsoft Excel, Microsoft Access, and other ODBC database programs, MathWorks MATLAB, and Intergraph SmartSketch. In order to use these applications as a component, you must have the application installed on your computer. If you save a Mathcad worksheet with a component, and then give the file to someone else to use, that person must have the same application installed on his or her computer. Some Application components embed the data in Mathcad. Other components provide a link to the data.

Application components are inserted into Mathcad by clicking **Component** from the **Insert** menu. This opens the Component Wizard dialog box. See Figure 21.1.

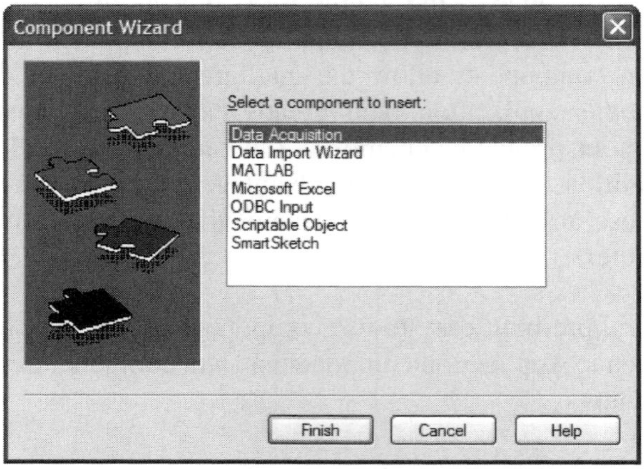

To open the Component Wizard dialog box, select Component from the Insert menu.

Figure 21.1 Component Wizard dialog box

Let's look at how to insert the Application components.

Microsoft Excel

Microsoft Excel is used widely in engineering project calculations. If you are like most engineers, you have multiple Excel spreadsheets that are used extensively. You can bring these Excel spreadsheets into Mathcad, have Mathcad provide the input to your spreadsheets, and then have Excel provide the results back to Mathcad.

Since Excel spreadsheets are so extensively used, Chapter 22 is dedicated to the use of Excel spreadsheets. We will delay our discussion of Excel until the next chapter.

Microsoft Access and other Open DataBase Connectivity (ODBC) components

The ODBC component allows you to read data from database programs such as dBASE, Microsoft Access, and Microsoft FoxPro. You can also read from Microsoft Excel. This component allows you to read from these programs, but you cannot write to them.

In order to use input from an ODBC component, Windows must establish a link to the database you want to use, prior to inserting the component. This is done through the Data Sources (ODBC) tool—found in Windows Administrative Tools. You can access Administrative tools from the Control Panel.

When you open the Data Sources from administrative tools, the ODBC Data Source Administrator dialog box opens. Once the box is open, select **Excel Files** and click the **Add** button. See Figure 21.2. For this example, we will use the sample database used in the Mathcad QuickSheet "Using an ODBC Read Component." After clicking the **Add** button, you should see the Create New Data Source dialog box. From this box, select the driver for the database you will be using. In our example, select **Microsoft Access Driver (*.mdb)**, then click Finish. This opens the ODBC Microsoft Access Setup dialog box.

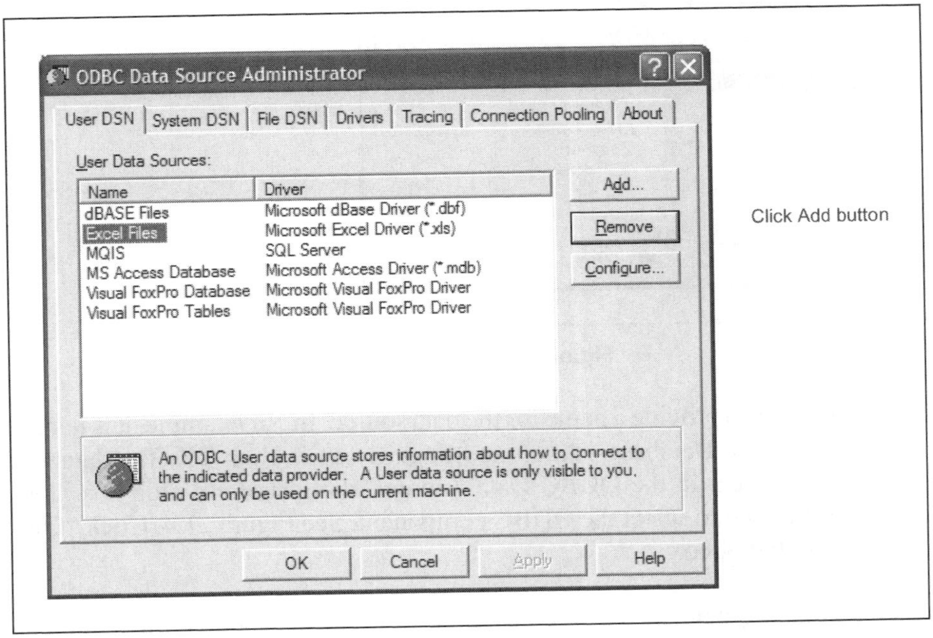

Figure 21.2 Configure data source in Windows

See Figure 21.4. From this dialog box in the **Database** section, click the **Select** button. This opens a Select Database dialog box. From this box, find the database file you want to access. The file we want to use is in the same directory as your Mathcad installation in the path: Mathcad 13\ qsheet\ samples\ ODBC. The file name is MCdb1.mdb. See Figure 21.3.

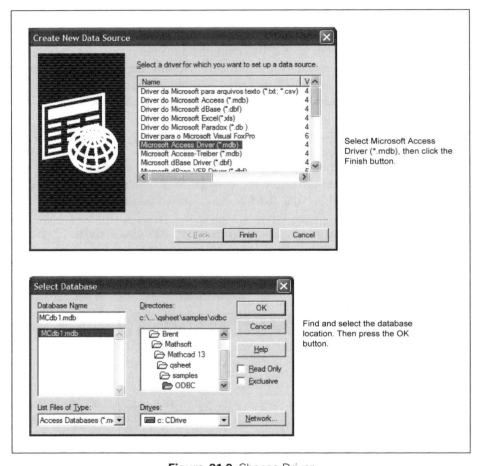

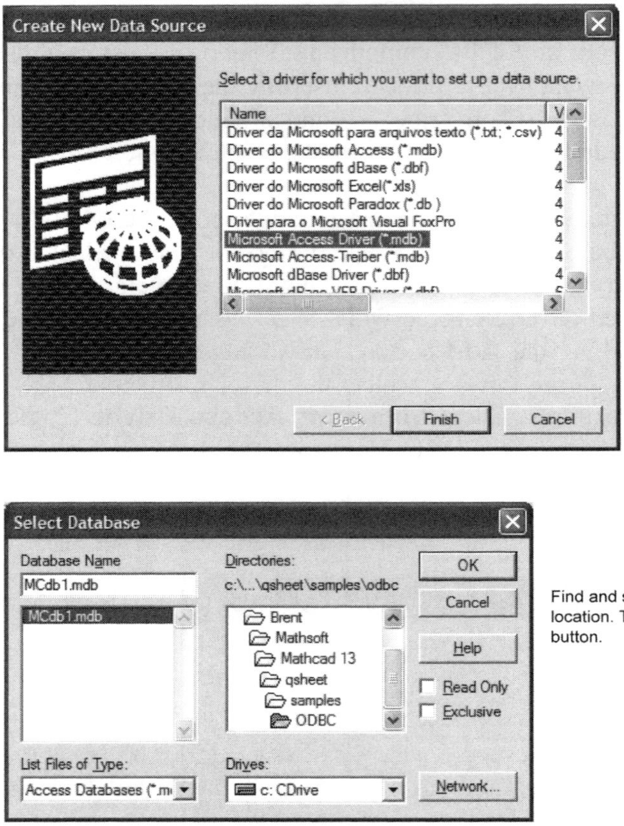

Figure 21.3 Choose Driver

The final step is to provide a name for the data source. In our example, it is named "Mathcad Sample Database." After inputting the name, click **OK**. The database should now appear in the ODBC Data Source Administrator dialog box. This name will be used to select the ODBC component. See Figure 21.4. Click "**OK**" to close the dialog box.

Once you have established a link in Windows to the data source, then you can insert the ODBC component. You can only bring in data from one table at

This database is located in the Mathcad program directory with the path ...Mathsoft\Mathcad 13\qsheet\samples\ODBC\MCdb1.mdb

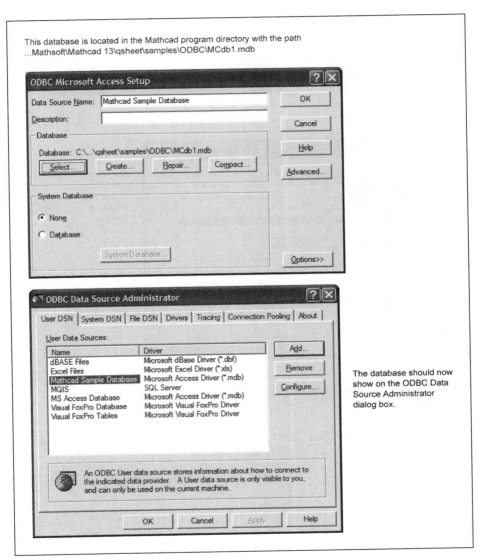

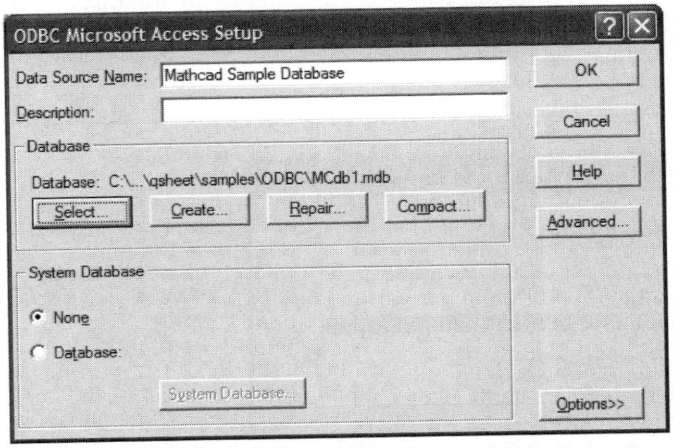

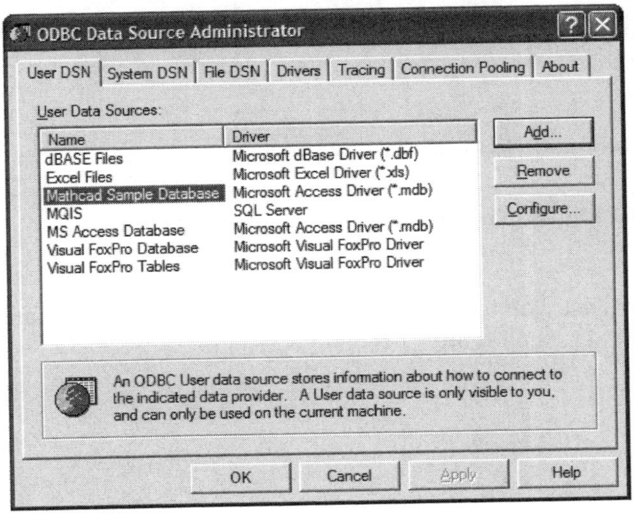

The database should now show on the ODBC Data Source Administrator dialog box.

Figure 21.4 Type a name for the data source

a time. If you want to import data from more than one table, you will need to insert multiple components. You are also able to choose which fields you want to import. If you know Structured Query Language (SQL), you can filter the data before bringing it into Mathcad.

Let's access data from the Mathcad Sample Database we just linked. Open the Component Wizard dialog box and select **ODBC Input**. This opens the ODBC Input Wizard dialog box. Find the Mathcad Sample Database from the drop-down list. See Figure 21.5.

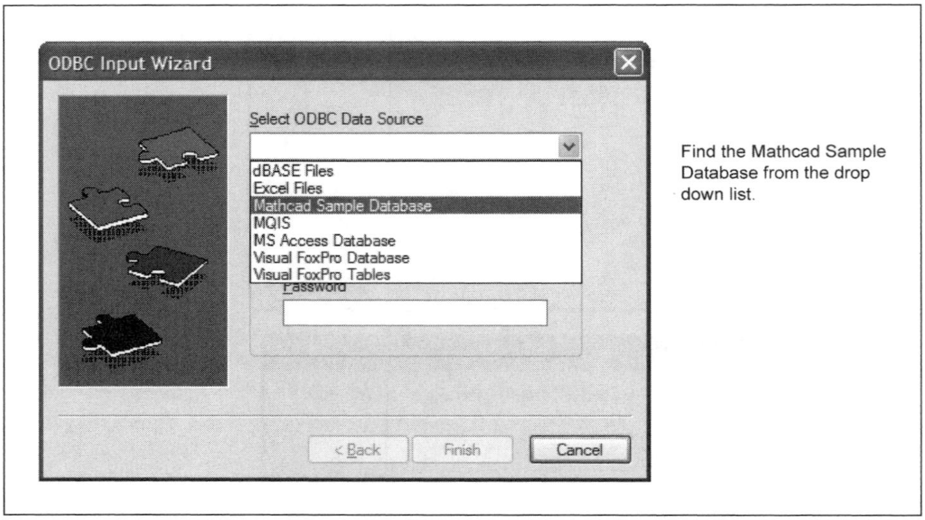

Figure 21.5 Inserting Microsoft Access component

The next dialog box will ask you to select a Table from the database file, and to select fields from the Table. See Figure 21.6.

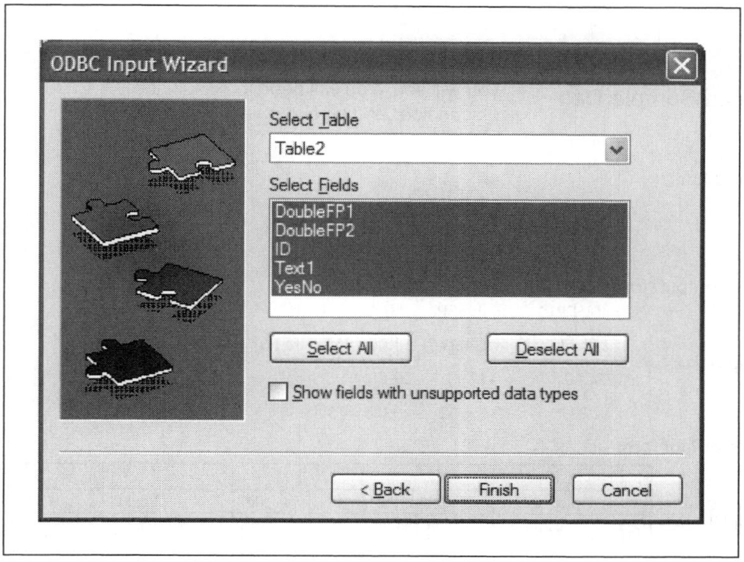

Figure 21.6 Select Table and Fields

After you select **Finish**, a region will appear with the name of the linked database. The placeholder is blank and needs to have a variable name added. When you display the value of the variable, a matrix appears, which is the data from the Access database. See Figure 21.7. The data is now usable in your Mathcad worksheet. If the Access database is changed, the data in Mathcad is also updated.

■ :=

ODBC
athcad Sample Dat:

After you close the dialog box, this is what you will see. You will need to click in the placeholder and type in a variable name.

SampleData :=

ODBC
Mathcad Sample Database

Matrix output form

$$\textbf{SampleData} = \begin{pmatrix} \text{"DoubleFP1"} & \text{"DoubleFP2"} & \text{"ID"} & \text{"Text1"} & \text{"YesNo"} \\ 1234.33 & 1234.57 & 1.00 & \text{"MyText1"} & 0.00 \\ 234.00 & 111.11 & 2.00 & \text{"23-45"} & 1.00 \\ 12.23 & 3424.00 & 3.00 & \text{"3.43.34"} & 0.00 \end{pmatrix}$$

Table output form

SampleData =

	1	2	3	4	5
1	"DoubleFP1"	"DoubleFP2"	"ID"	"Text1"	"YesNo"
2	1234.33	1234.57	1.00	"MyText1"	0.00
3	234.00	111.11	2.00	"23-45"	1.00
4	12.23	3424.00	3.00	"3.43.34"	0.00

You can now access individual elements of the matrix "SampleData"

$\textbf{SampleData}_{1,1} = \text{"DoubleFP1"}$

$\textbf{SampleData}_{1,2} = \text{"DoubleFP2"}$

$\textbf{SampleData}_{1,3} = \text{"ID"}$

$\textbf{SampleData}_{3,3} = 2.00$

The above matrix "SampleData" is dynamically linked to the Microsoft Access database file "MCdb1.mdb. If the data in the database file is changed, the data in the matrix "SampleData" will also be changed.

Figure 21.7 Database example

To configure the component, right-click on the component definition and select **Properties**. If you know SQL, you can filter the data prior to importing. See Figure 21.8.

To open this dialog box, right click on the component and choose Properties.

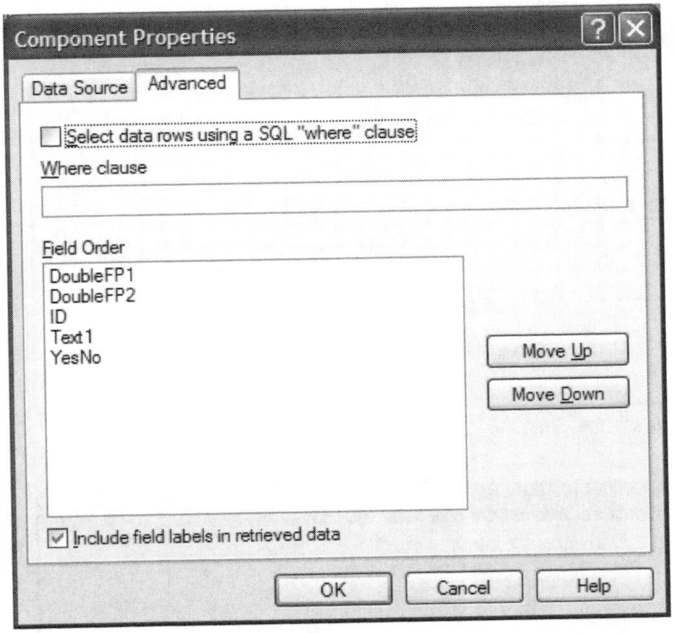

The field labels in the displayed array are turned on by checking the "Include field labels in retrieved data" box.

Figure 21.8 Configuring ODBC component

This is just a sample of what you can do with ODBC components. We are not able to give complete instructions on how to use all ODBC components. For more information, search for ODBC in Mathcad Help.

Intergraph SmartSketch®

The Intergraph SmartSketch component allows you to input drawing dimensions, angles, and other data in Mathcad, and have these plotted in the SmartSketch component.

To insert a SmartSketch component use the Component Wizard. See Figure 21.9.

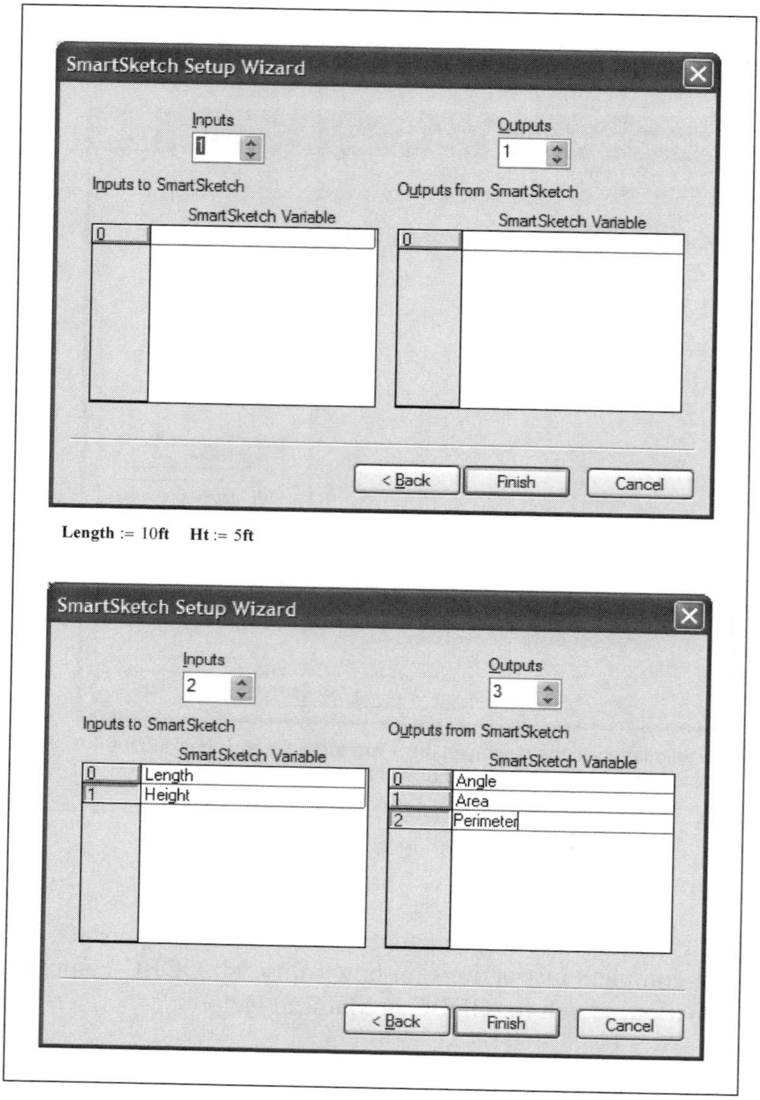

Figure 21.9 SmartSketch component

MathWorks MATLAB®

You can use the MATLAB component to execute MATLAB scripts using data from Mathcad, and then have MATLAB return the result back to Mathcad. The Mathcad component works with MATLAB .M files.

Unlike other application components, the MATLAB component does not have a wizard. When you select MATLAB from the Component wizard, you get a region similar to Figure 21.10. Right-click on this region to add input or output variables. You may have up to four input and four output variables. Right-click on the region and select **Properties** to edit the input and output variable names for MATLAB. See Figure 21.11. Right-click on the region and select **Edit Script** to open and edit the script for the MATLAB component. See Figure 21.12.

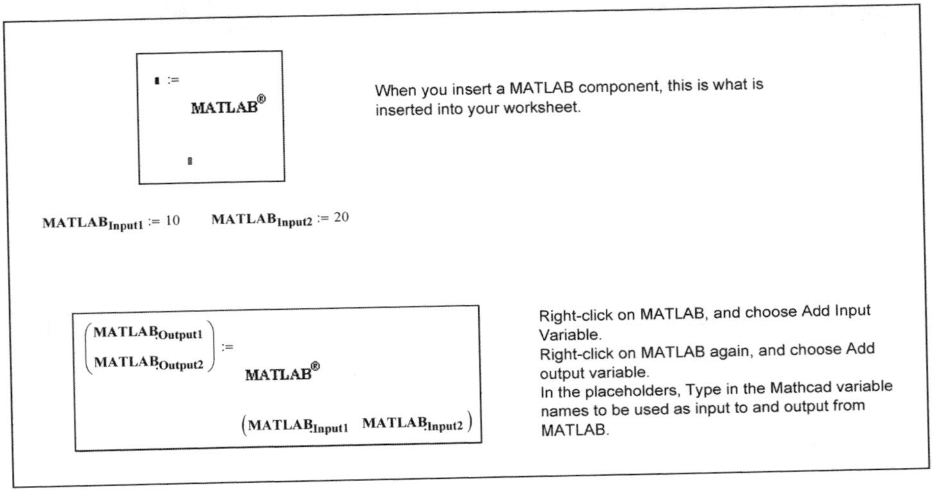

Figure 21.10 MATLAB component

$$\begin{pmatrix} \mathbf{MATLAB}_{\mathbf{Output1}} \\ \mathbf{MATLAB}_{\mathbf{Output2}} \end{pmatrix} := \mathbf{MATLAB}^{®} \left(\mathbf{MATLAB}_{\mathbf{Input1}} \quad \mathbf{MATLAB}_{\mathbf{Input2}} \right)$$

Right-click on MATLAB, and choose Properties. The default MATLAB variable names for input are in0, in1, in2, and in3. You are limited to four input variables. The default MATLAB variable names for output are out0, out1, out2, and out3. You are limited to four output variables. The value of in0 and in1 are taken from Mathcad variables $\text{MATLAB}_{\text{Input1}}$ and $\text{MATLAB}_{\text{Input2}}$. The MATLAB Variable out0 and out1 are transferred back to Mathcad variables $\text{MATLAB}_{\text{Ouput1}}$ and $\text{MATLAB}_{\text{Output2}}$.

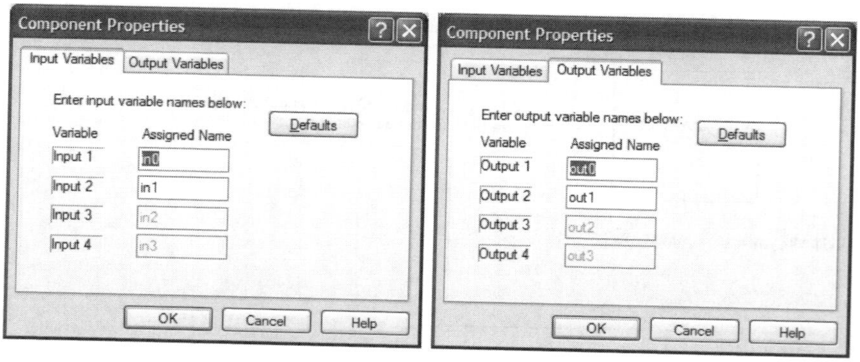

Figure 21.11 Naming MATLAB variables

$$\begin{pmatrix} \mathbf{MATLAB}_{\mathbf{Output1}} \\ \mathbf{MATLAB}_{\mathbf{Output2}} \end{pmatrix} := \mathbf{MATLAB}^{®} \left(\mathbf{MATLAB}_{\mathbf{Input1}} \quad \mathbf{MATLAB}_{\mathbf{Input2}} \right)$$

Right-click on MATLAB, and choose Edit Script. A blank MATLAB Script Editor opens, where you can type your MATLAB script.

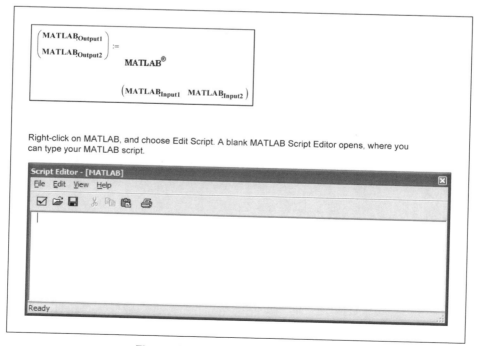

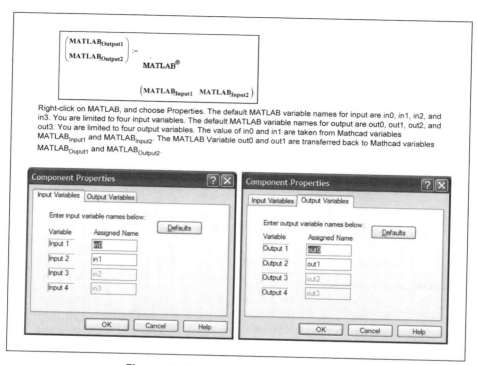

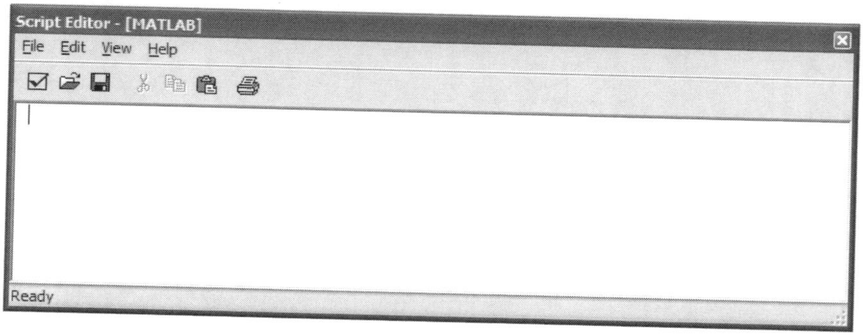

Figure 21.12 Writing MATLAB script

Data components

Data components allow you to read from or write to data files. Once imported into Mathcad, you have full use of the data. You select a matrix variable name, and all data is accessible by using the matrix name and array subscripts. The ODBC component allows you to bring in data from database programs. The data components allow you to access data from a much larger number of format types.

There are several ways to import data from other files. The first methods establish a link with the data files, so that the information brought into Mathcad is always current. The second method only brings in the data once. Once brought into Mathcad, the data does not change.

Data Import Wizard component

The Data Import Wizard is located in two places. You can access it by selecting **Components** from the **Insert** menu, or by selecting **Data** from the **Insert** menu. Both methods bring up the same dialog box. See Figure 21.13 for an example of the Data Import Wizard dialog box.

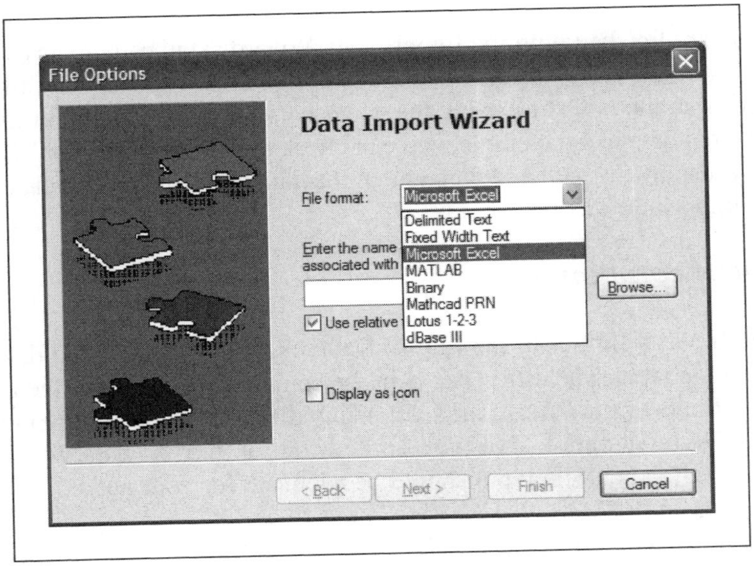

Figure 21.13 Data Import Wizard

The Data Import Wizard establishes a link between your Mathcad worksheet and the data file. Every time you recalculate your worksheet, Mathcad goes to the data file, gets the most up-to-date data, and imports it into the worksheet.

The Data Import Wizard allows you to access the following data types: Delimited text (you tell Mathcad the type of delimiter), Fixed-Width Text, Microsoft Excel, MATLAB, Binary, Mathcad PRN, and dBase. Once you select the file format and filename from the Data Import Wizard, Mathcad will begin asking you questions about the data file. You will be able to preview the data before you finish. You will be able to select things such as:

- Beginning row.
- Ending row.
- Rows to read.
- Columns to read.
- Delimiter type.
- What to do with blank rows.
- How to recognize text.
- What to do with unrecognized or missing data.
- Decimal symbol.
- Thousands separator.
- Named data range (Microsoft Excel).
- Worksheet name (Microsoft Excel).
- Binary data type including endian.
- Column widths (for fixed-width data).

Once you are finished with the Import Data Wizard, Mathcad creates an output table showing all or only a portion of the imported data. You can view all the data by clicking in the table and using the slider bars. If you do not want to display the imported data in a data table, you can check the "Display as icon" box. This will only display the path and filename of the data file. All data is still accessible from within Mathcad.

File Input component

The File Input is similar to the Import Data Wizard. The imported data is still linked and updated. The difference is in the control you have over the imported data. The Import Data Wizard asks you many different questions about the data file. The File Input option gives you limited control over the data. If you have a simple data file, then File Input should be adequate for your needs.

The File Input is used by selecting **Data** from the **Insert** menu and then selecting **File Input**. This opens the File Options dialog box. See Figure 21.14.

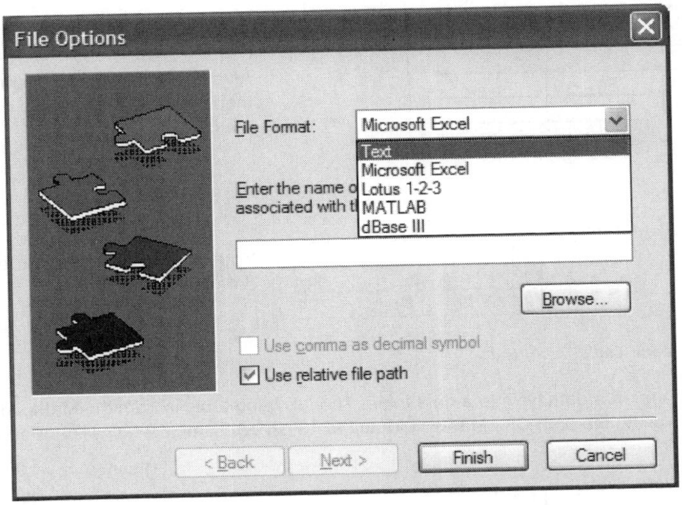

Figure 21.14 File Input component

Select a file type and a file name. When you click Next, Mathcad asks a much more abbreviated list of questions than the Import Data Wizard. The data is inserted into Mathcad as an icon. Once the icon is inserted, click the output placeholder and type in the variable name of the matrix. You can view the data by typing the variable name followed by the colon. You have the same access to the data as you did with the Import Data Wizard.

Data table

A data table is not a true component. It allows you to insert data from a file only once. The data is not updated, and it is not manipulated by another program. You can also use a data table to input your own data into a matrix rather than use the insert matrix dialog box discussed in Chapter 9.

To insert a data table into your worksheet, select **Data** from the **Insert** menu and then click **Table**. This inserts a blank data table. See Figure 21.15. Once the data table has been inserted into your worksheet, you can give the data table a variable name. You can now input data into the table, or you can import a data file into the data table. To import a data file, left-click in the data table, and then right-click and select **Import**. This opens the File Options dialog box. This is the same dialog box as from the File Input component, except now you are embedding the file instead of linking to it.

■ :=

	0	1
0	0	
1		

After the data table is inserted, then you type the variable name in the placeholder.

Sample_Data_Table$_1$:=

	1	2
1	25	"text"
2	2	42

You can enter numbers or text in a data table.

Sample_Data_Table$_{1_{1,2}}$ = "text"

You can import a data file into a data table. The following table is from the Mathcad program directory using the path C:\...Mathsoft\Mathcad 13\qsheet\samples\excel\Thermocouple.xls

Sample_Data_Table$_2$:=

	1	2	3	4	5
1	nd calculation"	0	0	0	0
2	"Coefficients"	0	temperature."	0	0
3	0.039	0	sured Voltage"	ulated Temp."	"Date"
4	25.892	0	1.94	47.706	"05/19/2004"
5	11.87	0	1.84	45.358	"05/19/2004"
6	1.939	0	1.95	0	0
7	0.132	0	2.065	0	0
8	3.152·10-3	0	2.11	0	0
9	0.488	0	0	0	0
10	0.087	0	0	0	0
11	6.795·10-3	0	0	0	0
12	2.1·10-4	0	0	0	0
13	8.635·10-7	0	0	0	0
14	-5.27·10-8	0	0	0	0
15	1.43·10-9	0	0	0	0

Sample_Data_Table$_{2_{1,1}}$ = "Type T Thermocouple calibration and calculation"

Figure 21.15 Inserting a data table

File Output component

The File Output component allows you to save a data table to a file. You can save the data in a variety of file formats. You can choose from: Formatted Text, Tab Delimited Text, Comma Separated Values, Microsoft Excel, Lotus 1-2-3, MATLAB, and dBase III. Using the File Output component is very similar to using the File Input component.

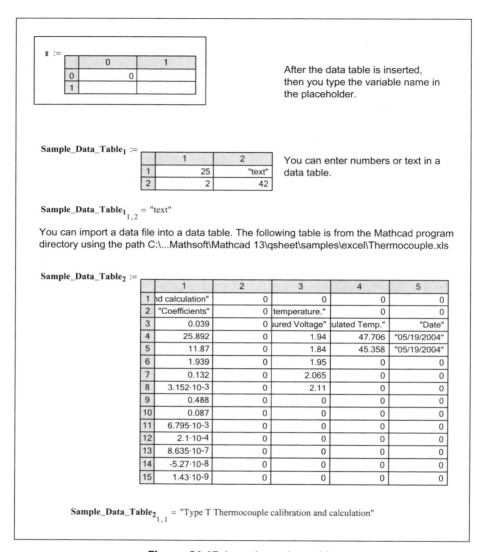

To insert a File Output component into your worksheet, hover your mouse over **Data** on the **Insert** menu and click **File Output**. This opens the File Options dialog box. See Figure 21.16. From this dialog box, choose the type of file format you want to use. Next, browse to the folder location where you want the file to be written, and then enter a file name.

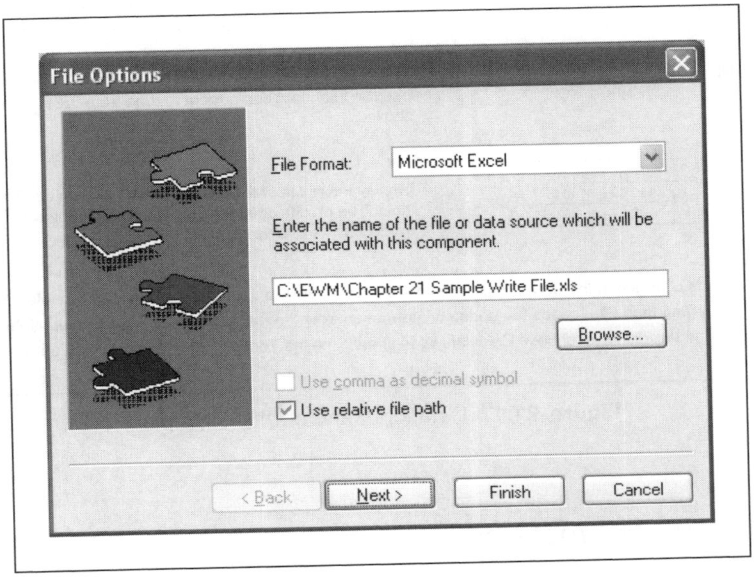

Figure 21.16 File Output component

Figure 21.17 provides an example of creating a data table, writing to a data file, reading from the data file and displaying the results. If the data in the table is updated, the file is updated, which is indicated by the variable $InputFile_1$.

In order to update the worksheet results type `CTRL+F9` or hover your mouse over **Calculate** on the **Tools** menu and click **Calculate Worksheet**.

$\mathbf{DataTable_1} :=$

	1	2
1	10	100
2	30	40

Create a simple data table. This is a static table. It only changes if you make a change.

...\Chapter 21 Sample Write File.xls

$\mathbf{DataTable_1}$

Use the File Output component to write the data table to a file. In this case we use Microsoft Excel .xls format. It is stored in the file, "Chapter 21 Sample Write File.xls".

$\mathbf{InputFile_1} :=$

...\Chapter 21 Sample Write File.xls

Use the File Input component to read from the file created above.

$\mathbf{InputFile_1} =$

	1	2
1	10.00	100.00
2	30.00	40.00

Display the result of the file just read. This is a dynamic table. The results change when the file, "Chapter 21 Sample Write File" is updated.

If a number is changed in the $DataTable_1$, then the change is written to the file "Chapter 21 Sample Write File". The region $InputFile_1$ reads the updated information from the file, which is displayed in $InputFile_1$. In order to refresh all displays, choose Calculate Worksheet from the Tools>Calculate menu, or type CTRL+F9.

Figure 21.17 Data input and output example

Read and write functions

Mathcad comes with numerous read and write functions that allow you to read from or write to various files in various file formats. These functions are beyond the scope of this book. If you are interested in learning about these functions, refer to Mathcad Help under Functions then File Access.

Data Acquisition

The Data Acquisition Component (DAC) allows you to read from and write to recognized measurement devices. With the DAC you can bring "real time" data into Mathcad and perform "real time" analysis of the data. Numerous settings can be set with the DAC. For more on the DAC, see Mathcad Help and the Mathcad Developer's Reference from the **Help** menu.

Scriptable Object component

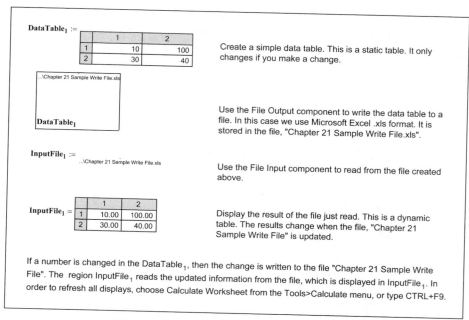

We have just discussed the pre-built Application components and Data components. These are relatively easy to use because they have been built and scripted for you. The Scriptable Object component can add increased functionality to your Mathcad worksheets, but you must have some knowledge of scripting and Visual Basic. The Scriptable Object component can exchange data between your Mathcad and any other application that supports OLE Automation. The component uses Microsoft's Active X scripting specification.

To get an idea of how many different applications support OLE Automation, open the Scripting Wizard. This is opened by clicking **Components** from the **Insert** menu and then selecting Scriptable Object. The Scripting Wizard lists the many different applications that can be scripted to communicate with Mathcad. See Figure 21.18.

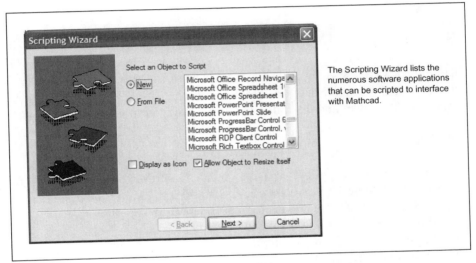

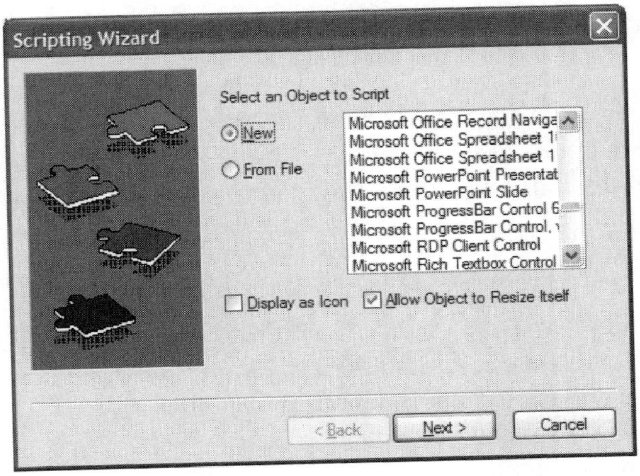

The Scripting Wizard lists the numerous software applications that can be scripted to interface with Mathcad.

Figure 21.18 Scripted Object component

If you are familiar with scripting and with Visual Basic, the Mathcad Help provides complete instructions on how to include a Scriptable Object component into your Mathcad worksheet.

Once a component has been scripted, it can be exported as a customized component and reused in other Mathcad worksheets. Information on this feature can be found in Mathcad Help under Export as Component.

Controls

Mathcad controls are a Scriptable Object component. These will be discussed in Chapter 23.

Inserting Mathcad into other applications

Mathcad has a built-in Application Programming Interface (API) that allows Mathcad to be used as an OLE Automation server from within another application. To use this interface, you need to have knowledge of scripting and Visual Basic. Information on this feature can be found in Mathcad Help or in the Mathcad Developer's Reference under the **Help** menu.

The Automation client interface has already been created for several applications. These are called Add-ins. To view and download these Add-ins go to www.mathcad.com/downloads.

Summary

Components are a wonderful feature of Mathcad because they let you communicate and share information with other files and applications. This allows much greater power and flexibility when creating your project calculations.

In Chapter 21 we:

- Discussed the Application components: Microsoft Excel, Microsoft Access, Intergraph SmartSketch, and MathWorks MATLAB.
- Described the different ways to import data from data files.
- Explained the Data Import Wizard component
- Explained the File Input component.
- Discussed data tables.
- Showed how to write data tables to a data file.
- Introduced the read and write functions.
- Introduced the concept of Data Acquisition.
- Introduced the topic of Scriptable Object components.
- Introduced the idea of inserting Mathcad into other software applications.

22
Microsoft Excel component

Because so many engineers have spent considerable time and money developing Microsoft Excel spreadsheets, this entire chapter will focus on integrating Microsoft Excel spreadsheets into Mathcad.

As discussed in earlier chapters, you can include your previously written spreadsheets within your Mathcad calculations. This chapter will show how to take a previously written Excel spreadsheet and have Mathcad provide the input to these spreadsheets, and how to have Excel pass the results back to Mathcad.

Chapter 22 will:

- Explore the use of Microsoft Excel as a Mathcad component.
- Discuss the advantages and disadvantages of Microsoft Excel.
- Provide several examples of incorporating input and output from existing spreadsheets into Mathcad.

Introduction

Mathcad communicates with the Excel component through input and output variables. You tell the component what the input variables are and which cells to put the values in. You also tell Mathcad which cells to look in for the output values, and what variable name to use for the output. Chapter 21 discussed the use of the Excel component to store and retrieve data. This chapter will focus on the use of Excel for its computational capability.

Inserting new Excel spreadsheets

To insert a new Excel spreadsheet component into your worksheet, click **Components** from the **Insert** menu, and then select **Microsoft Excel** and click the **Next** button. Select, **Create an empty Excel worksheet**, and click the **Next** button. This opens the Excel Setup Wizard dialog box. See Figure 22.1. From this dialog box, you tell Mathcad how many input variables and output variables you will have. You also tell Mathcad where to put the input variables in the Excel worksheet, and where to get the output variables. Both the input

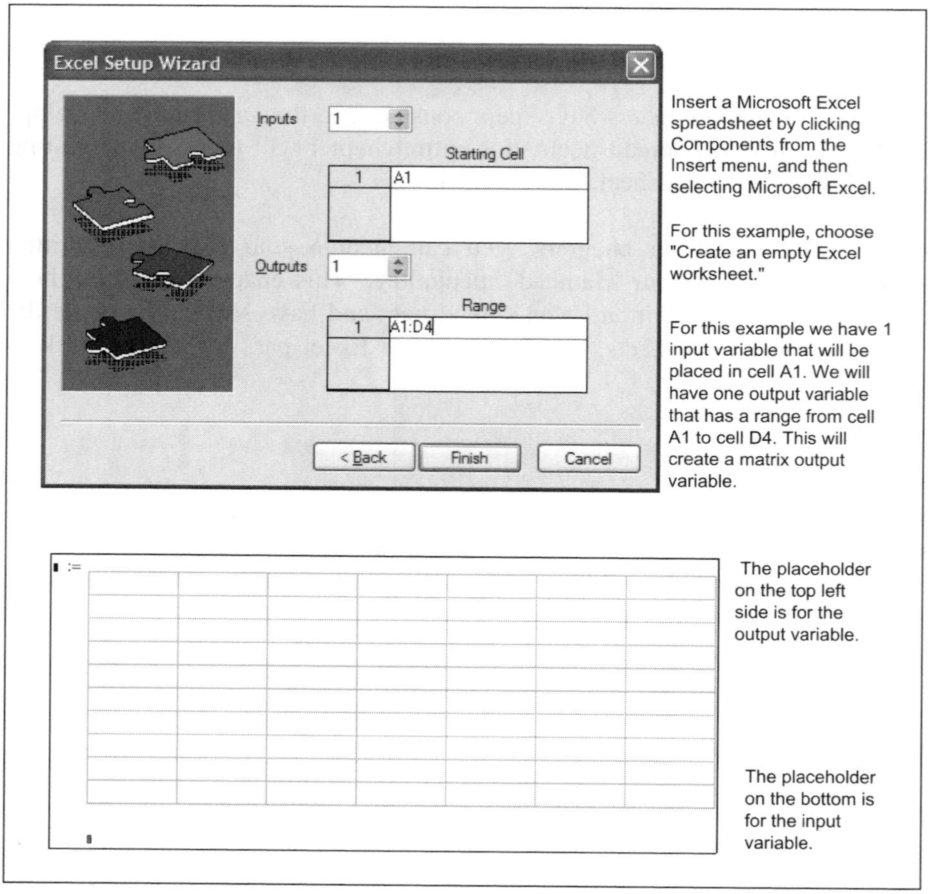

Figure 22.1 Inserting a blank Microsoft Excel worksheet as a component

and output variables can be arrays or matrices. If the input variable is a vector or matrix, you enter the starting cell and Mathcad will fill adjacent cells to the right and below the starting cell. If you select a range for the output variable, then the result will be a vector or a matrix. If you select more than one input or output, then you will need to select a starting cell and range for each input and output. The output range can be a single cell. The example in Figure 22.1 has cell A1 as the input cell and the range A1:D4 for the output range.

If you do not know what cells you will use for your input or output, then stay with the default. You will be able to change the inputs, outputs, and cell locations later. After you have entered the required cell locations, click the **Finish** button. This inserts a blank Excel spreadsheet into your Mathcad worksheet. If you want to reduce the number of cells shown in the Excel component, double-click on the component and drag the side and bottom handles. If you only drag the region, it makes the region smaller, but does not change the number of cells shown.

The placeholder on the bottom is for the input variable that Mathcad provides Excel. The placeholder on the top left side is for the output variable name for the data that Excel provides Mathcad.

Figure 22.2 gives an example of an Excel spreadsheet with input from Mathcad and output from Excel. If you change variable Excel_Input1 from 10 to 20, then the data within Excel changes, and the output from Excel also changes.

In order to change the location of the input cells and output cells, right-click on the component and select **Properties**. This opens the component Properties dialog box. From this dialog box, you can change the number of input or output variables, and change the location of the input starting cell and the range of the output cells. See Figure 22.3.

Figure 22.4 warns of the problems that can occur if you change the output range after you have begun using the output data.

$\textbf{Excel\_Input}_1 := 20$

This input will be placed in cell A1 of the Excel spreadsheet. If you change this value, the Excel table changes and the output to Mathcad changes.

$\textbf{Excel\_Output}_1 :=$

20	x*2	x^2	x^3
20	40	400	8000
21	42	441	9261
22	44	484	10648

$\textbf{Excel\_Input}_1$

The formulas entered into this worksheet are listed at the top of the columns.
The input for this Excel component is the variable Excel_Input_1.
The output range A1:D4 is output to the variable Excel_Output_1 as a matrix.

$$\textbf{Excel\_Output}_1 = \begin{pmatrix} 20.00 & \text{"x*2"} & \text{"x^2"} & \text{"x^3"} \\ 20.00 & 40.00 & 400.00 & 8000.00 \\ 21.00 & 42.00 & 441.00 & 9261.00 \\ 22.00 & 44.00 & 484.00 & 10648.00 \end{pmatrix}$$

Output can be in matrix form or table form.

$\textbf{Excel\_Output}_1 =$

	1	2	3	4
1	20.00	"x*2"	"x^2"	"x^3"
2	20.00	40.00	400.00	8000.00
3	21.00	42.00	441.00	9261.00
4	22.00	44.00	484.00	10648.00

Individual elements of the output are accessible by using the subscript operator.

$\textbf{Excel\_Output}_{1_{4,4}} = 10648.00$

$\textbf{Excel\_Output}_{1_{2,3}} = 400.00$

You can assign variable names to specific output values.

$\textbf{Test} := \textbf{Excel\_Output}_{1_{3,3}}$ $\qquad \textbf{Test} = 441.00$

Figure 22.2 Using a blank Excel worksheet

Refer to Figure 22.2 for Excel worksheet.

In Figure 22.2, if you do not want to have the formulas be a part of your output data, then you can change the output range from A1:D4 to A2:D4.

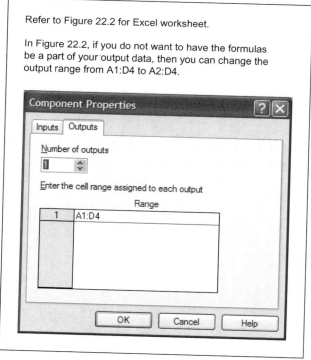

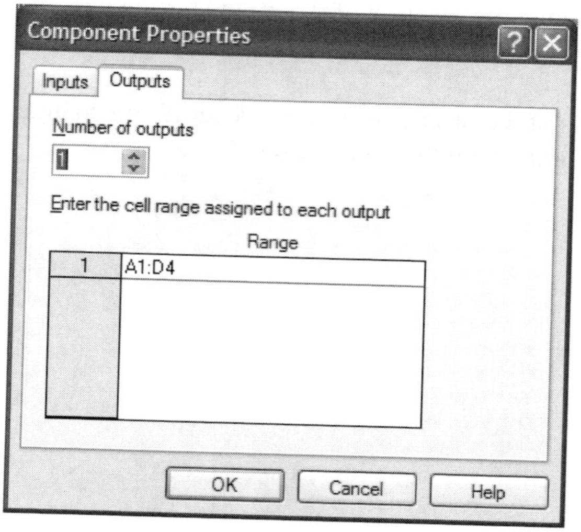

Figure 22.3 Modifying input and output properties

Be careful when changing the output range as this may have drastic impact on your worksheet.

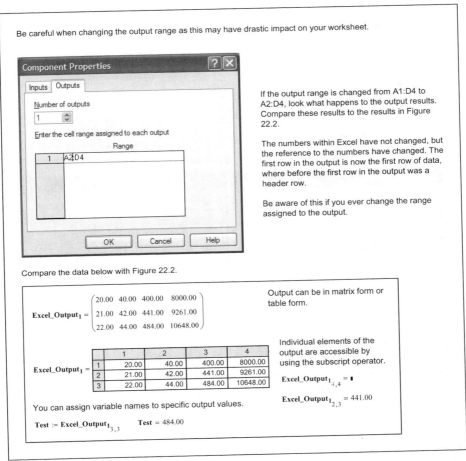

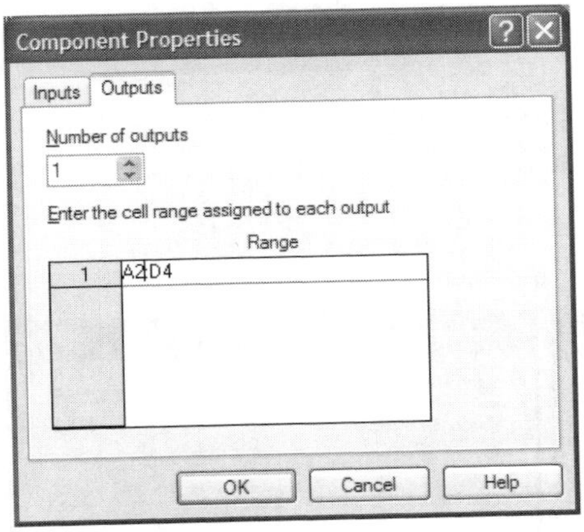

If the output range is changed from A1:D4 to A2:D4, look what happens to the output results. Compare these results to the results in Figure 22.2.

The numbers within Excel have not changed, but the reference to the numbers have changed. The first row in the output is now the first row of data, where before the first row in the output was a header row.

Be aware of this if you ever change the range assigned to the output.

Compare the data below with Figure 22.2.

$$\textbf{Excel\_Output}_1 = \begin{pmatrix} 20.00 & 40.00 & 400.00 & 8000.00 \\ 21.00 & 42.00 & 441.00 & 9261.00 \\ 22.00 & 44.00 & 484.00 & 10648.00 \end{pmatrix}$$

Output can be in matrix form or table form.

Excel_Output$_1$ =

	1	2	3	4
1	20.00	40.00	400.00	8000.00
2	21.00	42.00	441.00	9261.00
3	22.00	44.00	484.00	10648.00

Individual elements of the output are accessible by using the subscript operator.

$\textbf{Excel\_Output}_{1_{4,4}} =$ ▮

$\textbf{Excel\_Output}_{1_{2,3}} = 441.00$

You can assign variable names to specific output values.

$\textbf{Test} := \textbf{Excel\_Output}_{1_{3,3}}$ $\textbf{Test} = 484.00$

Figure 22.4 Beware of changing the output range

Multiple input and output

You can have an unlimited number of input and output variables in an Excel component. You are only limited by your system resources. You can add more input and output variables by right-clicking in the component and selecting **Properties**. Figure 22.5 gives an example of a component with four input variables and four output variables. Notice the Component Properties dialog boxes showing the cell assignments for the variables.

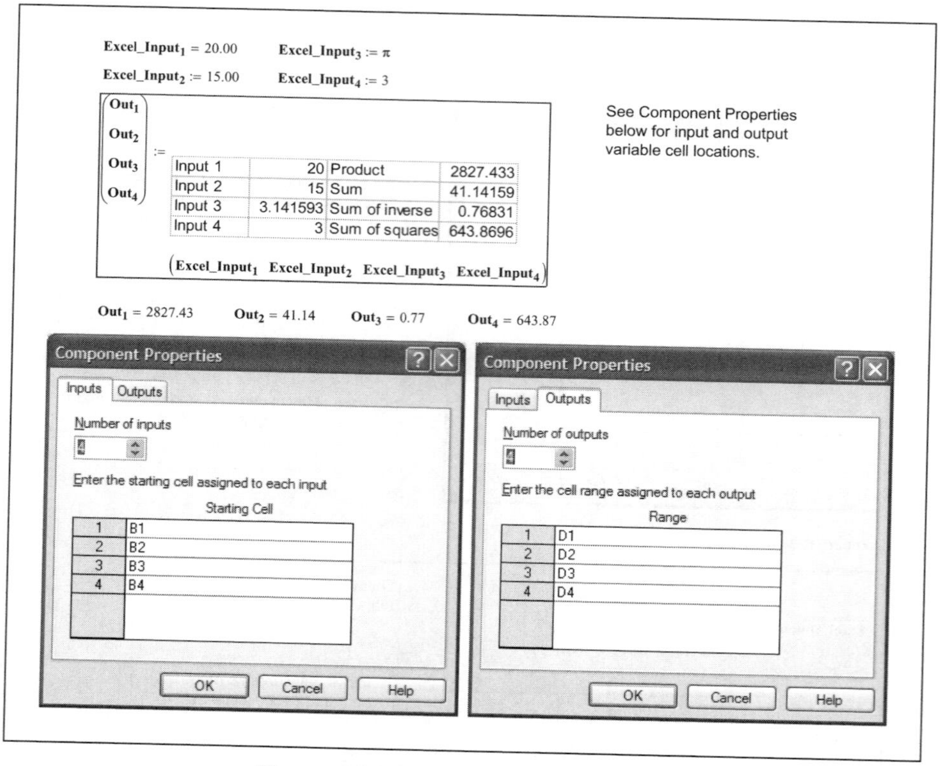

Figure 22.5 Multiple input and output

If you have multiple inputs, it may be easier to include all your input into a data table, a vector or a matrix. Figure 22.6 shows the same component, except with a single input variable and a single output variable. Compare the Component Properties dialog boxes with Figure 22.5.

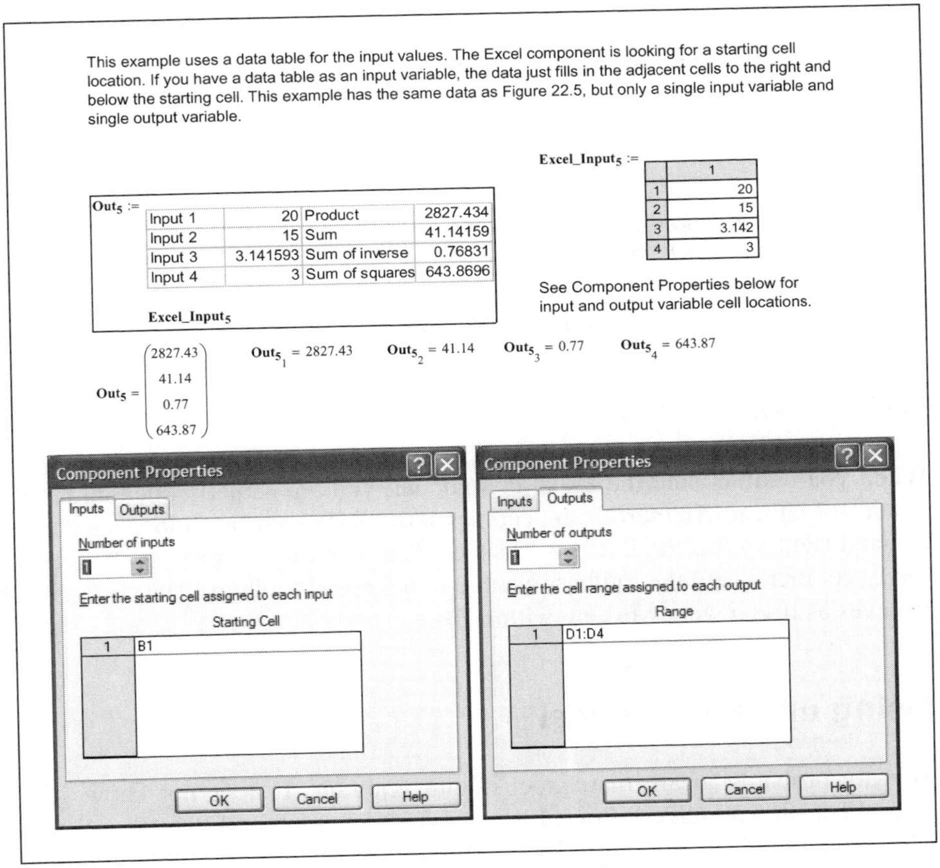

Figure 22.6 Multiple input and output

Hiding arguments

It is possible to hide the display of the input and output variable names so that only the Excel spreadsheet is visible. To do this, right-click in the component, and select **Hide Arguments**. See Figure 22.7.

Input 1	20	Product	2827.433
Input 2	15	Sum	41.14159
Input 3	3.141593	Sum of inverse	0.76831
Input 4	3	Sum of squares	643.8696

This is what the component from Figure 22.5 looks like with hidden arguments.

You can hide the display of the input and output variables. To do this right click in the component and select "Hide Arguments."

To show the input and output variables again, right click and select "Show Arguments."

Figure 22.7 Hiding arguments

Using Excel within Mathcad

When you double-click the Excel component, you are actually opening Excel inside of Mathcad. The scroll bar becomes active so that you can move up, down, left and right within the Excel worksheet. The menu bar at the top changes to the Excel menu, and the toolbars change to the Excel toolbars. The component behaves as if you were working within Excel.

Using units with Excel

You are probably aware that Excel is unit ignorant. It does not know 12 ft from 12 m. If you have followed the recommendations in this book, all your Mathcad worksheets use units. The use of units in Mathcad presents a problem when passing values to Excel. Figure 22.8 illustrates this problem. In the case of passing values to Excel, what you see is not what you get. Mathcad does not necessarily pass the displayed value to Excel. It passes the internally stored default value. This is illustrated in Figure 22.8.

Let's look at some simple examples of using units.

Example 1

$\mathbf{Unit\_In_1} := 5\mathbf{m}$

$\mathbf{Unit\_Out_1} :=$ [5] $\mathbf{Unit\_In_1}$

$\mathbf{Unit\_Out_1} = 5.00$

This example inputs into Excel, a variable (Unit_In$_1$) with units of length attached. The output uses the same cell as the input.

Microsoft Excel does not understand or recognize Mathcad units. The input into Excel is the number 5. The units are stripped off the number. The output is also unitless.

Example 2

$\mathbf{Unit\_In_2} := 5\mathbf{ft}$

$\mathbf{Unit\_Out_2} :=$ [1.524] $\mathbf{Unit\_In_2}$

$\mathbf{Unit\_Out_2} = 1.52$

Why does Excel not display the number 5?

$5\mathbf{ft} = 1.524\,\mathbf{m}$

Excel does not understand units. Mathcad does not give the displayed unit to Excel; Mathcad gives the default unit to Excel. This worksheet has SI unit preferences. Thus the unit of length defaults to meters (See example above). Mathcad gives this default value of length to Excel.

Example 3

$\mathbf{Unit\_In_3} := 5\mathbf{ft}$

$\mathbf{Unit\_Out_3} :=$ [5] $\mathbf{Unit\_In_3}$

$\mathbf{Unit\_Out_3} = 5.00$

$5\mathbf{ft} = 5.00\ \mathbf{ft}$

If the preferences of this sheet were set to US, then the default unit of length is feet and the number 5 would be passed to excel.

Example 4

$\mathbf{Unit\_In_4} := 5\mathbf{ft}$

$\mathbf{Unit\_Out_4} :=$ [60] $\mathbf{Unit\_In_4}$

$\mathbf{Unit\_Out_4} = 60.00$

$5\mathbf{ft} = 60.00\ \mathbf{in}$

If the unit preferences of this sheet were set to Custom based on US and the default unit of length was set to inches, then this would be the result.

Figure 22.8 Using units with the Excel component

It is important to understand what values you want to input into Excel, and it is equally important to understand how to get the proper values into Excel. The solution is similar to what you need to do for empirical equations and for plots. You divide the input variable by the units you want to be used in Excel. When the output comes back from Excel, you need to attach the proper units. See Figure 22.9.

Divide the input variable by the units expected by Excel. For example, if Excel is expecting kg, then divide the mass by kg. Multiply the output by the proper unit.

$\mathbf{Unit\_In_5} := 50\mathbf{kg}$

Example 5

$\mathbf{Unit\_Out_5} :=$ [50] $\dfrac{\mathbf{Unit\_In_5}}{\mathbf{kg}}$

$\mathbf{Unit\_Out_5} = 50.00$

$\mathbf{Out_{5a}} := \mathbf{Unit\_Out_5} \cdot \mathbf{kg}$ $\quad \mathbf{Out_{5a}} = 50.00\ \mathbf{kg}$

Even though this spreadsheet has SI unit default and mass defaults to kg, it is still a good idea to divide the input by kg. This way, if the worksheet unit defaults change, you still have the correct input into Excel.

Example 6

$\mathbf{Unit\_Out_6} :=$ [110.2311] $\dfrac{\mathbf{Unit\_In_5}}{\mathbf{lb}}$

$\mathbf{Unit\_Out_6} = 110.23$

$\mathbf{Out_6} := \mathbf{Unit\_Out_6} \cdot \mathbf{lb}$ $\quad \mathbf{Out_6} = 110.23\,\mathbf{lb}$ $\quad \mathbf{Out_6} = 50.00\ \mathbf{kg}$

To have the value be input as pounds mass, divide the input by lb.

$50\mathbf{kg} = 110.23\,\mathbf{lb}$

Example 7

$\mathbf{Unit\_In_5} = 50.00\ \mathbf{kg}$ $\quad \mathbf{Unit\_Out_7} :=$ [1763.698] $\dfrac{\mathbf{Unit\_In_5}}{\mathbf{oz}}$

$\mathbf{Unit\_Out_7} = 1763.70$

$\mathbf{Out_7} := \mathbf{Unit\_Out_7} \cdot \mathbf{oz}$ $\quad \mathbf{Out_7} = 1763.70\,\mathbf{oz}$ $\quad \mathbf{Out_7} = 50.00\ \mathbf{kg}$

To have the value be input as ounces, divide the input by oz.

$50\mathbf{kg} = 1763.70\,\mathbf{oz}$

Example 8

$\mathbf{Unit\_In_5} = 50.00\ \mathbf{kg}$ $\quad \mathbf{Unit\_Out_8} :=$ [50000] $\dfrac{\mathbf{Unit\_In_5}}{\mathbf{gm}}$

$\mathbf{Unit\_Out_8} = 50000.00$

$\mathbf{Out_8} := \mathbf{Unit\_Out_8} \cdot \mathbf{gm}$ $\quad \mathbf{Out_8} = 50000.00\,\mathbf{gm}$ $\quad \mathbf{Out_8} = 50.00\ \mathbf{kg}$

To have the value be input as grams, divide the input by gm.

$50\mathbf{kg} = 50000.00\,\mathbf{gm}$

Figure 22.9 Units continued

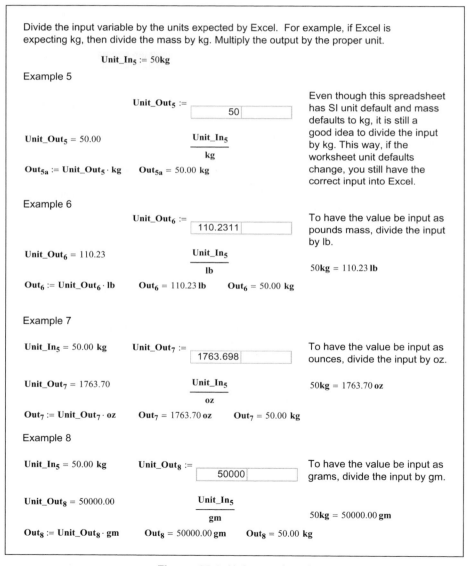

Figure 22.10 illustrates the need to understand what values to input into Excel. There are times when it does not matter what units are input as long as consistent units are used. There are other times when it does matter. The last two examples in Figure 22.10 illustrate this.

For this example, let's use Excel to calculate the velocity of an object given its acceleration (a), distance traveled (Dist) and initial velocity (V_0). (Yes, it is much easier to calculate this in Mathcad, but we are using this example to illustrate how important it is to understand units when using an existing Excel spreadsheet.)
Input values are in SI units. V_1 converts units to US units and Excel calculates the results in US units of ft/sec. V_2 uses SI units to calculate the velocity. Both results give the same final answer if the appropriate units are attached to the Excel output.

$a := g \quad a = 9.81\,\frac{m}{s^2} \qquad Dist := 100m \qquad V_0 := 50\frac{m}{s}$

$Velocity := \sqrt{V_0^{\,2} + 2 \cdot a \cdot Dist} \qquad Velocity = 66.79\,\frac{m}{s} \qquad Velocity = 219.14\frac{ft}{s} \qquad Velocity = 149.41\,mph$

$V_1 :=$

32.17405
328.084
164.042
219.1378

$V_1 = 219.14$

$\begin{pmatrix} \frac{a}{\frac{ft}{s^2}} & \frac{Dist}{ft} & \frac{V_0}{\frac{ft}{s}} \end{pmatrix}$

$V_2 :=$

9.80665
100
50
66.79319

The values V1 and V2 are different, but when the proper units are attached the end results are the same.

$V_2 = 66.79$

$\begin{pmatrix} \frac{a}{\frac{m}{s^2}} & \frac{Dist}{m} & \frac{V_0}{\frac{m}{s}} \end{pmatrix}$

$Velocity_1 := V_1 \cdot \frac{ft}{s} \qquad Velocity_1 = 219.14\frac{ft}{s} \qquad Velocity_1 = 66.79\,\frac{m}{s}$ Must attach ft/s units to V_1

$Velocity_2 := V_2 \cdot \frac{m}{s} \qquad Velocity_2 = 219.14\frac{ft}{s} \qquad Velocity_2 = 66.79\,\frac{m}{s}$ Must attach m/s units to V_2

These results work because the Excel formula is =SQRT(A3^2 +2*A1*A2) a formula that works for any consistent set of units.

If the Excel spreadsheet was written assuming units of feet and seconds in cell A3, and then converts the velocity to miles per hour (mph), then the results will be wrong if SI units are input into Excel.

$V_3 :=$

32.17405	
328.084	
164.042	
219.1378	ft/sec
149.4121	mph

$\begin{pmatrix} \frac{a}{\frac{ft}{s^2}} & \frac{Dist}{ft} & \frac{V_0}{\frac{ft}{s}} \end{pmatrix} \qquad V_3 = 149.41$

$V_4 :=$

9.80665	
100	
50	
66.79319	Excel expects this in ft/sec, but it is m/s
45.54081	mph is incorrect

$\begin{pmatrix} \frac{a}{\frac{m}{s^2}} & \frac{Dist}{m} & \frac{V_0}{\frac{m}{s}} \end{pmatrix} \qquad V_4 = 45.54$

$Velocity_3 := V_3 \cdot mph \qquad Velocity_3 = 149.41\,mph \qquad Velocity_3 = 66.79\,\frac{m}{s}$ Correct answer

$Velocity_4 := V_4 \cdot \frac{m}{s} \qquad Velocity_4 = 101.87\,mph \qquad Velocity_4 = 45.54\,\frac{m}{s}$ Incorrect answer, because the spreadsheet was not set-up for SI units.

Figure 22.10 Example of using units in Excel

Inserting existing Excel files

Mechanics

You have many existing Microsoft Excel spreadsheets. Let's learn how to insert these into Mathcad so that Mathcad can use the data. To insert an existing Excel spreadsheet into your worksheet, click **Components** from the **Insert** menu, and then select **Microsoft Excel**, and click the **Next** button. Select **Create from file**. Next, browse to the location of the file you want to insert, and click the **Next** button. This opens the Excel Setup Wizard dialog box. See Figure 22.1. When you insert an Excel component from a file, you may not know where to put the input values, and where to get the output values. If you do not know, then change both the input and output values to zero. You can modify the number and location of input and output values later using the Properties dialog box.

If you assign an input value to a specific cell, Mathcad will overwrite whatever was in that cell. If you accidentally assign the input value to a wrong cell, you could overwrite a very important cell in the spreadsheet. That is why it is a good idea to have no inputs until you know exactly where the input values need to go. Once you know where the input values need to go, open the Properties dialog box by right-clicking in the component and selecting **Properties**.

It is a good practice to turn on protection of your Excel worksheet prior to importing it into Mathcad. That way you will not accidentally overwrite critical cells. You can always turn off the protection after you bring the component into Mathcad.

Sizing the component

You can change how many cells are shown in the component by double-clicking on the component to open Excel, and then dragging the handles on the side and bottom to show fewer or more cells. If you click on the edge of the region, you can shrink or enlarge the size of the region by dragging the corner handles. This displays the same information, just smaller or larger.

Embedding versus linking

When you add an Excel component from an existing file, the original file becomes embedded in Mathcad. There is no more link to the original file. If you change the file later, the Mathcad component is not updated. If you change the Mathcad component, the file is not updated. It is not possible to link a file with the Excel component. If you want to keep a link to the existing file, then use one of the data components discussed in Chapter 21. However, this will not allow you to use the computing power of Excel. You can also add an OLE link, but this will not give Mathcad access to the data in the Excel spreadsheet.

It is possible to save the Excel component spreadsheet to an Excel file. To do this, single-click in the component, then right-click and select **Save As** from the pop-up menu. If you want to update the original file, then select the name of the file and choose to overwrite the file.

Printing the Excel component

Mathcad will only print what is visible within the component region. This has some drawbacks, especially if you want to print the entire spreadsheet. There is a way around this problem. Simply do a Save As to an Excel file (as explained above), then open the file in Excel and print it from Excel.

Getting Mathcad data from and into existing spreadsheets

The quickest way to use your existing Excel spreadsheets is to include it as a component, then double-click the component to open the component in Excel. This allows you to work in the spreadsheet just as you would in Excel. You can enter input and make changes just as you have always done. If you want to transfer data back to Excel, write down the cell addresses of the cells that you want to transfer back to Mathcad. Be sure to write down the units so that you can attach units to the output. Now you can use the Properties dialog box to add outputs and cell locations. After you close the Properties dialog box, you should have a column of placeholders on the left side of the component. This is where you add the Mathcad variable names for your output. Make sure that you have the variable names corresponding to the correct cell address. Unfortunately, you cannot attach units when you define the variable.

If you want Mathcad to provide the input information, then follow the same procedure as discussed above. Activate the spreadsheet; write down the input locations and units expected and use the Properties dialog box to add inputs and cell locations. This should add a row of placeholders below the component. You will now need to define the input variables above the Excel component (including units), and then include the variable names in the row of placeholders below the component. Be sure to divide each variable by the units expected by Excel. Also, be sure to put the variables in the same order that you used in the Properties dialog box.

Warning! Once you have used the Mathcad input feature to add data to Excel, do not go into the spreadsheet and manually change the input cells. As soon as you click outside the component, Mathcad will update the component with input values, and overwrite any changes you just made.

> Tip! I recommend protecting the Excel worksheet, except for the cells that require input. Do this prior to importing the spreadsheet into Mathcad. This way you do not accidentally overwrite a critical cell by providing a wrong cell address. You cannot protect the entire worksheet, because that would not allow Mathcad to change the input values.

Mathcad can only pass data to and retrieve data from the top Excel worksheet. If you have multiple sheets, make sure that the sheet you need to access is on top (on the far left side of the sheets listed at the bottom of your workbook). If you need to get data from another worksheet, then create a cell in the top worksheet that references the data on the other worksheet. If you need to input data into another worksheet, then have Mathcad place the data somewhere in the top worksheet. You will then need to activate the component, and go to the other worksheet, and add a reference to look on the top worksheet for input value. Make sure that you do not accidentally move the top sheet once you have set-up your Mathcad component.

Another way to input and output data to your Excel spreadsheet is to create a block of cells somewhere in your existing top worksheet. You can have an input block and an output block. Mathcad can then place the input values in consecutive cells. You can also have an output block, where you have referenced all the output values. If you use this method, you will need to rewrite your spreadsheet so that the input values are placed in the proper location in the spreadsheet.

Example

Figure 22.11 is an example of an existing Excel spreadsheet being brought in Mathcad. The spreadsheet calculates the uniform moment on a beam, and then calculates the bending stress in a solid beam. The first example returns incorrect results because the input values were not divided by the expected units. The second example returns the correct results. The input and output properties are shown in Figure 22.12.

DeadLoad := 300**plf**

LiveLoad := 500**plf**

BeamSpan := 15**ft**

BeamDepth := 1.5**ft**

BeamWidth := .5**ft**

The moment in the beam is calculated from the formula M=w*L^2/8. The beam stress is calculated from the formula Stress=M/S, where S= (1/6)*b*d^2.

The following component was input from the file Chapter 22/Excel Example - Beam Stress.xls. See Figure 22.12 for a list of the input and output properties.

$\mathbf{BeamStress_1} :=$

Calculate bending stress in a solid beam		
Dead Load lbf/ft	4378.171	lbf/ft
Live Load lbf/ft	7296.951	lbf/ft
Beam Span (ft)	4.572	ft
Beam depth (inches)	0.4572	inches
Beam width (inches)	0.1524	inches
DL Moment (W*L^2/8)	11439.71	ft*kip
LL Moment (W*L^2/8)	19066.19	ft*kip
Total Moment (DL+LL)	30505.9	ft*kip
Section Modulus b*d^2/6	0.005309	in^3
Bending Stress (M*12/6)	68947573	psi

(**DeadLoad** **LiveLoad** **BeamSpan** **BeamDepth** **BeamWidth**)

Notice how easy this is in Mathcad!

$$\mathbf{M} := \frac{(\mathbf{DeadLoad} + \mathbf{LiveLoad}) \cdot \mathbf{BeamSpan}^2}{8}$$

$$\mathbf{M} = 22500.00\,\mathbf{ft} \cdot \mathbf{lbf}$$

$$\mathbf{Section} := \frac{1}{6} \cdot \mathbf{BeamWidth} \cdot \mathbf{BeamDepth}^2$$

$$\mathbf{Section} = 324.00\,\mathbf{in}^3$$

$$\mathbf{F_b} := \frac{\mathbf{M}}{\mathbf{Section}} \qquad \mathbf{F_b} = 833.33\,\mathbf{psi}$$

$\mathbf{BeamStress_1} := \mathbf{BeamStress_1} \cdot \mathbf{psi}$ $\qquad \mathbf{BeamStress_1} = 6.89 \times 10^7\,\mathbf{psi}$

Results are incorrect because input values were not divided by the units that Excel was expecting.

$\mathbf{BeamStress_2} :=$

Calculate bending stress in a solid beam		
Dead Load lbf/ft	300	lbf/ft
Live Load lbf/ft	500	lbf/ft
Beam Span (ft)	15	ft
Beam depth (inches)	18	inches
Beam width (inches)	6	inches
DL Moment (W*L^2/8)	8437.5	ft*kip
LL Moment (W*L^2/8)	14062.5	ft*kip
Total Moment (DL+LL)	22500	ft*kip
Section Modulus b*d^2/6	324	in^3
Bending Stress (M*12/6)	833.3333	psi

Remember: If you change any of the input cells within Excel, Mathcad will overwrite them once you click outside of the Excel component.

$$\left(\frac{\mathbf{DeadLoad}}{\mathbf{plf}} \quad \frac{\mathbf{LiveLoad}}{\mathbf{plf}} \quad \frac{\mathbf{BeamSpan}}{\mathbf{ft}} \quad \frac{\mathbf{BeamDepth}}{\mathbf{in}} \quad \frac{\mathbf{BeamWidth}}{\mathbf{in}}\right)$$

$\mathbf{BeamStress_2} := \mathbf{BeamStress_2} \cdot \mathbf{psi}$ $\qquad \mathbf{BeamStress_2} = 833.33\,\mathbf{psi}$

Note that this result is correct even though beam depth and width were input in feet.

Figure 22.11 Inserting an Excel component from a file

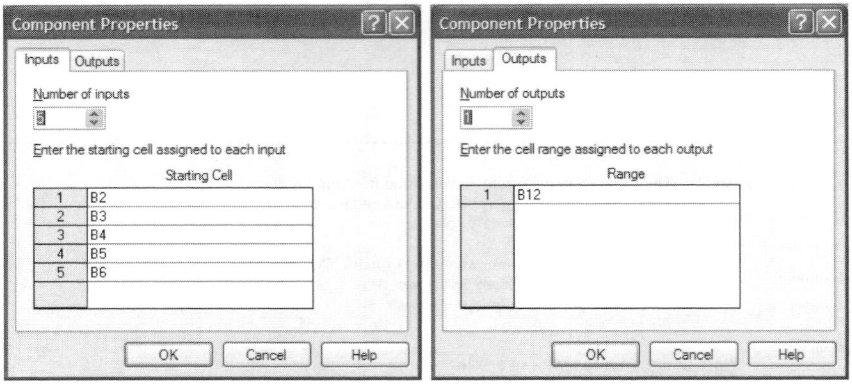

Figure 22.12 Component Properties from Figure 22.11

Summary

The Excel component makes the transition to Mathcad much easier because you do not have to start over. You can easily bring your existing Excel spreadsheets into your project calculations. You can also create new Excel spreadsheets in your project calculations to take advantage of the Excels strengths. The two programs can work together by exchanging data back and forth.

In Chapter 22 we:

- Learned how to insert a blank Excel worksheet into Mathcad.
- Learned how to reference the Excel cells to transfer information into Excel and extract information from Excel.
- Showed how to get the proper values into Excel when using Mathcad units.
- Discussed ways of attaching units to Excel output.
- Explained how to bring in your existing Excel files into Mathcad including:
    - How to size the region and worksheet
    - How to reference the correct cell addresses
    - How to save and print the component

- Encouraged you to protect your spreadsheet prior to inserting it into Mathcad.
- Warned about manually changing variables once you have added Mathcad inputs.
- Warned about mistakes that can occur if you do not divide your input by the proper units.
- Provided an example to illustrate the concepts.

Practice

1. From your field of study, write 5 Excel components. Add input from Mathcad, and provide output to Mathcad. Use units.
2. Bring 2 existing Excel worksheets into Mathcad. Be sure that Mathcad inserts information in the correct cells. List items that need to be considered when using units.

23
Inputs and outputs

It is important to have a simple way of organizing and inputting design criteria and assumptions into Mathcad. Calculations for engineering projects can become rather large. Having a clear way to identify input and output information is very useful. This chapter will explore different ways to include input information in Mathcad worksheets. It will also give suggestions on ways of making results standout.

Chapter 23 will:

- Discuss the need to highlight the calculation input variables to make them standout from other variables.
- Discuss various ways of inputting information.
- Discuss the various ways of making output information standout.
- Introduce the concept of Mathcad controls.
- Explain how to use Web Controls for calculation input.

Emphasizing input and output values

Input

In project calculations, some regions are more important than other regions. Regions that require user input can often get lost amid the surrounding regions. It is very helpful to have a consistent means of identifying the variable definitions that need to be input or changed by the user of the worksheet. Some variables are input at the beginning of a project and should not be changed. You should have a means of protecting these variables.

There are various means available to make variables standout. The easiest method is to highlight the region. This places colored shading inside of the region. The default color is yellow, but you can change the color. If each region that requires user input is shaded, then it makes it very easy to quickly scan a worksheet and review the input values. To highlight a region, right-click on the region, and select **Properties**. This opens the Properties dialog box. See Figure 23.1. Place a check mark in the "Highlight Region" check box. If you click the **OK** button, the default yellow color is used. If you click the **Choose Color** button, then the Color dialog box opens. You can now select one of the basic colors, or you can define your own custom color. You can change the default color by selecting **color>highlight** from the **Format** menu. See Figure 23.1.

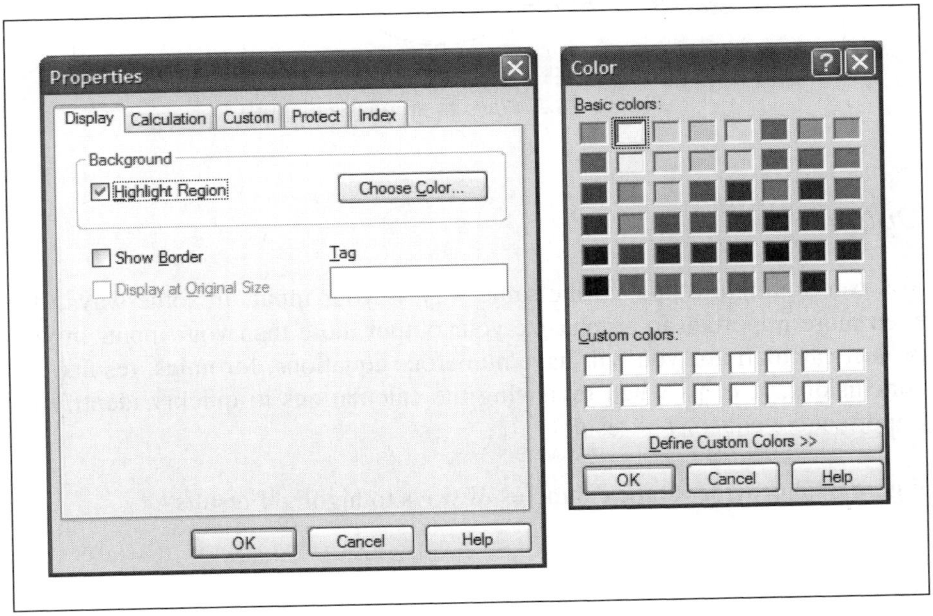

Figure 23.1 Highlight region

Another method of making an input region standout is to put a box around the region. To do this, right-click in the region, open the Properties dialog box and click the "Show Border box." This places a black border around the region.

You can also place a graphics symbol adjacent to regions requiring user input.

Figure 23.2 provides some examples of ways to highlight input.

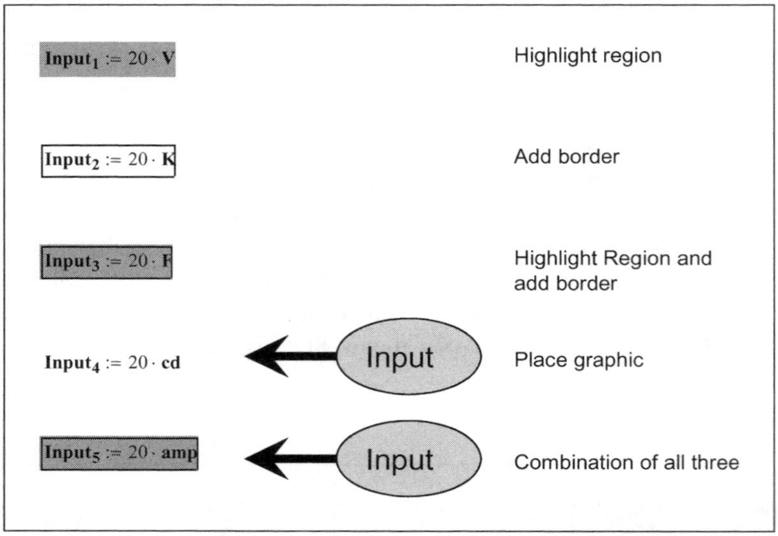

Figure 23.2 Ways to emphasize input

Output

You can highlight output in the same way as your input. In some ways, it is even more important to emphasize your output more than your input. In your project calculations, you will have numerous equations, formulas, results, and conclusions. It helps when reviewing the calculations to quickly identify the important conclusions or results.

Figure 23.3 provides some examples of ways to highlight results.

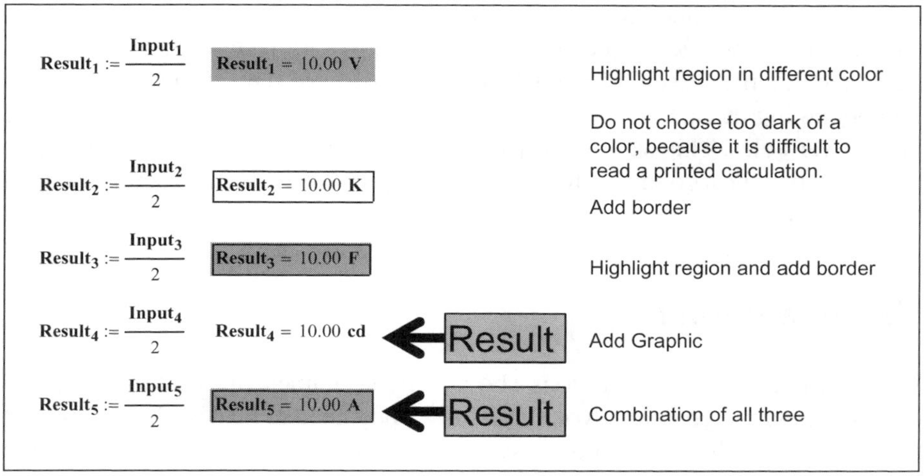

Figure 23.3 Ways to emphasize results

Project calculation input

There are many different ways to organize the input variables needed for your project calculations. For our discussion, let's establish two different types of variables: Project Variables and Specific Variables. Project Variables will be variables that will be used repeatedly throughout the project calculations. These variables include things such as: design criteria, material strengths, loading criteria, design parameters, etc. Specific Variables will be variables necessary to calculate a specific element in the calculations. These variables may be used in several calculations, but they are not applicable to the entire project.

The easiest method of inputting variables is to create a variable whenever and wherever it is needed in the project calculations. This does not require much planning or forethought. Just start working and whenever you need a variable input, just create it. This method works, but if your project calculations are very long, it makes it difficult to find the location where the variable was defined. It also makes it more likely that a variable will be redefined or defined again with a slightly different name.

If all of your Project Variables are placed at the top of your worksheet, then these will be easy to find and will be prominently displayed. They will not become hidden in your calculations. It also causes you to think about the organization of your project calculations.

If your project calculations will include several different Mathcad files, then create a specific project input file. All other project worksheets should reference this file. (See Chapter 19 for a discussion of the ***reference*** function.) The beginning of each of these worksheets should display, but not redefine the Project Variables. This way, all project worksheets will be using the same project criteria. It may be a good idea to protect the project input file from being changed accidentally.

Variable names

We have had several discussions about variable names. It is important to be consistent in how you name your variables. They need to be descriptive, but not too long. They also need to be easy to remember. The names need to be able to distinguish between very similar values. For example, if you have different strengths of steel then you will need to create a different variable name for each different strength of steel.

Literal subscripts are very useful to distinguish between closely related input variables. There are some cases where array subscripts will be even more useful. (Remember that array subscripts will create an array and are defined using the [key.) Figure 23.4 gives an example where array subscripts will be more useful than literal subscripts. By using array subscripts in this example, it is possible to sum all the variables and get a total.

Let's suppose you are inputting and calculating the loads on the roof of a structure.

Using Literal Subscripts

Roofing material	$\mathbf{DeadLoad_1 := 10psf}$	$\mathbf{DL_1 := 10psf}$
Ceiling weight	$\mathbf{DeadLoad_2 := 5psf}$	$\mathbf{DL_2 := 5psf}$
Piping	$\mathbf{DeadLoad_3 := 3psf}$	$\mathbf{DL_3 := 3psf}$
Roof equipment	$\mathbf{DeadLoad_4 := 7psf}$	$\mathbf{DL_4 := 7psf}$

Using Array Subscripts

$\mathbf{TotalWeight := DeadLoad_1 + DeadLoad_2 + DeadLoad_3 + DeadLoad_4}$

$\mathbf{TotalDL := \sum DL}$

The summation operator is on the Matrix menu.

$\mathbf{TotalWeight} = 25.00\,\mathbf{psf}$

$\mathbf{TotalDL} = 25.00\,\mathbf{psf}$

When using array subscripts, you are actually making the array variable DL larger each time you add a new variable. Let's illustrate using a new variable name.:

$\mathbf{Dead_1 := 10psf}$ $\qquad \mathbf{Dead} = (10.00)\ \mathbf{psf}$

$\mathbf{Dead_2 := 5psf}$ $\qquad \mathbf{Dead} = \begin{pmatrix} 10.00 \\ 5.00 \end{pmatrix} \mathbf{psf}$

$\mathbf{Dead_3 := 3psf}$ $\qquad \mathbf{Dead} = \begin{pmatrix} 10.00 \\ 5.00 \\ 3.00 \end{pmatrix} \mathbf{psf}$

$\mathbf{Dead_4 := 7psf}$ $\qquad \mathbf{Dead} = \begin{pmatrix} 10.00 \\ 5.00 \\ 3.00 \\ 7.00 \end{pmatrix} \mathbf{psf}$

Figure 23.4 Using array subscripts

Creating input for standard calculation worksheets

A standard calculation will be used repeatedly, so it is important to consider which variables will be changed when the worksheet is reused. This section discusses the different methods available for inputting information into your standard calculation worksheets.

We have discussed locating these types of variables at the top of the worksheet, so that they are all accessible in one place. This way, you do not need to scan the worksheet looking for variables to change. It also allows you to protect or lock the remainder of your worksheet to prevent unwanted changes.

Let's look at different methods you can use to create input for your standard calculation worksheets.

Inputting information from Mathcad variables

The most common way of inputting information into your Mathcad worksheet is by creating *variable definitions*. You can then click on the variable definition value and change the value. This has a big advantage over other methods, because you can input your units at the same time as you input your data. If your input has the units of length, you can attach units of meters, mm, feet, or inches. Mathcad does not care what unit of length you choose. If you use another method of input, as will be shown in the next section, it is critical to input the data with specific units of lengths. For example, if your data is input in a data table or in Microsoft Excel, then it is critical to know if you need to input meters, mm, feet, inches, or some other length unit. You will need to do the conversion before inputting the value.

Another advantage of inputting information from Mathcad is that it usually takes less space. Your definitions are already made. When you use a data table or Microsoft Excel, you may need to reassign the variable or redefine the variable in order to attach units to the variable. This adds more variable definitions to your worksheet.

A disadvantage of inputting information by individual variable definition is that it is sometimes much more time consuming. You will need to click every definition and change the value. If you use a data table or Microsoft Excel, you can quickly input large amounts of data without using your mouse to click on each input variable. If you have large amounts of data to input or change, you may want to consider using a data table or Excel.

Data tables

If you have considerable input that does not have units attached, or that has consistent units, a data table is a useful way to input and change many variables very easily. A data table creates an array, thus, every cell in the table has a unique variable name using array subscripts.

The data table allows for easy input because you can use the Enter key to move to the next input cell. You can also use the up and down arrows to scroll through the input. You do not need to click on every variable and change it, as you would have to do in Mathcad. You can also include a column of text to describe what values to input.

The biggest disadvantage to using a data table is that you cannot attach units to the input. All units are attached after the data is exported to Mathcad. This can lead to errors if the proper values are not entered. Another disadvantage to using a data table is that all the output is contained in a single variable matrix. There are ways of taking values from this matrix and assigning them to specific variables. Figure 23.5 gives two examples of how you can take the input data from the data table and assign it to variable names and attach units to the input.

$\mathbf{Table_1} :=$

	1	2
1	20	30
2	10	40

$$\mathbf{Table_1} = \begin{pmatrix} 20.00 & 30.00 \\ 10.00 & 40.00 \end{pmatrix}$$

$\mathbf{Resistance_1} := \mathbf{Table1}_{1,2} \cdot \Omega$ $\quad \mathbf{Resistance_1} = 30.00\ \Omega$

$\mathbf{Voltage_1} := \mathbf{Table1}_{2,2} \cdot \mathbf{V}$ $\quad \mathbf{Voltage_1} = 40.00\ \mathbf{V}$

After data is input, you can use Mathcad to reassign names and to add units.

Note: The data table below has font and number format changes made to the default data table.

$\mathbf{Table_2} :=$

	1	2
1	"Roofing Material"	30.0000
2	"Ceiling Weight"	5.0000
3	"Piping"	5.0000
4	"Roof Equipment"	10.0000
5		

$$\mathbf{Table_2} = \begin{pmatrix} \text{"Roofing Material"} & 30.00 \\ \text{"Ceiling Weight"} & 5.00 \\ \text{"Piping"} & 5.00 \\ \text{"Roof Equipment"} & 10.00 \end{pmatrix}$$

$\mathbf{DeadLoad} := \mathbf{Table_2}^{\langle 2 \rangle} \cdot \mathbf{psf}$

Use the matrix column operator [CTRL+6] to extract the column data from the matrix "Table_2." Also attach units to the data.

$$\mathbf{DeadLoad} = \begin{pmatrix} 30.00 \\ 5.00 \\ 5.00 \\ 10.00 \end{pmatrix} \mathbf{psf}$$

$\mathbf{TotalDeadLoad} := \sum \mathbf{DeadLoad}$

$\mathbf{TotalDeadLoad} = 50.00\ \mathbf{psf}$

Figure 23.5 Using data tables for input

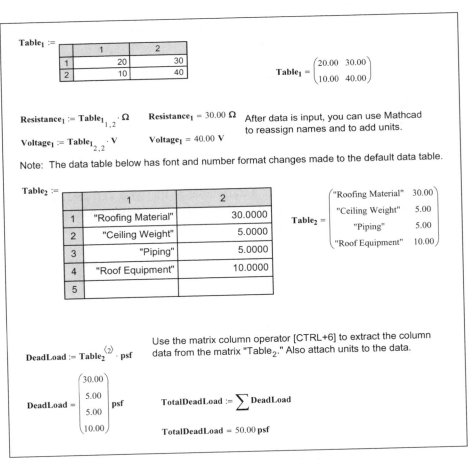

You can format the font, number format, trailing zeros, and tolerances of the values in a data table. To do this, single click on the table, then right-click and select **Properties**.

Another thing to be aware of with data tables is that if you enter a value in a cell, that cell will always retain a value. You can change the value in the cell, but you cannot delete the cell. Mathcad reads all cells with values (including zero) in a data table. Thus, if you add unnecessary cells in your data table, your input array will also contain unnecessary values. For this reason, it is usually best to use a Microsoft Excel component (if you have Excel loaded on your computer) rather than using a data table.

Microsoft Excel component

A Microsoft Excel component can also be used to input values into Mathcad. This is very similar to a data table, except that you now have the full functionality of Microsoft Excel. You can:

- Format and highlight individual cells.
- Protect specific cells.
- Use lookup tables.
- Use controls such as drop-down boxes.
- Adjust the width of each column.
- Place borders around specific cells.

Another advantage of Excel over a data table is that you can assign variable names to each cell in the Excel component rather than having a single variable name — as required for a data table.

Figure 23.6 gives an example of using Excel to input values. Notice the following in this example:

- Some cells are shaded.
- The input cells are highlighted.
- Different fonts are used.
- Bold is used in some columns.
- The column widths are different.
- A border is used.
- There are four outputs associated with the four input values.
- Units had to be attached outside of the component definition.

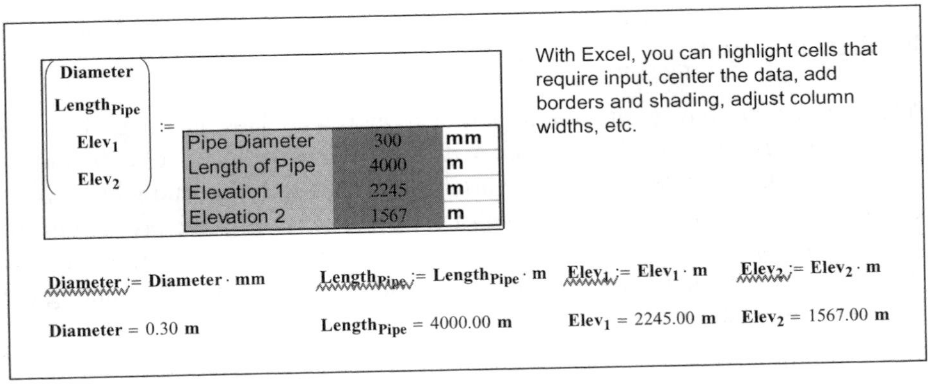

Figure 23.6 Using Microsoft Excel for input

In this example, the spreadsheet is also protected. You can only change the input values. The other cells are locked. It is also critical to input the proper units. For example, the pipe diameter must be input as mm, not meters.

Which method is best to use?

Which input method is the best method to use for standard calculation worksheets? It depends. Data tables are not a good option if you have Microsoft Excel available. There are advantages to each method. The answer to the question depends on how much input is necessary. Does the extra work of creating an Excel input component outweigh the extra time of clicking and changing individual variable definitions. The answer also depend on whether you want the flexibility of inputting data using different units, rather than being forced to use a specific unit as you would be required to do in Excel.

The examples used in Figures 23.5 and 23.6 are very short. Your input would need to be much more extensive before an Excel input component would be worthwhile. The examples are only for demonstration.

Summarizing output

If you have a long calculation worksheet, you may want to summarize the results at the end of the worksheet.

You can display each result again by typing the variable name followed by the equal sign. By listing all the variables, you will be able to see a summary of the results. You may also want to use one of the method discussed earlier to highlight the results.

You can also put the results into an Excel spreadsheet. This, however, is much more work. In order to do this, you will need to create an Excel component, list each result variable as an input variable and then tell Mathcad which Excel cell to put it into. You will also have to divide the result by the proper unit so that Excel displays the proper value. This is quite a bit of work, but there may be times when having the results in Excel outweighs the effort of creating the output component.

Controls

Controls are programmed boxes and buttons that you can add to your worksheets to help automate or limit input. They are best used in worksheets that will be developed by one person and used by another, or in worksheets that will be used repeatedly. Controls are useful in some standard calculation worksheets because you can limit input to just a few selected items. You can also use controls to have the user select if a specific condition is met.

The standard Mathcad controls use a scripted language. If you are familiar with VBScript or Jscript, then these controls will be easy for you to learn and to use. They are very powerful and useful. If you are not familiar with scripting, then the Mathcad Developer's Reference provides a good description of how to use the standard controls.

For those who do not want to use or learn scripting, Version 12 of Mathcad added what are called Web Controls. The Web Controls have been pre-scripted, and are easy to use. They were developed for web-based document servers such as the Mathsoft Application Server, but they work fine in Mathcad. They are limited in their application, but can still be useful to those who do not want to use scripted language.

Web Controls are added by selecting **Controls>Web Controls** from the **Insert** Menu. This opens the Web Control Setup Wizard. See Figure 23.7.

Web Controls do not require input values, and each control provides one output value that is assigned to a Mathcad variable. Thus, Web Controls give a means of providing input to your calculations. Let's look at each of the different Web Controls provided in Mathcad.

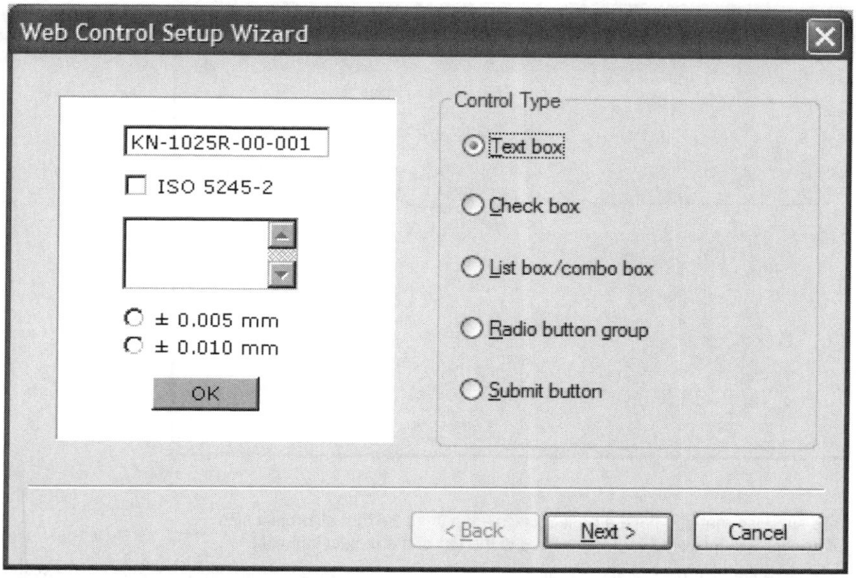

Figure 23.7 Web Controls

Text box

The Text box control provides a box where the user can input text. You can set the limit on how many characters are visible, and you can also limit how many characters can be entered. The Mathcad limit is 255 characters. The output from a text box is a text string. This text string is assigned to a variable.

When would a Text box be used?

- You could have the user input his/her name. This name could then be displayed as a text string anywhere in the calculations.
- You could have the user input today's date. This date could then be added to the calculations so that you know when the calculations were modified.
- If a teacher were creating a test, the text string could be used to ask the student to respond to a question.

Figure 23.8 gives an example of how to create a text string and how to apply it.

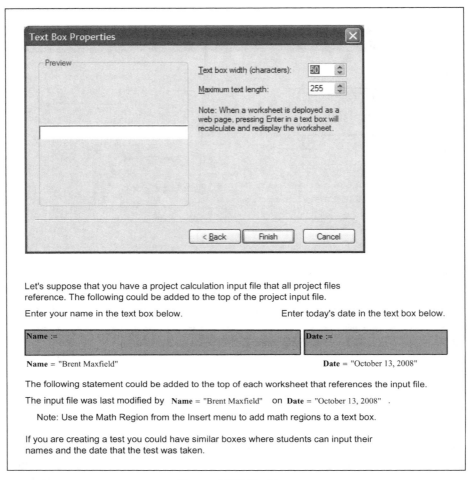

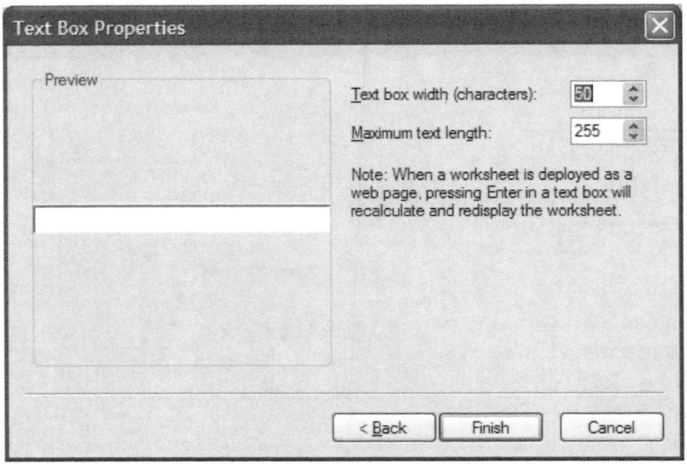

Let's suppose that you have a project calculation input file that all project files reference. The following could be added to the top of the project input file.

Enter your name in the text box below. Enter today's date in the text box below.

Name := **Date** :=

Name = "Brent Maxfield" **Date** = "October 13, 2008"

The following statement could be added to the top of each worksheet that references the input file.

The input file was last modified by **Name** = "Brent Maxfield" on **Date** = "October 13, 2008" .

Note: Use the Math Region from the Insert menu to add math regions to a text box.

If you are creating a test you could have similar boxes where students can input their names and the date that the test was taken.

Figure 23.8 Text box

Check box

The Web Controls check box only allows for a single check box in the control. The output is a 1 if the box is checked, and a 0 if the box is unchecked. This makes the box useful to use to ask if a specific condition is met. You can then use a logical program to determine the result.

See Figure 23.9 for an example.

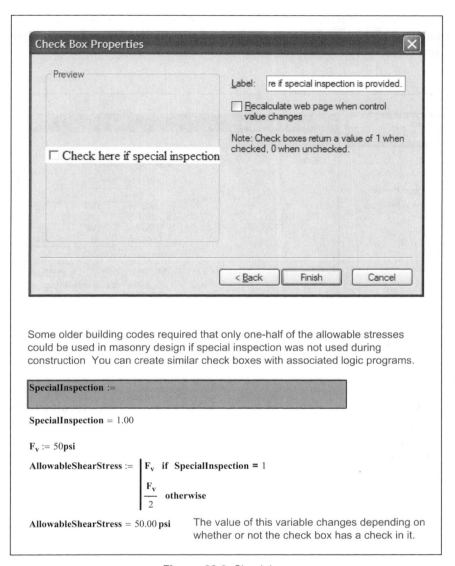

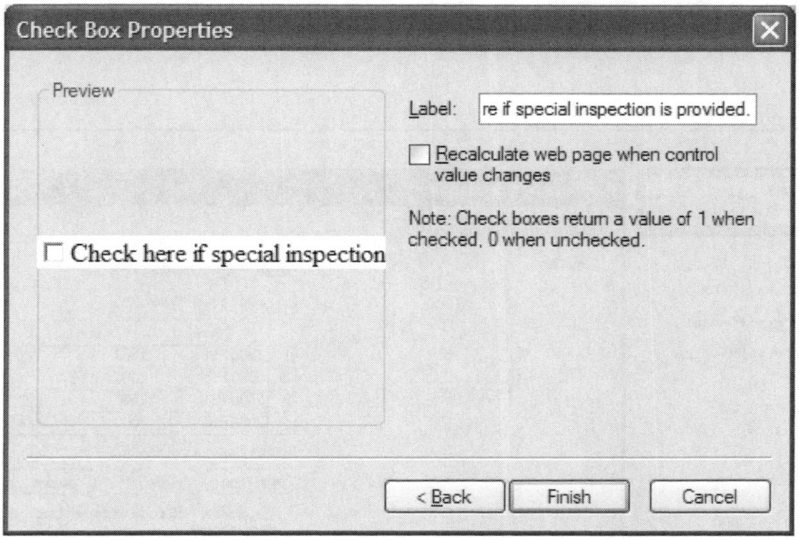

Some older building codes required that only one-half of the allowable stresses could be used in masonry design if special inspection was not used during construction You can create similar check boxes with associated logic programs.

SpecialInspection :=

SpecialInspection = 1.00

$\mathbf{F_v} := 50\mathbf{psi}$

$$\mathbf{AllowableShearStress} := \left| \begin{array}{ll} \mathbf{F_v} & \textbf{if SpecialInspection} = 1 \\ \dfrac{\mathbf{F_v}}{2} & \textbf{otherwise} \end{array} \right.$$

AllowableShearStress = 50.00 **psi** The value of this variable changes depending on whether or not the check box has a check in it.

Figure 23.9 Check box

List box

A List box allows the user to select from a list of items. The list can include up to 256 items. The selected item is returned to Mathcad as a variable.

The way a list box appears in the worksheet depends on how many rows are displayed. You can tell Mathcad how many rows to display. If the display is set to show one row, then the box displays as a standard combo box where all the items appear when you select the down-arrow. This condition is illustrated in Figure 23.10.

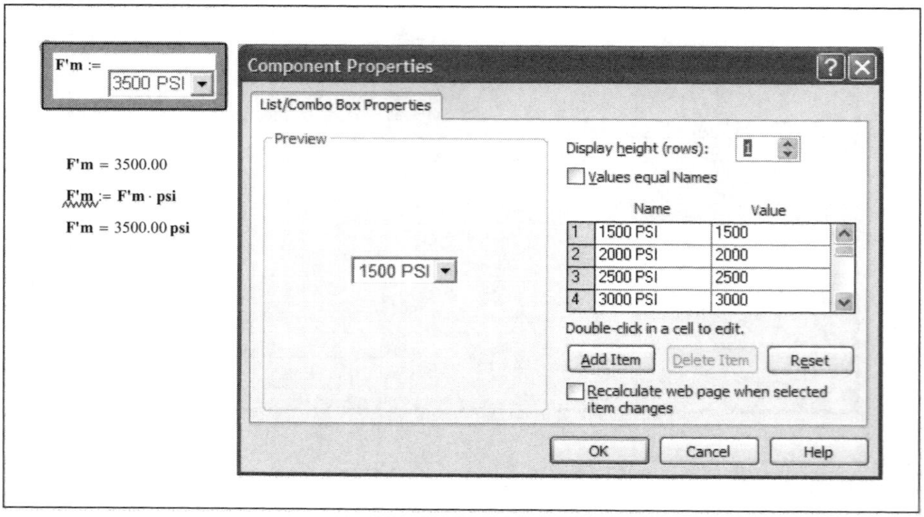

Figure 23.10 List box

If the display is set to 2 rows or higher, then the list appears as a scrolling list box if there are more items than can be displayed. If the display is set to show more rows than are in the list, then the box shows empty unselectable lines below the list. The scrolling list box is illustrated in Figure 23.11.

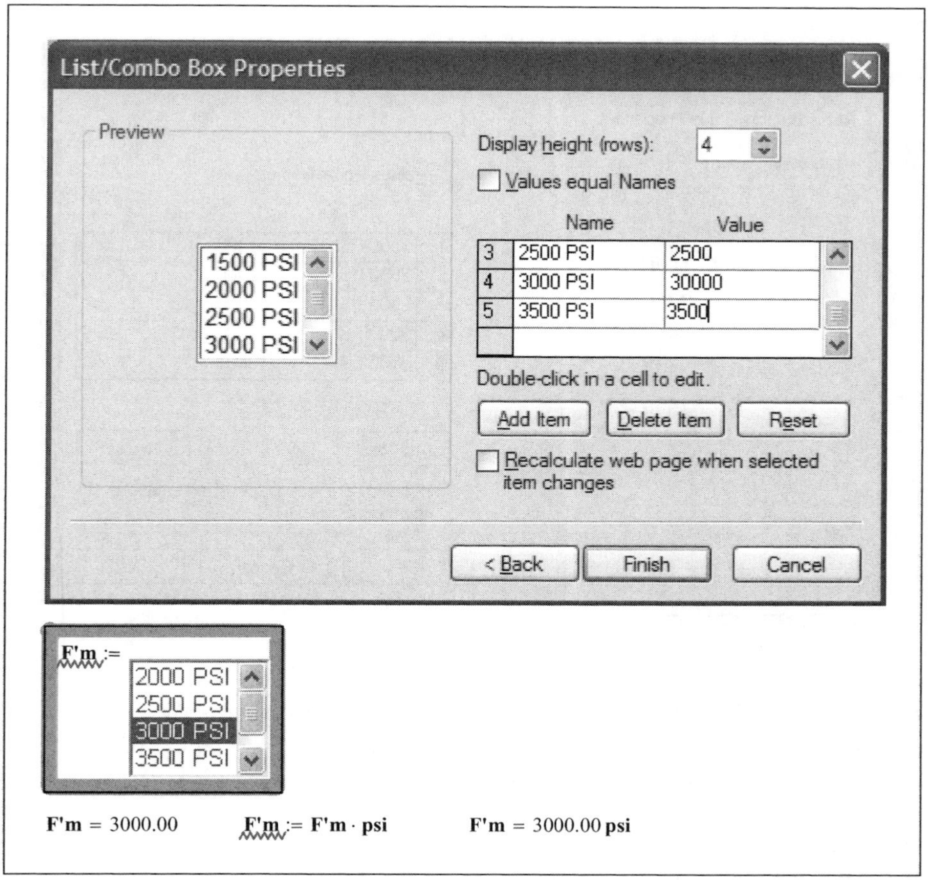

Figure 23.11 List/combo box

Radio box

A radio box is similar to a list/combo box. Both types of boxes select an input from a list. The list/combo box item is selected by selecting the item from a list. A radio box has small circles to the left of the listed items. You select an item by clicking in the circle adjacent to the item.

Figure 23.12 gives an example of a list box.

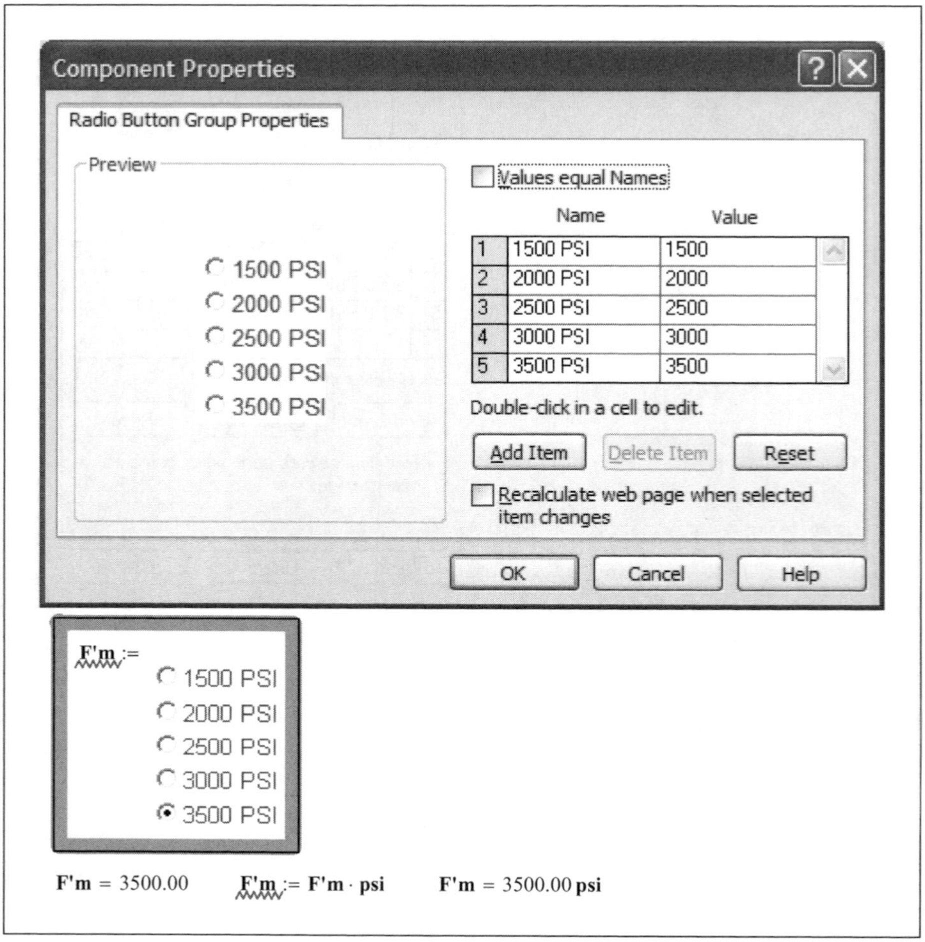

Figure 23.12 Radio button

Submit box

The submit button is used only when a worksheet is published on the web. It does not have any affect for normal worksheets. See Figure 23.13.

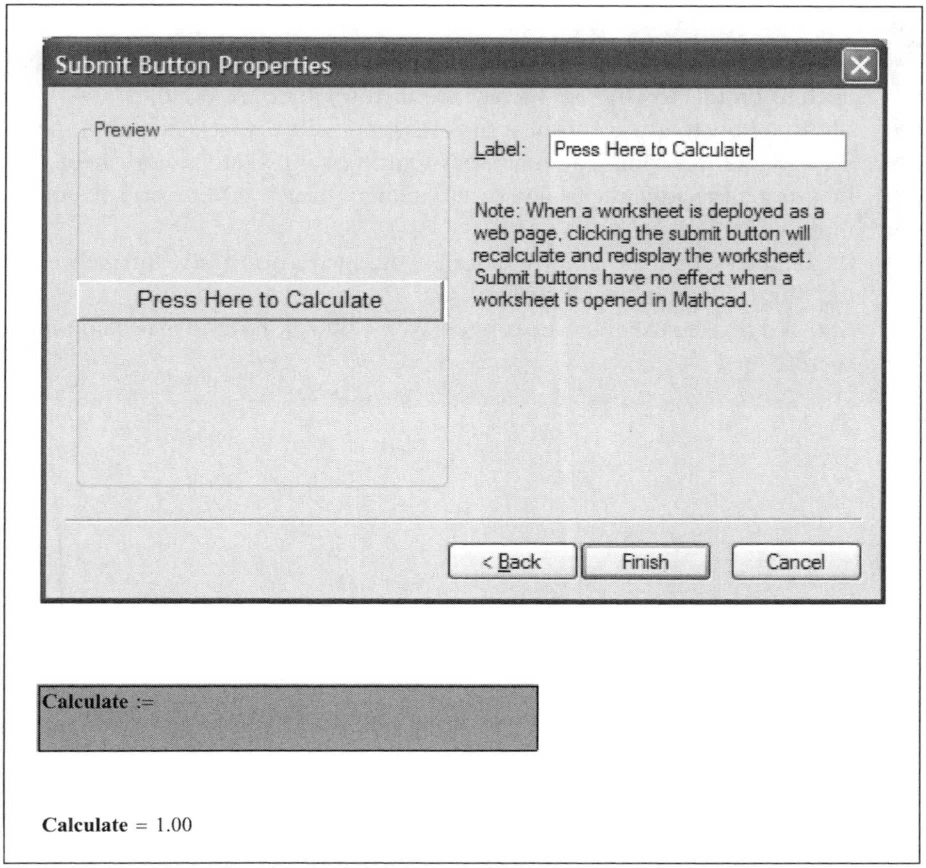

Figure 23.13 Submit button

Summary

This chapter discussed the various way of getting information into your Mathcad worksheets and how to make your critical values standout from the rest of your worksheet.

In Chapter 23 we:

- Showed how to highlight input variables so that they will be easy to see.
- Gave other ways to identify input variables, such as using borders or placing a graphic adjacent to the variables.
- Suggested using different means to highlight results so that the critical results are clearly identified.

- Discussed having a project calculation input file, and emphasized the need to protect the file, and reference it from other project files.
- Showed how to use array subscripts for input values that will be summed.
- Explained the various methods for inputting values into a worksheet.
- Discussed the advantages and disadvantages of using Microsoft Excel to input variables.
- Introduced Mathcad Controls and encouraged you to research more in the Mathcad Developer's Reference.
- Showed how to use the pre-scripted Web Control in standard calculation worksheets.

24 Hyperlinks and Table of Contents

A hyperlink is a special region (text or graphic) that when double-clicked, opens the document and/or location referenced by the hyperlink. You can create hyperlinks that will take you to regions within the current worksheet, regions within other worksheets, a Web page on the internet or even a file created by another program.

By using hyperlinks, you can refer the user of your worksheet to reference material, other worksheets or locations within your current worksheet. By using hyperlinks, you are able to create a Table of Contents and link all files in your project calculations. This will be illustrated shortly.

Chapter 24 will:

- Define hyperlinks.
- Show how hyperlinks are created.
- Discuss the use of hyperlinks in text regions and non-text regions.
- Encourage the use of hyperlinks only in text regions.
- Explain the use of "relative path."
- Tell how to create a Table of Contents.
- Explain Author's Resources.
- Introduce how to create a calculation E-book.

Hyperlinks

A hyperlink is created from selected text in a text box or from a graphic image. You can select the entire content of a text box or only a single character. Once you have selected the text or graphic, which you want to become your hyperlink, then select **Hyperlink** from the **Insert** menu. You can also right-click and select **Hyperlink** from the pop-up menu. This opens the Insert Hyperlink dialog box. In Figure 24.1, the link is to the Mathcad Web site. Be sure to include the http:// before the www. You can add a text message that will appear at the bottom of your screen in the status line when your mouse is placed over the hypertext. When you click **OK** to close the box, the selected text becomes bold and underlined. When you place your cursor over the hyperlink text, the cursor changes to a hand symbol. You need to double-click to activate the hyperlink. If you used a graphic as your hyperlink, there is no change in the appearance of the graphic, but if you place your cursor over the graphic, the cursor changes to a hand icon.

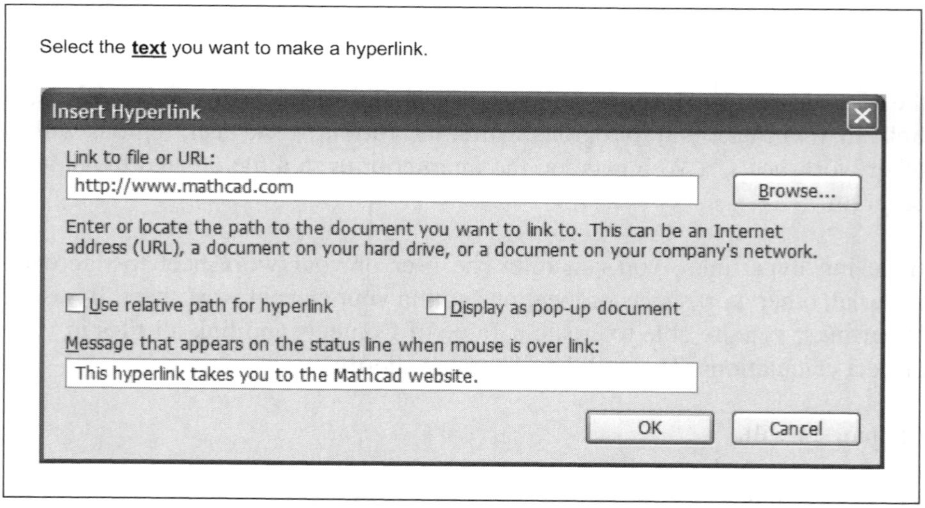

Figure 24.1 Inserting a hyperlink

Tip! You may want to highlight your graphic regions with a color, so that you can easily identify the graphics with hyperlinks attached.

The hyperlink can be to any location where you can navigate on your computer. If you link to a file that has a software program assigned to the file type, then the file will open in the associated program. If you link to a Mathcad worksheet, the worksheet will open and your cursor will be placed at the top of the worksheet.

Linking to regions in your current worksheet

You can also link to a region in your current worksheet or to a region in another worksheet. In order to link to a specific region, you need to assign a tag to the region to which you want to link. This is similar to attaching a bookmark in a word processing program. To do this, right-click on the region and select **Properties**. On the Display tab, add a name or phrase in the "Tag" box. A hyperlink will refer to this tag name. The tag name cannot have a period in it such as "Voltage.1."

> **Tip!** There is no way to browse to a region, so you must remember the exact name of a tag. For this reason, keep your tag names simple and easy to remember.

Once you have a tag name assigned to a region, you can now create a hyperlink to that region. To do this, highlight the chosen text, or select a graphic, and open the Insert Hyperlink dialog box. In the "Link to file or URL:" box, type the # symbol followed by the name of the tag. See Figure 24.2 for an example.

Create a tag in the following region. Right-click the region and add a name in the Tag box. Do not use a period "." in the tag name.

$\mathbf{Voltage_1 := 900amp}$

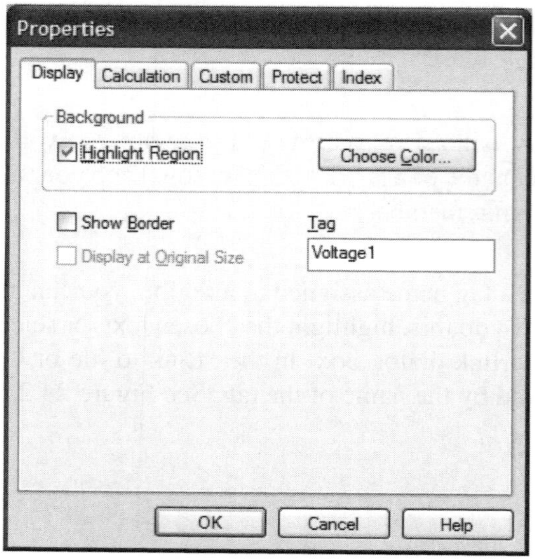

The above cell region now has a tag called, "Voltage1." You can now hyperlink to this region.

Hyperlink to **Voltage1**

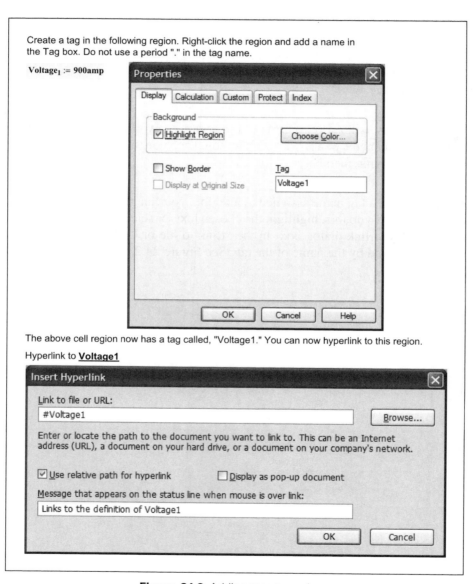

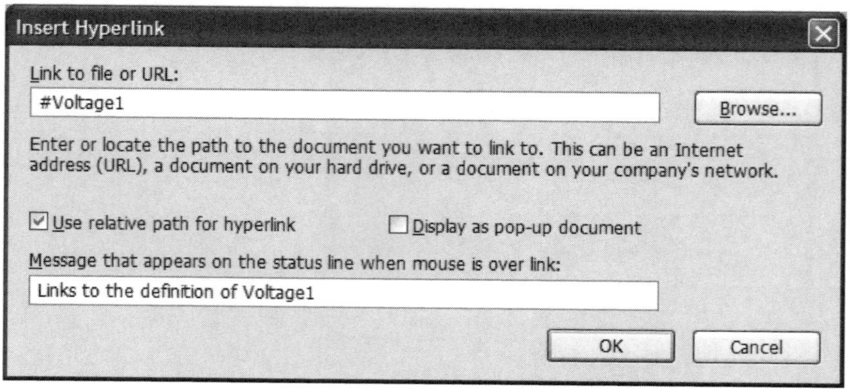

Figure 24.2 Adding tags to regions

Linking to region in another worksheet

If you want to link to a region in another worksheet, then open the Insert Hyperlink dialog box. The "Use relative path for hyperlink" box is checked by default. Keep this box checked unless the file you are referring to is a permanent location. If you intend to copy or move the current worksheet and the hyperlinked worksheet, then you will want to use the relative path. Click the **Browse** button to browse to the worksheet, and then click **Open**. If you have the "Use relative path for hyperlink" box checked, and the file is in the same folder, then the path name will not be added. If the file is in a different folder, or the relative path box is not checked, then a path will be added. Once the filename is added, then type a # and the name of the tag. See Figure 24.3.

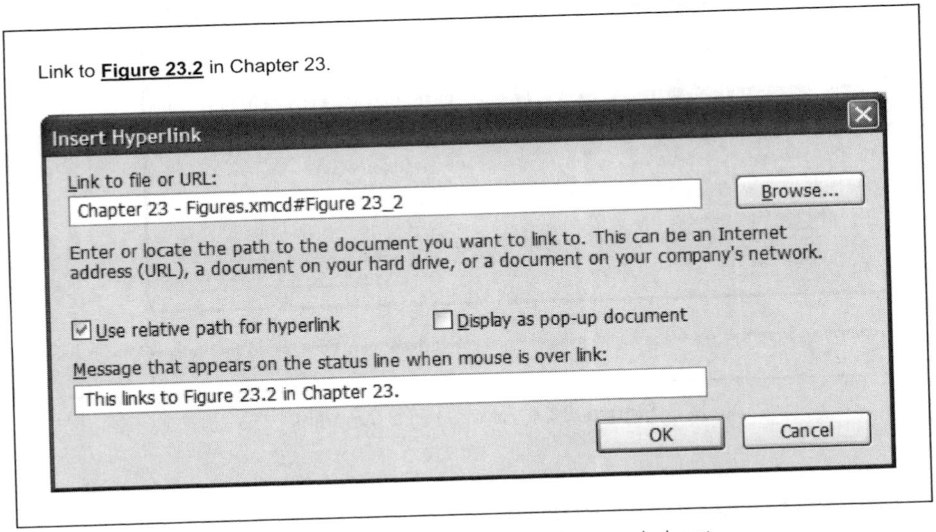

Figure 24.3 Hyperlink to another worksheet

When you double-click on the hyperlink, the worksheet will be opened and the linked region will appear at the top of your screen.

Notes about hyperlinks

It is important to note that when you link to a region, the screen moves to the region, but the cursor does not. If you link to a region in your current worksheet, the cursor remains in the hyperlink. If you link to a region in another worksheet, the cursor is moved to the top of the worksheet, even though the region is visible.

It is possible to create a hyperlink from a math region or from a plot region. Do not do this. It is much better to use a text box or a graphic. For example, if you create a hyperlink from a plot, when you double-click the plot, it opens the link associated with the hyperlink, rather than opening the Plot Format dialog box. A hyperlink in a math or plot region does not appear any different, so it is also difficult to distinguish the hyperlink. When you move your cursor over a math or plot region that has a hyperlink assigned, the cursor changes to a hand symbol.

To remove a hyperlink, right-click on the hyperlink, select **Hyperlink**, and click on the **Remove Link** button. This will remove the link and associated bold and underlined text. See Figure 24.4.

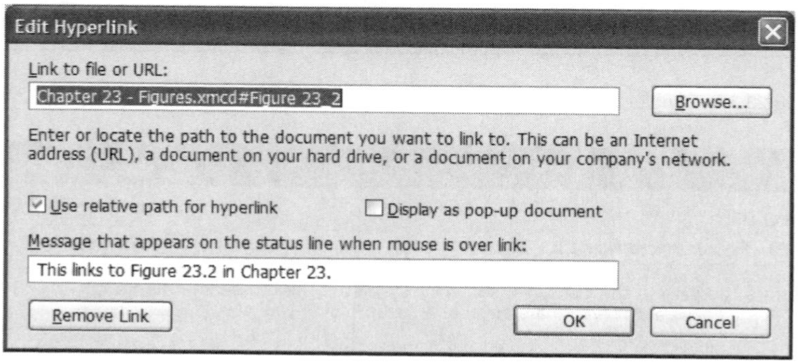

To remove a hyperlink, right-click on the hyperlink and select Hyperlink. Then click on the "Remove Link" button.

Figure 24.4 Removing a hyperlink

Creating a pop-up document

A pop-up document is one that appears in a separate smaller window. You can create a pop-up document for Mathcad worksheets, but not for a region in a worksheet. To create a pop-up document, click the "Display as pop-up" checkbox in the Insert Hyperlink dialog box. See Figure 24.5. It is best to have the linked pop-up worksheets be small with minimal regions. It is possible to have large worksheets appear in a pop-up window, but you would need to scroll through the window to see all of the information.

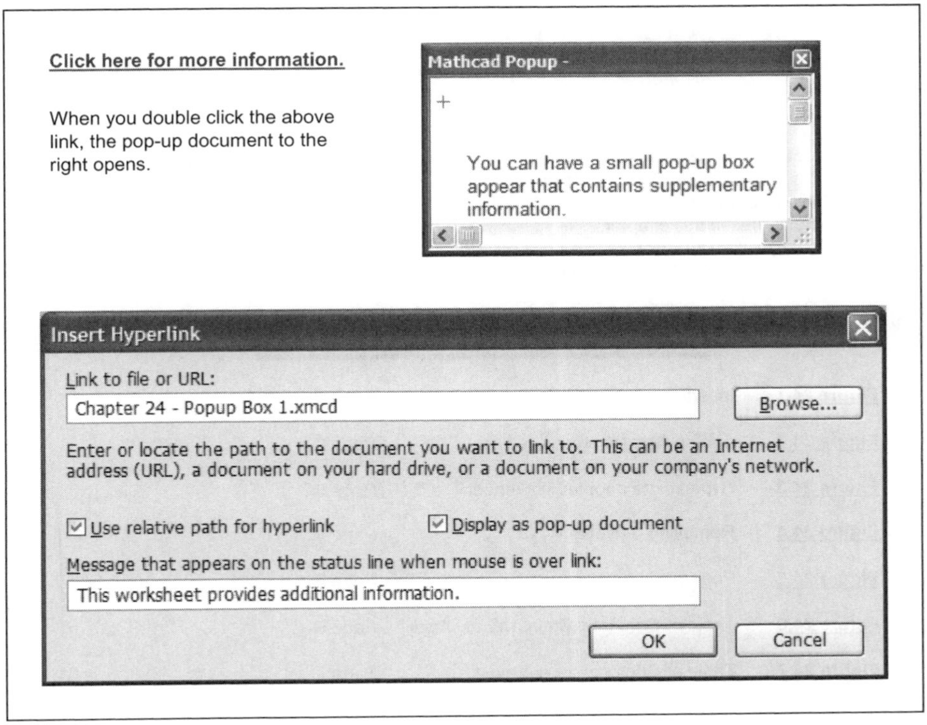

Figure 24.5 Pop-up windows

Table of Contents

Now that you understand how to link to regions in your worksheets, it is possible to create a Table of Contents. We will discuss two types. The first will be a Table of Contents for the data within your current worksheet. The second type will be a Table of Contents that links to many different files in your project calculations.

To create a Table of Contents to your current worksheet, first examine your worksheet and determine where you would like to have links added. In each of these regions, add a tag to the region. Be sure to write down the tag name for each region.

You can then go to the top of your document and create a text region. The entire Table of Contents can be in a single text region. Type the text of the hyperlinks, and then go back and add the region hyperlink to each item. See Figure 24.6 for an example.

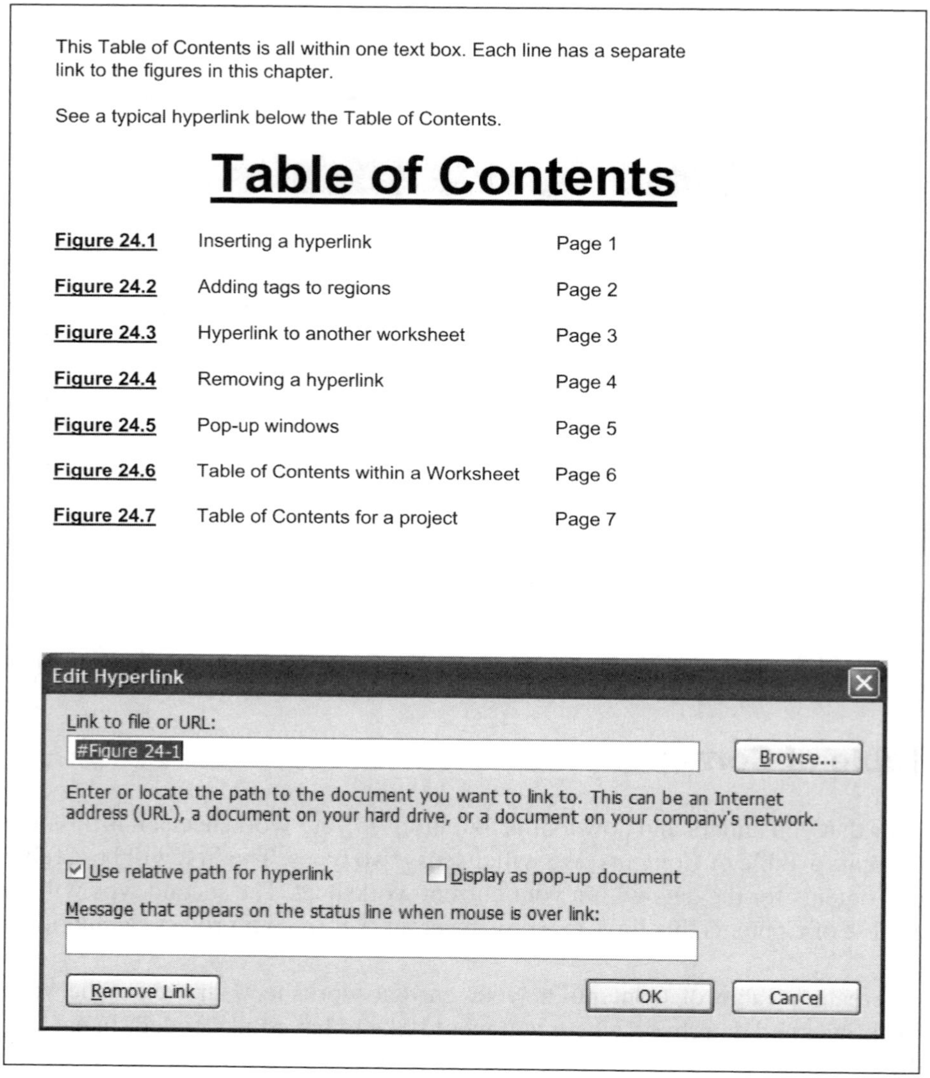

Figure 24.6 Table of Contents within a worksheet

If you have a project that includes many different worksheet files, then it is very useful to have a Table of Contents file that will link to each of the different files. The Table of Contents will be very similar to Figure 24.6. The only difference is that each hyperlink will reference a different file. The project Table of Contents will list each file in the order that it should be viewed or printed. It will also provide a hyperlink to each file. See Figure 24.7.

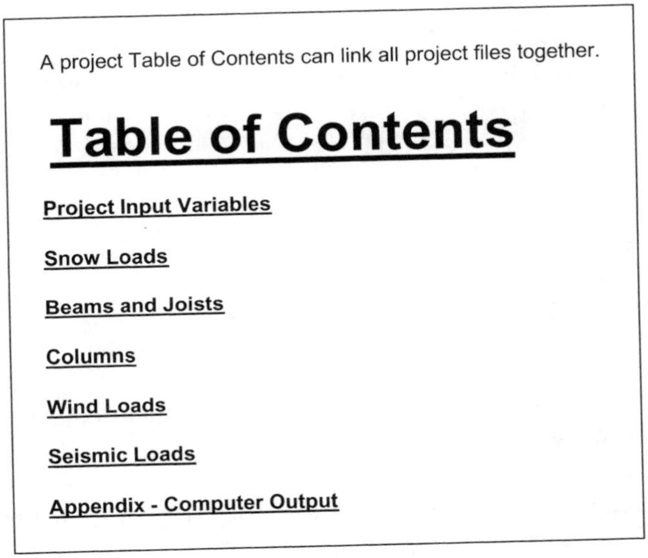

Figure 24.7 Project Table of Contents

Mathcad calculation E-book

A Mathcad E-book has all the information of a printed book, plus all the interactive math tools available in Mathcad. These books can be distributed to anyone with the same or higher version of Mathcad.

An E-book is a series of Mathcad files electronically bound together. You can annotate the book, and you can change the math regions to experiment with results. The original copy of the book remains unchanged, but you can save a copy with your annotations and changes. It is also possible to drag and drop regions from the E-book to other Mathcad worksheets.

It takes some effort to create an E-book, but if your calculations are going to be widely distributed, the effort may be worthwhile. The Mathcad Help and Tutorial were created from an electronic E-book.

The details for creating a Mathcad E-book are beyond the scope of this book. The Author's Resources on the **Help** menu has detailed instructions for creating a Mathcad E-book.

Summary

Hyperlinks are a wonderful tool to use in creating project calculations because they allow you to link to other Mathcad calculation files and also to regions in you current worksheet.

In Chapter 24 we:

- Introduced hyperlinks.
- Discussed how to create a hyperlink.
- Showed how to link to files and Web sites.
- Showed how to create a tag in a region by using the Properties dialog box.
- Demonstrated how to link to a region in a current or other worksheet by using the # symbol following the path.
- Told about pop-up windows.
- Learned about creating a Table of Contents for your local worksheet.
- Learned about creating a Table of Contents for a project calculation.
- Briefly introduced the concept of Mathcad E-books.

25 Conclusion

I hope that you have enjoyed learning about the benefits of using Mathcad to create and organize your engineering calculations. I hope even more, that you have followed along with your own version of Mathcad, and have practiced what you have learned. There is no substitute for hands-on practice and application.

Advantages of Mathcad

You may have already known about the powerful mathematical calculation power of Mathcad. After reading this book, you should now see how useful Mathcad is to do everyday engineering calculations. You can now use Mathcad not only for its powerful mathematical abilities, but also for its ability to help you create and organize your engineering calculations. You can now use Mathcad as the primary tool for your project calculations.

Remember some of the key advantages to using Mathcad as the primary tool for your project calculations:

- Units! Mathcad alerts you when you are using inconsistent units. You can easily get results in SI or U.S. units. You can work in both systems, simultaneously displaying the results in both systems.
- Your formulas and equations are visible and can be easily checked. They are not hidden in cells where only the results are visible.
- You can change an input variable, and your results are immediately updated.
- You can create standard calculations worksheets that can be use over and over again.
- Your calculations can be reused in other projects.

- Using Mathcad's metadata and Provenance, you can trace where regions were copied from.
- You can continue using the programs you are comfortable working with, and then store and present the results within Mathcad.
- You can archive the entire project.
- Using the Mathcad Application Server from the Mathcad Calculation Management Suite, you can share your calculations via the internet.
- Using Designate from the Mathcad Calculation Management Suite, you can create a library of Mathcad worksheets that can be checked-out and copied to other projects. Designate keeps track of which projects have used calculations from the library so that you can trace where the calculations went. You can also search the library for keywords and results from files stored in the library.

Creating project calculations

Let's review some of the key topics essential to creating project calculations with Mathcad.

- Use the tools in your Mathcad toolbox. Keep them sharp, by reviewing the topics covered in Parts I, II and III.
- Keep using your existing spreadsheets. Learn how to incorporate them as components in Mathcad so that Mathcad can provide input and receive output from them.
- Use the software programs that make the most sense. Each program has its strengths and weaknesses. Mathcad is not the ideal software program for all applications.
- Keep using your specialty software programs. Link or embed the results in Mathcad. Embed the input and output files in Mathcad.
- Use Mathcad to calculate input and to use the output from your other software programs.
- Use Mathcad to collect and organize the many different software files into one comprehensive project calculation.
- Link all your files together with a Table of Contents file, which contains hyperlinks to each file in the project calculations.
- Use a Table of Contents at the top of each file so that specific areas in the worksheet can be easily found.

Additional resources

We have just scratched the surface. There is so much more that could have been discussed. Our intent was to point out some useful features, and to get you started using Mathcad for all your project calculations.

Here is a list of things that you can do to increase your knowledge and Mathcad skills:

- Open the Mathcad Tutorials from the **Help** menu and review each tutorial.
- Open the Mathcad QuickSheets from the **Help** menu and review each sheet. There are many hidden treasures in the QuickSheets.
- Open the Reference Tables from the **Help** menu and explore the wealth of information contained in the tables.
- Review the Mathcad functions from the Insert Function dialog box. Research additional functions that are not discussed in this book.
- Visit the Mathcad User Forums on the internet. There is a link on the **Help** menu. Explore the many different topics. Ask a question. Look for answers.
- Search for topics at http://support.mathsoft.com.
- Read articles from old issues of the Mathcad Advisor Newsletter, available from the Resources section of www.mathcad.com.
- Attend a Mathcad on-line webinar (Web seminar). Visit www.mathcad.com for topics and time.
- Review the list of E-books and Mathcad files available in the Resources section of www.mathcad.com.

Conclusion

I wish you well, as you begin to use Mathcad in your engineering and technical calculations. I encourage a continued effort to add more tools to your Mathcad toolbox. There are many tools to add that we have not discussed.

It is also important to keep your tools sharp. Tools that are not used very often get rusty. I encourage you to review the chapters in this book and try to incorporate some of the features in your calculations. Review the chapters from time-to-time and polish your tools so that they will be sharp and ready to use when occasion permits.

I wish you the best of success as you enjoy the wonderful world of Mathcad.

Index

CD instructions

This CD contains the following:

1. A 120 day Academic Evaluation version of Mathcad 13.1.
2. The Mathcad files used to create the figures in "Engineering with Mathcad." These files are included so that you can follow along with the book. The files contain most of the screen shots used in the book, as well as the Mathcad expressions. In order to create the screen shots for the book, it was necessary to insert some graphic screenshot images into the Mathcad worksheets. So be aware that some of the information in the files is actually graphics and not active Mathcad regions.
3. A Mathcad electronic book (E-book) that also contains the same figures.

Installing the Mathcad Software:

If the Mathcad software does not install automatically, double-click the setup.exe file.

When the installation process asks for a product code, leave it blank and click Next. This will install the evaluation version, which will expire 120 days after installation.

When starting Mathcad for the first time, you will be asked to activate Mathcad. Choose the "Activate Later" option.

Engineering with Mathcad files:

The Mathcad E-book opens from within Mathcad. In order to use the E-book you need to copy some files from the CD to the Mathcad installation directory.

Copy the Handbook folder and its contents to the following Mathcad directory: C:\Program Files\Mathsoft\Mathcad 13\ (or another path where you installed Mathcad). If the Handbook folder already exists, then just copy the contents of the Handbook folder on the CD into the Handbook folder in the Mathcad 13 directory. Once you have done this, open Mathcad and select E-books> Engineering with Mathcad Figures from the Help menu. The E-book contains all the Mathcad examples, but does not contain all of the functionality that the actual Mathcad files have. For example some of the error messages, warnings, and styles are not included. All the chapters are hyperlinked from the main Table of Contents of the E-book.

If you would like to open the figures without using the E-book, you may open them directly in Mathcad. The Mathcad files are contained in the Handbook folder with the E-book. You may also copy the files from CD (EWM Files\Handbook\EWM) to a location on your hard drive. If you try to open the Mathcad files directly from the CD, you will get an error.

The Mathcad files will have the error messages, warnings, and styles available. A TOC.xmcd file provides a hyperlink to each chapter. These files will be copied to the Handbook folder in the Mathcad directory if you copy the Handbook folder as noted above.

Use the Mathcad files and E-book as you practice the concepts taught in Engineering with Mathcad. You can change the Mathcad values and immediately see the effects of the change. It will also allow you to see the error messages shown in the figures.